Historical-Analytical Studies on Nature, Mind and Action

Volume 11

Historical-Analytical Studies on Nature, Mind and Action provides a forum for integrative, multidisciplinary, analytic studies in the areas of philosophy of nature, philosophical anthropology, and the philosophy of mind and action in their social setting. Tackling these subject areas from both a historical and contemporary systematic perspective, this approach allows for various "paradigm-straddlers" to come together under a common umbrella. Digging down to the conceptual-historical roots of contemporary problems, one will inevitably find common strands which have since branched out into isolated disciplines. This series seeks to fill the void for studies that reach beyond their own strictly defined boundaries not only synchronically (reaching out to contemporary disciplines), but also diachronically, by investigating the unquestioned contemporary presumptions of their own discipline by taking a look at the historical development of those presumptions and the key concepts they involve. This series, providing a common forum for this sort of research in a wide range of disciplines, is designed to work against the well-known phenomenon of disciplinary isolation by seeking answers to our fundamental questions of the human condition: What is there? – What can we know about it? – What should we do about it? – indicated by the three key-words in the series title: Nature, Mind and Action. This series will publish monographs, edited volumes, and commented translations.

Marek Gensler • Monika Mansfeld
Monika Michałowska
Editors

The Embodied Soul

Aristotelian Psychology and Physiology in Medieval Europe between 1200 and 1420

Springer

Editors
Marek Gensler
University of Łódź
Łódź, Poland

Monika Mansfeld
University of Łódź
Łódź, Poland

Monika Michałowska
Medical University of Łódź
Łódź, Poland

Akademie Věd České Republiky
Narodowe Centrum Nauki
This work was supported by the Polish National Science Centre (NCN) grant n. UMO-2016/23/B/HS1/00430. The Academy of Sciences of the Czech Republic has funded the Open Access publication of the chapter "Buridan Wycliffised? The Nature of the Intellect in Late Medieval Prague University Disputations" by Lukáš Lička.

ISSN 2509-4793 ISSN 2509-4807 (electronic)
Historical-Analytical Studies on Nature, Mind and Action
ISBN 978-3-030-99452-5 ISBN 978-3-030-99453-2 (eBook)
https://doi.org/10.1007/978-3-030-99453-2

This Springer imprint is published by the registered company Springer Nature Switzerland AG
The registered company address is: Gewerbestrasse 11, 6330 Cham, Switzerland

Preface

The Embodied Soul: Aristotelian Psychology and Physiology in Medieval Europe Between 1200 and 1420 explores a wide range of issues of philosophical anthropology in the later Middle Ages. The purpose of this volume is to acquaint a wider audience with the dynamic of the reception of Aristotle-inspired views on psychology and physiology and the mind-body problem, which increased vigorously after the translation of Aristotle's natural philosophy treatises into Latin. It sheds light on the writings of both the best-known masters of that time and authors whose works and ideas have undeservedly fallen into oblivion and still await critical editions and investigation.

The collection evolved from the papers read at the conference Mind and Body: Aspects of Medieval Natural Philosophy, which was held at the University of Łódź, Poland, on 13–14 June 2019, to conclude the project entitled The Problem of Mind-Body Relation in the Philosophy of Walter Burley.

This book came into being during a very disturbing time, when the COVID-19 pandemic was announced and lockdowns were imposed all over the world. We are deeply grateful to the authors with whom we had the pleasure to work on this volume for their determination and engagement in the production of the volume. We would like to thank all those who encouraged us during this time, and especially our families for their love and unfailing support.

Łódź, Poland Marek Gensler

Łódź, Poland Monika Mansfeld

Łódź, Poland Monika Michałowska

Contents

Chapter 1
The Development of Aristotelian Psychology and Physiology in Medieval Europe Between 1200 and 1420: Introduction

Marek Gensler, Monika Mansfeld, and Monika Michałowska

Abstract The chapter is devoted to various aspects of philosophical anthropology in the views of Latin authors of the Later Middle Ages. The boundaries of the period are set by two important historical occurrences: the translation of Aristotelian natural philosophy treatises into Latin at the turn of the thirteenth century and the early signs of crisis in scholastic philosophy precipitating the rejection of the pre-modern vision of human being as a psychosomatic unity by early modern science. It presents, in a chronological order, the changes and developments in the understanding of mutual relations between the soul and the body. It takes into account not only the areas of interest (reproduction, nutrition, growth, motion, perception, cognition, etc.) but also the forms of the analyzed works (literal commentaries, paraphrases, questions, treatises). Moreover, it pays attention to the way in which those works formed an interlocking chain through two and a half centuries of history with several recurrently crossing trails, some of which were frequented throughout the period, while others were of a more ephemeral character.

Keywords Medieval philosophy · Philosophy of nature · Philosophical anthropology · Mind-body problem · Aristotelianism · History of physiology · History of psychology

This work was supported by the Polish National Science Centre (NCN) grant n. UMO-2016/23/B/HS1/00430.

M. Gensler (✉) · M. Mansfeld
University of Łódź, Łódź, Poland
e-mail: marek.gensler@uni.lodz.pl; monika.mansfeld@uni.lodz.pl

M. Michałowska
Medical University of Łódź, Łódź, Poland
e-mail: monika.michalowska@umed.lodz.pl

M. Gensler et al. (eds.), *The Embodied Soul*, Historical-Analytical Studies on Nature, Mind and Action 11, https://doi.org/10.1007/978-3-030-99453-2_1

1.1 Introduction

The volume entitled *The Embodied Soul. Aristotelian Psychology and Physiology in Medieval Europe between 1200 and 1420* is devoted to various aspects of philosophical anthropology in the views of Latin authors of the Later Middle Ages. The boundaries of the period are set by two important historical occurrences: the translation of Aristotelian natural philosophy treatises into Latin at the turn of the thirteenth century and the early signs of crisis in scholastic philosophy precipitating the rejection of the pre-modern vision of man as a psychosomatic unity by early modern science, which apparently rendered the scholastic vision of the world obsolete. The period included between those dates was by no means uniform in its doctrinal character, since the dynamic of reception and accommodation of new ideas resulted in different forms of presenting issues belonging to that area of speculation. The volume sheds light on some aspects of that dynamic in the writings of both the best-known masters of that time and those whose works and ideas have been neglected so far and still await critical editions and investigation. For this reason, texts included in this volume address both general and particular problems.

An important criterion of the volume's structure and of the historical-doctrinal analysis of those works is the time of their composition. Accordingly, the editors have decided to arrange them primarily in a chronological order. However, two further criteria have also been taken into consideration: the form of the analyzed works (literal commentaries, paraphrases, questions, treatises) and the area of interest (reproduction, nutrition, growth, motion, perception, cognition, etc.). As a result, the order of the texts in the volume has been coordinated in such a way that they form an interlocking chain of studies leading the reader through two and a half centuries of history along several recurrently crossing trails, some of which were frequented throughout the period, while others were of a more ephemeral character or soon ran into dead ends.

1.2 The First Phase: Aristotle's *De anima* and Avicenna's *Liber sextus naturalium*

The scholastic commentary tradition of Aristotle's treatises on psychology and physiology started with the Latin translation of two texts, viz. Aristotle's *De anima* and Avicenna's *Liber de anima* (*Sextus de naturalibus*), in the second half of the twelfth century.[1] Their appearance in the Western institutions of learning coincided in time but it was not a mere coincidence: the reception of each of them was significantly influenced by the reception of the other. The reason for it was that the novel ideas from Aristotle had to be juxtaposed and coordinated with the established psychological doctrines of neo-Platonic origin. Avicenna's neo-Platonic

[1] See Hasse (2000).

interpretation of Aristotle provided a key that made such coordination possible. The earliest known attempt at such a synthesis of Aristotelian and neo-Platonic psychological views, with Avicenna serving as a catalyst, was *Tractatus de anima* by John Blund, a philosopher active in Oxford and Paris the beginning of the thirteenth century.[2] Riccardo Fedriga devotes his text to Blund's role as the trendsetter in this field. He points out astounding erudition of the English master, who knew well not only the earlier Latin tradition (Augustine, Boethius, Anselm of Canterbury and Peter Lombard) and the new authorities, Aristotle and Avicenna, but also a number of natural philosophy texts freshly translated from Arabic (al-Ghazali, pseudo-Alexander of Aphrodisias, Costa ben Luca). The mix of authors quoted by Blund can be seen as a deliberate attempt to integrate the strictly philosophical issues concerning the relation between the body and the soul with some of those that traditionally belonged to the realm of theology, leaving only the supernatural aspects of the soul as the sole preserve of theology. Fedriga argues that Blund, by skilfully combining various philosophical and theological traditions, succeeded in presenting the body-mind problem from an original perspective. This originality is shown on the example of the analysis of human sensory cognition by two of the so-called lower senses: touch and taste. The cognitive character of an act of perception, according to Blund, is possible only thanks to the soul's acts of voluntary attention. This allows Fedriga to conclude that for Blund the boundary between the body and the soul "is drawn between an externalist ontology of things and sense organs" (the Aristotelian component) "and an internalist ontology self-based on the intentional status of soul" (the Augustinian – neo-Platonic component).

Avicenna had more than one role to play in extending the scope of anthropological studies of the scholastics. He was not only a philosopher but also a physician and his medical views were closely related, in the spirit of Galen, with his philosophical ones. This Galenic approach, present in Avicenna's *Canon of Medicine*, encouraged Latin masters to read Galen too, not only for the sake of medical studies but also for the sake of philosophical ones.[3] Such an approach can be seen in the works of Gilbert the Englishman.[4] In her paper, Agnieszka Kijewska returns to the pioneering studies by Marian Kurdziałek, who saw Gilbert as one of the earliest medicine-oriented philosophers in the Latin West, paving the way for a more advanced study of anthropology that included both psychology and physiology.[5] She sets Kurdziałek's interpretation of Gilbert's anthropological views against the recent works of Michael McVaugh arguing that Gilbert was enthusiastically involved in medical debates as well as in the reception of Avicennian medicine at that time.[6] By pointing out the key elements of Gilbert's anthropology, Kijewska shows that he was not only a skilful philosopher but also a studious reader of both Aristotle and

[2] See Burnett (1996, 37).

[3] See McVaugh (2019, 384).

[4] See Esteban-Segura (2012, 18–19).

[5] See Kurdziałek (1950).

[6] See McVaugh (2010).

Avicenna, which allowed him to entwine their concepts and formulate his own theories. This blend is most clearly visible in Gilbert's concept of the soul as well as of the role of the intellect in the structure of the soul. Kijewska's chapter underlines a significant feature of Gilbert's writings, which she defines as "philosophical eclecticism, in which Aristotelian philosophy and science was interpreted through the prism of Avicenna's reinterpretation and often in a neo-Platonic spirit."

1.3 The Second Phase: *Parva Naturalia*'s *Translatio Vetus* and Averroes's Commentaries

Once the interest in physiology was ignited in the West, its study proceeded at an increasing pace. It took some time, however, before it gathered momentum. An important factor in its acceleration was the translation of Aristotle's short treatises in psychology and physiology, later known as *Parva naturalia*, which began at the turn of the thirteenth century.[7] Their reception, however, did not start immediately with the first translations of them, similarly to what seems to have been the case with the *De anima*. In the beginning, they were treated as a specific supplement to the Aristotelian theory of the soul, found in the *De anima*, addressing psychological issues from the point of view of the body animated by the soul and thus devoting a particular interest to the physiological processes in bodily organs corresponding to the actions of the soul. It took two generations for the Latin masters to get interested in the texts in their own right; actually, less than a half of them had any traceable reception by the 1260s.[8] Still further time passed before even those that had been read and commented upon were introduced to the university curricula.[9] Again, some help in interpreting the texts was provided by yet another Arabic commentator—Averroes, whose epitomes of *De sensu et sensato*, *De memoria et reminiscentia*, *De somno et vigilia*, and *De longitudine et brevitate vitae* (usually mislabelled as *De morte et vita*) were translated around 1230.[10] The authority of the Commentator helped to establish those four treatises as the standard set for commentators on the *Parva naturalia*, even though other texts were added to it later. The first author known to have commented on the *translatio vetus* of the original set was Adam of Buckfield, who wrote expositions to them between ca. 1240 and 1250.[11] The issues

[7] See Galle (2008, 9–11).

[8] See De Leemans (2010, 917–918).

[9] See Callus (1943, 4).

[10] See Shields and Blumberg (1949, xiii–xiv); Callus (1943, 38).

[11] There might have been some earlier attempts to comment on at least some of the treatises from the *Parva naturalia* set. According to an entry from a seventeenth-century catalogue of the Cathedral Library in Beauvais, there was a manuscript containing, among others, glosses on *De somno et vigilia* and *De morte et vita* (again, presumably *De longitudine et brevitate vitae*) by Alfred of Sareshel. This copy, however, did not survive and, if there are some others still extant, they have not been discovered yet. See Omont (1916, 48, n. 143); Callus (1943, 10–11).

related to early Oxford reception of the problems found there are analyzed by Christina Thomsen Thörnqvist. Her chapter is a survey of approaches to a single *aporia*, found on the pages of Aristotle's *De insomniis*. Aristotle characterizes the receptivity of our sensory apparatus using the metaphor of a mirror reflecting the images of objects placed in front of it. Attractive as it was, the metaphor had a drawback, which was important from the point of view of later commentators: in a passing remark, Aristotle states that the sight of menstruating women leaves a rusty stain on the surfaces of mirrors they use. This remark apparently undermines Aristotle's idea of purely passive character of the senses by pointing to some active element in human perception affecting its object. Thörnqvist's analysis offers an account of how medieval commentators of Aristotle's treatises on sleep and dreams—starting from Adam of Buckfield, through Albert the Great, Thomas Aquinas, Geoffrey of Aspall, Radulphus Brito, to Simon of Faversham and Walter Burley—approached the challenge of reconciling Aristotle's claim that the organ of sight acts upon a sensible object in such a way with their overall understanding of Aristotle's theory of perception. Additionally, Thörnqvist shows that in the earliest period of the reception of the *Parva naturalia* the concept of the active role of the subject of cognition and its influence on its object was closely related to the concept of *fascinatio*, the best-known example of which is the belief in the 'evil eye'.

In 1260s, Albert the Great undertook the task of rendering the whole of the Aristotelian corpus intelligible to the Latin scholars by the method of paraphrase. His project included paraphrasing both the *De anima* and all the treatises of the 'standard' *Parva naturalia* set: *De morte et vita*, *De sensu et sensato* together with *De memoria et reminiscentia*, which he treated as the second part of the same work, *De somno et vigilia* together with *De insomniis* and *De divinatione per somnium*. To these texts he added paraphrases of Aristotle's other works on similar topics, such as *De iuventute et senectute*, *De spiritu et respiratione*, and *De motu animalium*—the oldest surviving Latin commentary on that work[12]—and *De homine*, in which he compiled anthropological knowledge coming from various sources. Albert the Great was probably the most famous and most influential author in the earlier period of reception of Aristotelian psychology and physiology. His paraphrases and independent treatises emphasized the 'experimental' aspect of the new teaching, i.e., they described, analyzed and strove to generalize various recorded individual cases, with a special predilection for *curiositates* for which he tried to find rational explanations.[13] Albert's original and attractive but relatively little researched contributions to this field are the subject of four studies included in the volume, by Evelina Miteva, Mario Loconsole, Marilena Panarelli and Paloma Hernández-Rubio.

The description of the world requires adopting some definitions, classifications and categorizations, with the help of which one can understand not only the nature of various things but also introduce some order to the variety and diversity of the world. This explains why medieval masters, for whom Albert the Great was the

[12] See De Leemans (2000, 272–360, esp. 275–276).

[13] See Wöllmer (2013, 250–251).

model, were so fascinated by everything that was bizarre and escaped easy classification. Referring to the problem of what human nature actually means, Miteva analyzes two cases in which human nature ranges on the borderline. Her focus is on Albert the Great's concepts of pygmies and twins with reference to his definition of a monster. While the generation of twins and twins as such fall under the category of monstrosity, pygmies, although very similar to humans, are not defined as monsters. They are not categorized as monsters since, in Albert's view, nature did its job equipping them with 'normal human functioning', to paraphrase Norman Daniels. They have the potency of reason, yet due to the damaging role of corporeal matter they cannot develop it. Examining Albert's attempts at providing a description of 'normal species functioning', Miteva reveals a rather distinctive feature embedded in Albert's theories, namely a problem that in our times has been eloquently formulated by Georges Canguilhem as 'normalcy vs normativity'.[14] Canguilhem points out the twofold meaning of the term normalcy: normalcy understood as an average state of the human condition as well as an ideal one which the human species should aspire to. He also underlines that some states called 'pathological' and understood as chaotic and norm-less aberrations have their own regularity and their own norms, and in this sense, they are also normal. The classification provided by Albert reflects that tension. On the one hand, Albert identifies what is normal with that what is normative, on the other he tries to describe some 'abnormal' cases by ascribing certain common and regular features to them. Any definition of normalcy refers—*explicite* or *implicite*—to the concept of human nature whose description, as Albert's examples show, is complex and difficult to explicate. Miteva's chapter depicts an important characteristic of Albert's concept of human nature, which, to some extent, is a social-cultural construct determined by the prevalent values of a medieval society and reflecting what was desirable and undesirable.[15]

Loconsole, who writes on human reproduction, focusing on the puzzling issue of generation of foolish children by wise men, analyzes not only the writings of Albert the Great but also those of Peter of Spain and James of Viterbo. Examining the decisive role of male semen in reproduction these authors state that its action can be hindered not only by physical circumstances but also by the mental conditions in which the man finds himself during reproductive practices, Loconsole also refers to the idea of 'specific' and 'normal' functions of humans. Since, as he rightly points out, medieval masters did not have "a general theory of biological hereditariness," they made various attempts at explaining how intellectual capacities are transmitted, if at all, using quite often contradictory concepts provided by both medical theory and natural philosophy.

The chapters by Miteva and Loconsole demonstrate a conspicuous tendency in medieval natural philosophy to evaluate any deviation or variation from normalcy negatively. For a modern reader, it is worth adding that this observation also applies to some contemporary concepts, for instance those formulated by Daniels in the

---

[14] See *Canguilhem* (1991).

[15] See Freidson (1970).

enhancement debate, in which enhancement is treated "as departures from species-typical normal functioning"[16] which, in his view, is a sufficient argument to consider it morally wrong. Miteva's and Loconsole's chapters eloquently illustrate that medieval writings were constructed in various languages: not only descriptive and explanatory ones but also, to paraphrase Hugo Tristram Engelhardt Jr., an evaluative one.[17]

The intersections of natural philosophy and medicine are discussed also by Panarelli who concentrates on the variety of meanings ascribed to the term 'spiritus', beginning with Aristotle's and Galen's concepts, through its interpretations by Costa ben Luca and Avicenna, ending with the approach adopted by Albert the Great. She demonstrates how Albert, drawing upon different traditions, developed his own theory of how spirit interacts with the soul. She shows how biological explanations, for example an anatomical division of the structure of the brain or of the digestive processes, served as a starting point to explain the functions of the animal spirit and allowed for formulating complex theories of spirit as a "vehicle of the soul"—the soul's instrument with which it can entertain a variety and diversity of functions ranging from the basic vital ones to the most complex and sophisticated ones, such as thinking. Panarelli's chapter tackles a profound shift in the perception of a body and the mind-body relationship in medieval natural philosophy and medicine. From a mere object of various medical practices, the body became an independent subject of investigation. As she argues, this new approach allowed of explaining "the polyvalence of the phenomena of the living."

Yet another aspect of Albert's natural philosophy is discussed by Hernández-Rubio. She devotes her chapter to Albert's characteristics of human nature focusing on sense perception. She shows how Albert reformulated key Avicennian terminology and developed a new original theory of sense perception influenced by Avicenna's inner-sense theory and a rejection of Averroes's distinction between animal senses and human senses.[18] Guiding us through meticulous distinctions and subtle differences between the terms Albert employed, Hernández-Rubio argues that Albert successfully overcame the problems Averroes posed, yet left unanswered, and explained how the senses can grasp the individual thing as individual without cognizing its substance. As a result, Albert's concept of perception turns out to be more universal than Averroes's, namely the former claims that all sentient beings can apprehend individuals in a meaningful way, since both rational and non-rational animals have identical sensitive apparatuses suitable for perceiving individuality as such.[19] Hernández-Rubio concludes that since "individuality is always grasped as a potential subject of the properties of the apprehended thing, [...] the sensorial apparatus shared by rational and non-rational animals grasps the same kind of mental object: a *suppositum*."

[16] See Daniels (2000, 318).

[17] See Engelhardt (1996).

[18] See Black (1991, 1996, 2000).

[19] See Tellkamp (2016).

1.4 The Third Phase: The Reception of the Last Wave of Translations

In the second half of the thirteenth century, one can see a significant growth of interest in commenting on both the *De anima* and the *Parva naturalia*, mainly for didactic purposes. The new commentaries are styled according to changing literary fashions. Earlier in the thirteenth century, the most frequently used commentary form was the literal exposition, which became more and more elaborate over time.[20] Yet another literary genre was added to the commentary tradition after 1250: it was the commentary *per modum quaestionis*, which soon became more popular than literal expositions and paraphrases. Questions appeared first at Oxford and only later gained popularity at Paris and other centres of learning. The first known question-commentaries written at Oxford were the ones by Geoffrey of Aspall on *De sensu et sensato*, *De longitudine et brevitate vitae*, and *De somno et vigilia*[21] in the early 1260s.[22] At the turn of the fourteenth century, Simon of Faversham composed questions on almost the whole set of the *Parva naturalia*.[23] The issues related to early Oxford reception of problems found there are analyzed by Michael Dunne.

Geoffrey of Aspall, to whose commentary on *De longitudine et brevitate vitae* Dunne devotes his chapter, may have been a student of Adam of Buckfield, the Oxford pioneer of *Parva naturalia* commentaries. He is regarded as having been part of the third and last phase of the reception of Aristotle at Oxford.[24] His surprisingly long set of questions on *De longitudine* (Dunne adds the critical edition of the fourth question from this set as an appendix to his chapter) forms a part of the collection of his commentaries on Aristotle's (mostly) natural philosophy works. One of the issues he discusses is the well-known problem of the immortality of the soul, analyzed separately for the intellectual, the sensitive and vegetative soul. Aspall does not surprise his readers with any unexpected solutions but provides some new, interesting arguments, e.g., drawn from Grosseteste's concept of light propagation, or the discussion concerning the first and the last instants of existence, which later became characteristic for Oxford philosophers, such as Walter Burley or Richard Kilvington.[25] Dunne concludes that "Aspall clearly distinguishes between 'reditio' and 'resurrectio', the return of an individual specimen of a species but not of the individual itself, but he has moved the discussion on from the renewal and return of

---

[20] See Dunne (2003, 320–336).

[21] See Ebbesen (2014, 257–341).

[22] See Lohr (2013, 127–128). For the list of questions of *De somno* et *De sensu* commentaries, see Ebbesen et al. (2015, 66, 96).

[23] See Ebbesen (2013, 90–145); Christensen (2015, 93–128). It was followed by a number of anonymous questions originating in the same academic environment. Some of them have been critically edited, see Ebbesen (2015, 128–161).

[24] Callus (1943, 38–39 and 45–47).

[25] See Marone (1983, 196).

an element such as water which transforms into air and then back into water to the death and return of a living thing, in a presumably likewise cyclical manner."

Already in the times of Blund, the new texts were analyzed not only from the philosophical but also from the theological perspective. The borderline between natural philosophy and theology remained an important area, in which the new scholarly approach was expected to achieve consonance with the already existing tone of teaching.[26] The doctrine of the immortality of the soul and the dominant Augustinian vision of it as an independent substance superior to the body in all respects were taken as given in all attempts to accommodate Aristotelian psychological teaching to the requirements of faith. A special place on the long list of medieval Latin scholars who tried to Christianize Aristotle is occupied by Thomas Aquinas.[27] These attempts are the subject of a study by Cecilia Trifogli, who discusses it on a broader historical background. Martin Pickavé, by contrast, documents the resistance offered by some thinkers (in his case, James of Viterbo) against the increasing dominance of Aquinas-inspired interpretation of Aristotelianism.

Trifogli addresses the quintessentially Aristotelian question of how deep the difference between human beings and the other animals is, first examining the reply given to it by Thomas Aquinas and then discussing the criticism of Aquinas's solution presented by Giles of Rome and Thomas Wylton. The difference rests on the assumption that a human being possesses an intellectual soul, which makes her the kind of being she is.[28] According to Trifogli, this assumption invites these questions: "What is the subject of thought? Is it the composite of body and the intellectual soul, or is it the intellectual soul alone? If the subject of thought is the intellectual soul alone, does this confer a special status on the intellectual soul compared to the souls of other animals? Does this make the intellectual soul itself somehow independent from the body?" Aquinas's reply affirms the intellectual soul's independence from the body, even though he admits the dependence of thought on phantasms. Trifogli shows that Aquinas was aware of the weakness of inferences leading to the conclusion that thinking is an action independent from the body and tried to strengthen his argument, yet "[d]espite Aquinas's authority as an Aristotelian commentator, his reading of the dependence of thought on imagination was scarcely convincing" for Giles of Rome and Thomas Wylton. The former argues that it was Aristotle's view that all the affections of the soul, including thought, belong to the composite and not to the soul alone. Moreover, Giles states, somewhat unfairly according to Trifogli, that Aquinas's position leads to the statement, opposed by the Church, that it is the soul that thinks, not the man. For Wylton "the Catholic view is the true one but he believes it by faith alone, since he cannot find any rational argument in its support." Consequently, he does not point out any inconsistency in the notion of self-subsisting form and dismisses Aquinas's argument trying to bridge the gap between the teaching of Aristotle and the Church.

[26] See Flint and Rea (2009).

[27] See Del Soldato (2020).

[28] See McCord Adams and Trifogli (2012).

Pickavé devotes his chapter to the epistemological views of James of Viterbo, who was one of the few later medieval philosophers and theologians explicitly arguing for innate cognition, even though he did not reject the basic tenets of Aristotelian empiricism.[29] His defence of innatism is sophisticated showing a great awareness of the anti-innatist arguments of his contemporaries and predecessors, such as Thomas Aquinas. After a review of the arguments adduced against innatism by Aquinas, most of all that in all of its forms it is ultimately based on a mistaken understanding of natural agency and causality, Pickavé goes on to show how James responds to Aquinas's concerns. According to James, innate active forms in the intellect, the will, and in matter are what makes genuine agency possible in the first place rather than undermining it. He does not see them as definite ideas that can be recovered in an act of recalling them from memory in the fashion of Plato but as aptitudes or propensities for forms pre-existing in matter that are being extracted from it and move from potency to act. At the same time Pickavé states that "he maintains that the object is indeed a real cause which does something: namely, it excites the soul. He even calls it an efficient cause," but its action is concomitant with the innate aptitude. Pickavé points out that James was convinced that his view could be reconciled with Aristotle and referred to Aristotle's concept of innate virtue as an argument in favour of his opinion.

In the generations after Aquinas, we can see that a variety of available sources stimulated new inquiry into the problems belonging to the borderland between natural philosophy and medicine going beyond the achievements of the thirteenth-century scholars engaged in those issues. In the early fourteenth century it was further facilitated by the appearance of Avicenna's *Canon of Medicine* and the works of Latin theoreticians of medicine, mostly from Montpellier and Salerno, in the curricula of medical faculties across Europe.[30] Those medical works were soon taken as additional sources by natural philosophers in creating overall, synthetic approaches to the issue of human cognition, e.g., in the ideas of Walter Burley, examined by Monika Mansfeld, and used in solving specific scientific problems, such as the nature of touch, studied by Chiara Beneduce.

Burley was a natural philosopher with some interest in medical books. In two short treatises, apparently written for the general public, *De potentiis animae* and *De sensibus* (which constitutes a kind of appendix to the former work), he tries to give a concise presentation of the complex issue of sensory cognition, analyzed at length in his commentaries on Aristotle.[31] Burley retraces the process step by step, first analysing the perceptions of the external senses, then the actions of the internal senses, which he arranges in a sequence, to the point when sensory data reach the intellect. The whole process of sensory cognition is shown as one of growing abstraction and dematerialisation of sensations. Reconstructing the process serves the purpose of presenting an ideal model as well as that of pointing out the possible

---

[29] Solère (2018).

[30] See O'Boyle (1998, 5–7).

[31] See Kitchel (1971, 1977).

sources of cognitive errors. Mansfeld reveals Burley's own contributions to creating a uniform interpretation of not entirely concordant authorities on the subject and assesses the extent to which his attempt was successful (e.g., in the case of the physiology of the external senses).

Beneduce is not the first author writing about the relationship between natural philosophy and medicine in the medieval tradition, as shown above, but she devotes her chapter to a later phase of those contacts, analysing the issue of the unicity or plurality of touch as formulated by John Buridan and Tommaso del Garbo.[32] Examining how the issue was understood in natural philosophy and medicine, she shows the common ground at their intersections.[33] By going back to the Aristotelian and Avicennian traditions of understanding the role and nature of touch, Beneduce deals with the diversity of a 'touch problem' that troubled the minds of medieval masters, and in particular a) whether animals have only one sense of touch responsible for all sense perceptions; or b) whether they have various senses of touch, and c) what is the proper object of touch. Beneduce's chapter provides a detailed introduction to understanding the importance of the problem as examined by both philosophers and physicians and shows how the concepts from the field of natural philosophy and medical theories intersected providing a platform for mutual inspirations.

The fifteenth century, commonly thought of as the beginning of the decline of scholastic philosophy, brought also a significant decline in independent studies devoted to psychological and physiological issues. They were incorporated as arguments into the robust debates of the period, such as the realist–nominalist controversy.[34] This lack of independence resulted in their reduction to forms and character proper to philosophical exercises practiced in lecture rooms or at university debates.[35] Lukáš Lička analyzes such debates, organized at Prague University, to show that in spite of their rigid and schematic forms of exchanging arguments of authority for a proposition and against it, they nevertheless allowed of a lively philosophical discourse. Lička's chapter refers to the most problematic and, consequently, vividly debated topics of forming a definition of the human being as well as indicating its distinctive and uniquely human features. Lička concentrates on the nature of the human intellect as discussed by Prague masters from around 1390 to 1420, stressing the tension between the proponents of materialism on the one hand and those of personal immortality on the other. Analyzing concepts formulated in critically unedited sources, he guides us through various approaches adopted by Prague masters to ease that tension: some of them followed "the Buridanian distinction between the rationally demonstrable materialist tendency and the indemonstrable 'catholic truth' about the intellect both inhering in and separable from the body," while others assumed an "immortal spirit hypostatically united to each

[32] See Beneduce (2019); De Asúa (1997).

[33] See Miteva (2018).

[34] See Hoenen (2011).

[35] See Pavlíček (2018); Lawn (1993).

human being." Lička's chapter provides a well-presented historical background of the *quodlibetal* disputations and descriptions of various textual sources of that time and also gives some insight into the doctrinal content of these texts and the various strategies employed by Prague masters to identify and solve 'the intellect problem'.

Taking a broader perspective, the material presented in this volume allows the reader to recognize certain common tendencies and traits, some of which reflect phenomena characteristic for the development of medieval philosophy in general, while others are peculiar for that particular period and field. Its authors show both the unity and the diversity of ideas associated with the fascinating area of psychology and physiology in this important, albeit still somewhat obscure period. One of the main assets of the volume is its focus on analyzes of source material frequently taken not only from fresh critical editions but also from unpublished material preserved in manuscripts and incunabula.

Several recurrently appearing issues deserve highlighting here. One has to do with the milieu in which the problems concerning the relations between body and soul were discussed. It was among the medicine-oriented philosophers and philosophy-oriented physicians that the new texts, especially the ones belonging to the *Parva naturalia* set or its commentaries, were discussed. It was at the intersections of natural philosophy and medicine that the most interesting and most inspiring problems can be found: the problem of the intellect and its relation to the other powers of the soul (especially to the senses), the numerous problems concerning the particular senses, their relations to one another, to the sense organs, to objects of sensory perception, to list just a few of those discussed in the volume. Fittingly for a field at the crossroads of philosophy and medicine, the sources come from both disciplines with two important authorities—Galen, a philosophizing physician, and Avicenna, a philosopher with a knack for medicine—both becoming central figures and points of reference.

The other issue concerns an intersection of a more general kind: as it was with many other areas of philosophical study in the Middle Ages, the most fruitful debates were the ones that set the neo-Platonic/Augustinian and the Aristotelian traditions against one another. A variety of approaches can be seen here—some prefer to adhere to the established tradition, some want to discard it in favour of novel ideas, while a great majority prefer to accommodate the new to the old somehow. It is this 'somehow' that is most interesting, because it allows medieval authors to test various hypotheses, discuss them, correct them or discard them in favour of a better solution, either found in a newly translated text or presented by a more talented, or sometimes merely more articulate colleague.

The variety and diversity of approaches adopted to examine the mind-body problem debated so intensively in the Middle Ages reveal that medieval masters took a genuine interest in finding conciliatory solutions; at the same time, however, they point out certain tensions between the various theoretical approaches. That diversity naturally continued in the history of philosophy and in the history of medicine, eventually leading to the development of new disciplines. The mind-body problem dealing with how to place human subjective experience in the objective description of the world has since developed into separate disciplines, such as psychology or neuroscience, trying to answer the most profound questions of human consciousness that already troubled the minds of medieval masters. It ought to be pointed out that the mutual inspirations of philosophy and medicine are nowadays acknowledged in both bioethics and clinical practice and understood as one of the most important factors contributing to understanding the human nature. One of its most recent and model examples is the 'avatar therapy' used in treating auditory hallucinations that draws on the achievements of psychological approaches as well as on Emmanuel Levinas's[36] concept of experiencing the other and face-to-face encounter (*le visage*). Despite numerous attempts to solve the troubling relation between intellect/intelligence and bodily perceptions, it still remains a field open for new research and discoveries both in medical and philosophical domains. The quest to find correlations between mental states and brain activity still goes on—we are on a path that started centuries ago and being aware of this makes the challenges look even more attractive.

References

Beneduce, Chiara. 2019. La fisiologia del tatto nel XIV secolo. Il caso di Giovanni Buridano. In *Rappresentazioni della natura nel Medioevo*, ed. Onorato Grassi and Giovanni Catapano, 207–220. Firenze: Sismel Edizioni del Galluzzo.

Black, Deborah. 1991. Estimation (Wahm) in Avicenna: The Logical and Psychological Dimensions. *Dialogue* 32 (02): 219–258.

———. 1996. Memory, Individuals, and the Past in Averroes' Psychology. *Medieval Philosophy and Theology* 5: 161–187.

———. 2000. Imagination and Estimation: Arabic Paradigms and Western Transformation. *Topoi* 19: 59–75.

Burnett, Charles. 1996. The Introduction of Aristotle's Natural Philosophy into Great Britain: A Preliminary Survey on the Manuscript Evidence. In *Aristotle in Britain during the Middle Ages*, ed. John Marenbon, 21–50. Turnhout: Brepols.

Callus, Daniel Angelo. 1943. Introduction of Aristotelian Learning to Oxford. In *Proceedings of the British Academy*, vol. 29, 3–55. London: Humphrey Milford.

Canguilhem, Georges. 1991. *The Normal and the Pathological*. New York: Zone Books.

Christensen, Michael Stenskjær. 2015. Simon of Faversham "Quaestiones super De motu animalium". A Partial Edition and Doctrinal Study. *Cahiers de l'Institut du Moyen-Âge Grec et Latin* 84: 93–128.

Daniels, Norman. 2000. Normal Functioning and the Treatment Enhancement Distinction. *Cambridge Quarterly of Healthcare Ethics* 9: 309–322.

De Asúa, Miguel. 1997. Pedro de Abano y la controversia entre médicos y filósofos. *Patristica et Mediaevalia* 18: 49–66.

De Leemans, Pieter. 2010. Parva naturalia, Commentaries on Aristotle's. In *Encyclopaedia of Medieval Philosophy between 500 and 1500*, ed. Henrik Lagerlund, 917–918. Dordrecht/New York: Springer.

———. 2000. Medieval Latin Commentaries on Aristotle's "De motu animalium". *Recherches de théologie et philosophie médiévales* 67 (2): 272–360.

[36] See Levinas (1961).

Del Soldato, Eva. 2020. *Early Modern Aristotle*. Philadelphia: University of Pennsylvania Press.
Dunne, Michael. 2003. Thirteenth and Fourteenth-Century Commentaries on the "De Longitudine et Brevitate Vitae". *Early Science and Medicine* 8 (4): 320–336.
Ebbesen, Sten. 2013. Simon of Faversham *Quaestiones super librum De somno et vigilia*: An Edition. *Cahiers de l'Institut du Moyen-Âge Grec et Latin* 82: 90–145.
———. 2014. Geoffrey of Aspall *Quaestiones super librum De somno et vigilia*: An Edition. *Cahiers de l'Institut du Moyen-Âge Grec et Latin* 83: 257–341.
———. 2015. Anonymous Orielensis 33 on *De memoria*: An Edition. *Cahiers de l'Institut du Moyen-Âge Grec et Latin* 85: 128–161.
Ebbesen, Sten, Christina Thomsen Thörnqvist, and Véronique Decaix. 2015. Questions on "De sensu et sensato", "De memoria et reminiscentia" and "De somno et vigilia". A Catalogue. *Bulletin de philosophie médiévale* 57: 59–115.
Engelhardt, Tristram H., Jr. 1996. *The Foundations of Bioethics*. Oxford: Oxford University Press.
Esteban-Segura, Laura. 2012. System of Physics (GUL MS Hunter 509, ff. 1r–167v). In *A Compendium of Mediaeval Medicine Including the Middle English Gilbertus Anglicus*. Bern–Berlin–Bruxelles–Frankfurt–New York–Oxford–Wien: Peter Lang.
Flint, Thomas, and Michael Rea. 2009. *The Oxford Handbook of Philosophical Theology*. Oxford: Oxford University Press.
Freidson, Eliot. 1970. *Profession of Medicine: A Study of the Sociology of Applied Knowledge*. New York: Harper and Row.
Galle, Griet. 2008. The Dating and Earliest Reception of the *Translatio vetus* of Aristotle's "De sensu". *Medioevo* 33: 7–96.
Hasse, Dag Nicolaus. 2000. *Avicenna's "De anima" in the Latin West. The Formation of a Peripatetic Philosophy of the Soul 1160–1300*. London/Turin: The Warburg Institute.
Hoenen, Maarten J.F.M. 2011. Nominalism in Cologne: The Student Notebook of the Dominican Servatius Fanckel. In *Crossing Boundaries at Medieval Universities*, ed. Spencer E. Young, 85–144. Leiden/Boston: Brill.
Kitchel, Mary Joan. 1977. Walter Burley's Doctrine of the Soul: Another View. *Mediaeval Studies* 39: 387–401.
———. 1971. The "De potentiis animae" of Walter Burley. *Mediaeval Studies* 33: 85–113.
Kurdziałek, Marian. 1950. *Gilbert Anglik i dygresje psychologiczne w jego "Compendium medicinae"*. Dissertation. Lublin: Katolicki Uniwersytet Lubelski.
Lawn, Brian. 1993. *The Rise and Decline of the Scholastic "Quaestio disputata". With Special Emphasis on its Use in the Teaching of Medicine and Science*. Leiden/New York/Köln: Brill.
Levinas, Emmanuel. 1961. *Totalité et infini: essais sur l'extériorité*. The Hague: Martinus Nijhoff.
Lohr, Charles H. 2013. *Latin Aristotle Commenaries I.1. Medieval Authors A–L*. Firenze: Sismel Edizioni del Galluzzo.
Marone, Stephen P. 1983. *William of Auvergne and Robert Grosseteste: New Ideas of Truth in Early Thirteenth Century*. Princeton: Princeton University Press.
McCord Adams, Marilyn, and Cecilia Trifogli. 2012. Whose Thought Is It? The Soul and the Subject of Action in Some Thirteen and Fourteenth Century Aristotelians. *Philosophy and Phenomenological Research* 85 (3): 624–647.
McVaugh, Michael. 2010. Who was Gilbert the Englishman? In *The Study of Medieval Manuscripts of England: Festschrift in Honor of Richard W. Pfaff*, ed. George Hardin Brown and Linda Ehrsam Voigts, 295–324. Turnhout: Brepols.
———. 2019. Galen in the Medieval Universities, 1200–1400. In *Brill's Companion to the Reception of Galen*, ed. Petros Bouras-Vallianatos and Barbara Zipser, 381–392. Leiden/Boston: Brill.
Miteva, Evelina. 2018. "Iam ergo patet veritas eius quod dixit Aristoteles, et causa deceptionis Galieni". Philosophers vs. Medics in Albertus Magnus' Account on Conception. In *Irrtum. Error. Erreur*, ed. Andreas Speer and Maxime Mauriège, 107–122. Berlin/Boston: Walter de Gruyter.

O'Boyle, Cornelius. 1998. *The Art of Medicine: Medical Teaching at the University of Paris, 1250–1400*. Leiden/Boston/Köln: Brill.

Omont, Henri. 1916. Recherches sur la Bibliothèque de l'Église cathédrale de Beauvais. *Mémoires de l'Académie des Inscriptions et Belles-Lettres* 40: 1–93.

Pavlíček, Ota. 2018. Parisian and Prague "Quodlibeta" Compared: The Transfer of the Quodlibetal Disputation between the Faculties and Jerome of Prague's Struggle against the Thematic Limitations Imposed on the Faculty of Arts. In *What is New in the New Universities? Learning in Central Europe in the Late Middle Ages (1348–1500)*, ed. Elżbieta Jung, 325–356. Warszawa: Wydawnictwo IFiS PAN.

Shields, Aemilia Ledyard, and Henricus Blumberg. 1949. Praefatio. In *Averrois Cordubensis Compendia librorum qui Parva naturalia vocantur*, ed. Aemilia Ledyard Shields and Henricus Blumberg, vii–viii. Cambridge, MA: Mediaeval Academy of America.

Solère, Jean-Luc. 2018. James of Viterbo's Innatist Theory of Cognition. In *A Companion to James of Viterbo*, ed. Antoine Côté and Martin Pickavé, 168–217. Leiden: Brill.

Tellkamp, Jörg A. 2016. Aping the Logic? Albert the Great on Animal Mind and Action. In *Subjectivity and Selfhood in Medieval and Early Modern Philosophy, Studies in the History of Philosophy of Mind*, ed. Jari Kaukua and Tomas Ekenberg, vol. 16, 109–123. Cham: Springer.

Wöllmer, Gilla. 2013. Albert the Great and His Botany. In *A Companion to Albert the Great. Theology, Philosophy, and the Sciences*, ed. Irven M. Resnick, 221–268. Leiden/Boston: Brill.

Chapter 2
Physiology of Taste and Intentionality in John Blund's *Tractatus De Anima*

Riccardo Fedriga

Abstract In the beginning of the thirteenth century, the circulation of Aristotle's natural philosophy books deeply changed the ways in which the relationships between soul and body, mind, as well as the physical world were perceived. The discussions concerning cognition focused on how man, characterized by bodily nature, can engage in the cognitive process until the stage in which it becomes immaterial. Blund firmly claims the right of natural philosophy (and Masters of Arts) to deal with the nature of the soul, from both physical and metaphysical points of view: in order to clarify its accidental relationship to the body regarding the former; in order to elucidate its aspect as a separate, individual substance with respect to the latter. At the same time, Blund assigns to theologians the task of delving into the soul's supernatural ends. His analysis proves to be highly original in its description of the relationships and boundaries between soul and body. This is especially true of Blund's analysis of the two apparently secondary senses, touch and taste, both involved in the physiology and perception of flavours. In his analysis of the senses, Blund distinguishes: (i) external bodies; (ii) the production of sensitive images; (iii) their reception and transmission; and (iv) the perception of such bodily modification. In particular, this last point depends on the intentional activity of the soul and, mostly, it is at this level that the boundary is drawn between an externalist ontology of things and sense organs and an internalist ontology self-based on the intentional status of soul.

Keywords Intentionality · Philosophy of perception · Subject · Physiology of taste · John Blund · De Anima

---

R. Fedriga (✉)
Università di Bologna, Bologna, Italy
e-mail: riccardo.fedriga@unibo.it

M. Gensler et al. (eds.), *The Embodied Soul*, Historical-Analytical Studies on Nature, Mind and Action 11, https://doi.org/10.1007/978-3-030-99453-2_2

2.1 On Blund's *De anima*

John Blund's *Tractatus de anima*[1] is the earliest medieval commentary we possess on one of Aristotle's natural philosophy books. It dates back to the early thirteenth century (ca. 1204) and Blund was one of the first *magistri* to give lectures about *novus Aristoteles* in both Oxford and Paris in the first years of the thirteenth century.[2] His knowledge of Aristotle's *De anima* was based on James of Venice's Latin translation, dating back to the mid-twelfth century.[3] Although Blund's thought was largely independent, his reflection about the soul, the subject and the body, was greatly influenced by many other sources: primarily by the *Shifā'* (and particularly by the Kitab-al-Nafs, known to the Latins as *Liber sextus de naturalibus*),[4] then also by al-Ĝhazālī, pseudo-Alexander of Aphrodisias and Costa ben Luca (but he rejects Avicebron's theory of hylomorphism); among the Latin sources we can find Augustine, Boethius, Anselm of Canterbury and, to a lesser degree, Peter Lombard's *Sentences*.[5]

Blund firmly claims the right of natural philosophers (and of masters of Arts) to deal with the nature of the soul, from both a physical and a metaphysical point of view: the former in order to clarify its accidental relationship to the body; the latter in order to elucidate its aspect as a separate, individual substance. At the same time, Blund assigns to theologians only the task of delving into the soul's supernatural ends: the necessity to practise a scientific working method leads Blund to ignore the theological issue. Besides Avicenna, Aristotle's *De anima* is obviously the main

[1] Edited by D.A. Callus and R.W. Hunt and published for the British Academy by Oxford U.P. in 1970. For further observations on this edition, see below, n. 9.

[2] Born in England around 1175, John Blund probably studied in Oxford in the final years of the twelfth century, later completing his studies in Paris, where he remained until 1230. He was Master of Arts in Oxford and Paris when Aristotle's thought was first starting to spread. After 1230, Blund returned to England, where he held prestigious positions, eventually becoming Archbishop of Canterbury in 1232. Cf. McEvoy (1998, 563). Cf. Gutas (1999). See also Knuuttila (2018).

[3] See Dunne (2013, xi).

[4] For Avicenna, sensible experience is a *condicio necessaria* of human knowledge, but not sufficient. Sensible images (*phantasmata*) do not generate concepts in the intellect, but only dispose it to receive the influence of the separate agent intellect, which has an illuminating function on our souls. Cf. Avicenna, *Kitāb al-najāt*, II.6.15 (trans. Rahman, 1952, 64–68). In this sense, sensible experience, which Avicenna studies referring to a medical paradigm of the physiological aspects of sense-data, as his theory of emotions shows, does not have a properly cognitive value; it is accidental and instrumental with respect to the substance of the individual subject, which does not need sensitive perception (contrary to Aristotle's "orthodox" theory). For Avicenna, the essence of man is not the body, but the human self, the individual and immaterial soul, capable of conceiving itself distinctly without the help of the body. This individual self, which for Avicenna is a sort of self-knowledge, is the essence. It does not presuppose existence. It follows that a quiddity can exist externally and visibly, either as a universal concept in the imagination, or again, as something the existence of which is possible without there being any relationship with concrete reality or with the imagination. Cf. Werner (2005, 79), cf. also Bernardini (2002, 24 ff.); Callus (1943, 242); Minio-Paluello (1969, 462). See also Saccenti (2016).

[5] About Blund's sources, see Van Steenberghen (1991, 160); Burnett (1994, 45); Werner (2005, 17–31); Hasse (2000, 18 ff.); Corrado (2013, 21); Dunne (2013, x–xiii).

philosophical source for him, and he comments on it extensively.[6] Blund relies on Aristotle's natural philosophy when dealing with the nature of the soul (in particular to clarify the relationship with the body); however, when it comes to the immaterial, "internal" and supernatural ontological status of the soul, he believes that we must refer to metaphysics.

2.2 Body, Soul and the Senses

If we move from words to deeds, Blund's analysis reveals itself as highly original in its description of the relationships and boundaries between soul and body. This is especially true of Blund's analysis of two apparently secondary senses, touch and taste, both involved in the physiology and perception of flavours. In his analysis of sense perception, Blund distinguishes: i) external bodies; ii) the production of sensitive images; iii) their reception and transmission; and iv) the perception of such bodily modification. In particular, the last point depends on the intentional activity of the soul and, mostly, it is at this level that the boundary is drawn between i) an externalist ontology of things and sense organs and ii) an intentionality self-based on the status of soul.[7] Indeed, as we shall see, perception derives from a specific activity of the soul.

Only the soul is the true subject of any intentional act of perception. In other words, to become true knowledge, any perceptual content should be verified through the intentional activity of the subject. In this context, it is the will (*voluntas*) that functions as a truth-maker, since *voluntas* i) orients the soul towards one external object or another, and therefore ii) it focuses its attention (*intentio*) on specific contents, determining qualities (such as flavour in the case of tasting). The will is an intentional act that is rightly directed toward some content and fixes the attention (*intentio*) upon one content or another. In this sense, intentionality is very close to an act that selects physiological contents through attention. Finally, the Aristotelian category of quality draws the boundary between external and internal objects (like sensible images and concepts) abstracted by the senses and other faculties of the soul. The constant reference to Avicenna allows Blund to frame Aristotle's doctrines in strongly Neoplatonic terms, as it is immediately evident from the opening discussion on the definition of soul. Blund does not agree with the interpretation of the soul as the form of the body and he follows Aristotle's *De anima* translating correctly *entelecheia* with *perfectio* but failing to report that, according to Aristotle, *perfectio* is the "first *entelecheia*": in his *Treatise*, Blund wrote, in fact, only

[6] John Blund, *Tractatus de anima*, II, *Cuius artificis sit speculatio de anima*, § 2 (ed. Callus, Hunt, 1970, 7). For the relevance of Avicenna's theory of the soul in Blund, see Hasse (2000, 22).

[7] John Blund, *Tractatus de anima*, VIII, *De viribus animae sensibilis apprehensivis*, § 83 (ed. Callus, Hunt, 1970, 22): "Sequitur de viribus animae sensibilis apprehensivis. Primo dicendum est de vi apprehensiva deforis, ut de sensu. Primo igitur inquirendum est quid sit sensus et quae sunt species sensus."

perfectio and not *perfectio prima*. This elision allows for a semantic shift of *perfectio* from the Aristotelian cognitive paradigm to that of Avicenna, according to whom:

> We further say that each form is a perfection, but not every perfection is a form. It is agreed therefore that the essence of the soul is not the body, but that animal or vegetative part that is its form or its quasi form or its quasi perfection [...] The soul is the form of the organic body that has life in potency.[8]

2.3 What the Soul Is

Blund's use of the term "soul", thus, indicates a substance that is in a relation to a body insofar as the latter is animated by it. Blund combines Avicenna's doctrine with Aristotle's ontology starting precisely from this active and intentional role of the soul. For in his view, the role of the soul as *forma* (or *perfectio*) *corporis* does not define the soul in its essence, as it does for Aristotle, but it is only a predication *per accidens* or as *officium*, according to which the soul in general (and the sensitive soul in the case of perception) is the animating, intentional subject and life-giving principle which does not require for its awareness and subsistence any reference to the definition of being the essential form of the body.

> [...] From Aristotle we have that the soul exists, we must see what the soul is. From Aristotle we have that the soul is the perfection of an organized body potentially having life.
>
> However, it is objected. It is the form which gives existence, and matter in itself is incomplete, so all perfection is from form. But no form is a thing existing in itself separate from a substance. Therefore, since the soul is a form, the soul should not be said to be a thing existing in itself separate from a substance. Therefore, the soul cannot be separated from the body but it perishes with the body.
>
> With regard to this, it should be said that this word "soul" refers to its objects in a concrete manner. For it connotes a substance subsumed under a certain accident in relation to an organized body, inasmuch as the latter is animated and vivified by the soul, and thanks to this accident it is said to be the perfection of the body, namely because the soul animates the body.[9]

---

[8] John Blund, *Tractatus de anima*, I, *De essentia animae*, § 3 (ed. Callus, Hunt, 1970, 18–19): "Constat ergo quod essentia animae non est corpus, sed est pars animalis aut vegetabilis, quae est ei forma aut quasi forma aut quasi perfectio. [...] Dicimus quod omnis forma est perfectio, sed non omnis perfectio est forma [...] Anima est corporis organici vitam habentis in potentia [...]."

[9] The English translations of Blund's treatise are those by Michael Dunne, *Treatise on the Soul* (Dunne, 2013), with some modifications (particularly from n. 30 onwards). It is important to recall here the difficulties with the Latin text, which have been known since the review by Marvin L. Colker (1972), who evidenced the problems of the Callus-Hunt edition. This is why I have quoted Blund's chapter index (*Index Capitolorum*), name of chapter, number of paragraph and number of paragraphs of the Callus-Hunt edition. As Michael Dunne pointed out to me in private communication, the deficiencies of the existing Latin edition carried over into his translation, which he did not have an opportunity to correct. Ideally, both the Latin edition and the translation should be redone; in the meantime, we can but follow our best version to date, i.e., that of Dunne. I take the opportunity to thank Michael Dunne for his help on this issue.

Actual knowledge is internal to the soul and happens according to an Avicennian-style ontology, insofar as it proceeds from a progressive abstraction thanks to the operations of the faculties of the sensitive soul (*vis apprehensiva deforis*, *sensus communis*, *imaginatio*, *memoria* and *aestimatio*) and of the rational soul (*intellectus*). Blund then asks where perception comes from, i.e., whether the *apprehensio* of inputs is only singular or can be both singular and universal. He answers the question by asserting that a sense, which is commonly considered to engage with particulars only, can also grasp universals:

> Sense is the apprehensive power of a present thing insofar as the thing itself is present through the modifications coming from outside received in the bodily organ from outside. One may then ask what the sensation is of, that is, whether it has as its primary object the singular alone, or the singular and the universal. The fact that only the singular is sensed is attested by all authors who have dealt with the sense. But it seems, however, that it can be shown that the sense also apprehends universals.[10]

For the object known by the five senses is particular in relation to its material and peculiar characteristics, but presents in itself also the universal, even though perceived from sensible determinations. Blund can therefore conclude that also the universal, in a certain way, falls within the scope of sensual perception. For example, when we perceive a bright object, our senses grasp in its particularity and specificity (it can be large or small, more or less bright, it can also produce differently coloured light, etc.). But the universal "light" is already present in it, and this is why the universal, in a certain way, does fall within the scope of sensual perception.[11]

While sense grasps the particular properties of the particular object, the latter also contains the universal, which is present in the object beyond all its possible particular determinations; thus, the universal, being present ever since the intentional knowledge of that which is perceived, falls within the scope of the sense, even though a sense is unable to properly know it, since its capacity is limited to sensible knowledge, and it is the intellect that ultimately grasps it. Blund can thus conclude, referring to Boethius's commentary on Porphyry's *Isagoge*:

[10] John Blund, *Tractatus de anima*, VIII, *De viribus animae sensibilis apprehensivis*, § 84 (ed. Callus, Hunt, 1970, 23): "Sensus est vis apprehensiva rei praesentis in quantum ipsa est praesens per immutationes receptas in instrumento corporeo ab extrinseco advenientes. Potest autem in primis quaeri quorum sit sensus, utrum scilicet solum singulare habeat sensum apprehendendi vel et singulare et universale. Quod solum singulare sentiatur testantur omnes auctores de sensu loquentes. Sed quod sensus apprehendat universalia videtur posse ostendi." Cf. Aristoteles, *De anima*, II.5, 417b20–25: "The cause of this is that perception in actuality has singulars as its objects, while science has universals, and the latter are somehow present in the soul itself. Therefore, thinking depends on the subject, whenever he wills to do so, while perceiving does not depend on him, since the presence of sensibles is necessary." Cf. also Avicenna, *Liber de anima seu sextus de naturalibus*, pars I, cap. 2 (ed. Van Riet, 1972, 18–19).

[11] John Blund, *Tractatus de anima*, VIII, *De viribus animae sensibilis apprehensivis*, § 86 (ed. Callus, Hunt, 1970, 23): "Praeterea. Lucidum per se et proprie videtur. Sed lucidum, secundum quod lucidum, est universale. Ergo universale per se et proprie videtur."

> However, because Boethius says in his second commentary on Porphyry that the particular is the same as the universal, particular when sensed, universal when thought, it can be conceded that that which is universal is sensed but non the universal as such.[12]

2.4 Touch and Taste

To better clarify these theoretical assumptions, let us see how they apply to the differences and boundaries between touch and taste. I will show (1) that the mechanism that makes the quality of a taste manifested to the sense has an internal basis and is governed by the *intentio animae*; (2) that this happens through an ordered principle that Blund calls *virtus ordinata* and which is grounded in the neural netwok (*nervus expansus*); (3) that it is not the case that the sensory nerve passively receives the external changes produced by taste, but it is rather the soul that moves in order to actualize the gustative power and orient it toward its proper object; (4) that this mechanism hinges on an intentional act that should be conceived as a voluntary attention directing the sense's capacity (*virtus*) towards the right object; (5) finally, that this perceptual activity belongs to the individual soul, and only this subject can produce "true knowledge". The relationship between externalism and internalism thus appears as firmly grounded in a unique, organic and external reality and as: a) based on two ontologies, Aristotelian and Avicennian; b) oriented by the soul; and c) hierarchically ordered according to the Neoplatonic principle of the *perfectio animae*.

At the beginning of chapter 15 of his treatise, devoted to taste, Blund explicitly declares that he will be treating this topic in accordance with Avicenna's teachings, as we can see from comparing the two texts:

Blund	Avicenna
Having spoken of the sense of smell, now we should talk about/discuss the sense of taste. Taste is described by Avicenna in the following way: taste is a power located in a nerve, which is spread out, aimed at acquiring flavours *coming from the outside*.[13]	Taste is a power located in the nerve, which is spread out *on the top of the tongue*, aimed at acquiring flavours *released by the bodies that touch the taste receptor when they get mixed with the tongue's viscous humour, which gets changed by being mixed*.[14]

[12] John Blund, *Tractatus de anima*, VIII, *De viribus animae sensibilis apprehensivis*, § 87 (ed. Callus, Hunt, 1970, 23): "Sed quia dicit Boethius in secunda editione *Super Porphirium*, quod idem est singulare quod universale, singulare cum sentitur, universale cum cogitatur, potest concedi quod id quod est universale sentitur, non tamen universale." For Boethius, see *In Isagogen Porphyrii Commenta* (ed. Brandt, 1906, 167).

[13] John Blund, *Tractatus de anima*, XV, *De gustu*, § 208 (ed. Callus, Hunt, 1970, 56): "Dicto de olfactu dicendum est de gustu. Gustus autem sic describitur ab Avicenna. Gustus est vis ordinata in nervo expanso ad apprehendendum sapores *ab extrinseco advenientes*."

[14] Avicenna, *Liber de anima seu sextus de naturalibus*, pars I, cap. 5 (ed. Van Riet, 1972, 84—my translation): "Gustus [...] est vis ordinata in nervo expanso *super corpus linguae*, ad apprehendendum sapores *solutos ex corporibus contingentibus ipsum, cum permiscentur humori unctuoso linguae permixtione permutanti*." In this passage, Avicenna refers to the argument on saliva that Blund will resume at paragraph 216 of chapter XV, 116. According to Avicenna, since external objects touch the taste receptor, taste is a particular, inferior and mediated form of touch. For Aristotle's account, see *De anima*, II.10 (ed. Ross 1961, 422a17–19).

Taste is defined here as an arranged or ordered power (*virtus ordinata*), i.e., a power located in a nerve whose main function is to apprehend, that is, to perceive flavours intrinsic to external objects. After expounding this definition, which is more concise than Avicenna's,[15] Blund immediately presents a first objection to it:

> However, it is objected: a nerve which is spread out is a corporeal thing, but a corporeal thing does not perceive, for only the soul does. Therefore, the aforementioned description is false, since it is neither proper to the power of the nerve to perceive, nor even to the nerve itself.[16]

The objection points out the contradiction implied in asserting that something material and bodily, such as a nerve, can hold the power or capacity to apprehend something. For according to his sources, the true perceptive subject is the soul, not the body.[17]

Blund then continues his argument:

> Further, it seems that taste is a kind of touch, and, for this reason, there is in effect no taste except through contact between the thing which has the taste and the organ. But touch happens in this way; and so taste is touch. Perhaps, someone might say that taste is distinguished from touch inasmuch as taste has to grasp flavours, whereas touch has to grasp hot, cold, moist, dry, heavy, and light.[18]

In these lines, Blund presents the well-known problem of the distinction between taste and touch, a problem stressed by all authors who have dealt with the five senses: namely, that of distinguishing between the two senses which are both exercised by contact. Here, Blund refers to the classic Aristotelian account that admits taste to be a sort of touch, since it is exercised by contact, but also distinguishes it from touch by virtue of its having a different object. The objects of taste are flavours; the objects of touch are the qualities of hot, cold, dry, wet, hard and soft. This distinction too, however, does not seem to be satisfactory:

[15] On the nature of the text, i.e., *per modum quaestionis* (and its use from the beginning of the thirteenth century) or *disputatio* see Gardinali (1984, 385, n. 25); Weijers (1995, 40, n. 38); Hasse (2000, 20); Dunne (2013, xiii).

[16] John Blund, *Tractatus de anima*, XV, *De gustu*, § 209 (ed. Callus, Hunt, 1970, 56): "Sed obicitur. Nervus expansus est res corporea; sed rei corporeae non est apprehendere, sed solius animae; ergo falsa est superior descriptio, cum nec virtutis ipsius nervi sit apprehendere, nec etiam ipsius nervi."

[17] For Avicenna, according to the Neoplatonic tradition, not only *apprehensio*, but also *sentire est natura animae*. Cf. Avicenna, *Liber de anima seu sextus de naturalibus*, pars II, cap. 3 (ed. Van Riet, 1972, 131). The soul's typical nature is that of perceiving, even though it does not operate directly on the external object. For Aristotle, a perception was a "passion" suffered by the soul; for Blund, it was first recognized and then apprehended by the soul; for Avicenna, in turn, the perceptive capacity is the soul's direct prerogative.

[18] John Blund, *Tractatus de anima*, XV, *De gustu*, § 210 (ed. Callus, Hunt, 1970, 56): "Item. Videtur quod gustus sit species tactus, et hac ratione. Non est gustus in effectu nisi per contactum rei habentis saporem et ipsius instrumenti. Sed ita fit tactus; et ita gustus est tactus. Forte dicet [aliquis] quod gustus distinguitur a tactu per hoc quod gustus habet apprehendere sapores, tactus habet apprehendere calidum, frigidum, humidum, siccum, grave, leve."

> And again, against this: let touch insofar as it is perceptive of hot and cold be called A; and let touch insofar as it is perceptive of rough and smooth be called B. In this way, both A and B have being in act through touch, just like taste. Consequently, if you say that B is a separate kind of touch, for the same reason you should say that taste is a separate kind of touch; if not, why not? For just as taste is not one of the primary qualities, similarly, neither rough nor smooth nor viscous belong to the primary qualities; rather, they belong to the secondary qualities.[19]

Hot and cold, rough and smooth are two pairs of opposites both subject to touch. However, there is a substantial difference between the two: for hot and cold are primary qualities of elements, while rough and smooth are secondary qualities.[20] But, Blunt affirms, rough and smooth are like flavours then, for they too are secondary, and not primary, qualities. The conclusion necessarily follows: if, whether rightly or wrongly, one admits that the pair of secondary qualities rough and smooth are the objects of a sort of touch, then one should also admit that taste too is a sort of touch, given that it too has secondary qualities as its objects.

> Moreover, the organ of touch is made up of a temperate complexion. Thus, since no animal is of such a temperate complexion as a human being, touch does not thrive in any animal as much as it does in a human being. Thus, since touch exists according to the complexion received from the primary qualities, touch does not leave an animal. Again touch is perceptive of qualities which are from the primary qualities, namely, hot, cold, dry, moist, and it happens in accordance with a temperate complexion. Moreover, touch is in a human being in order to keep the subject in existence, since we perceive both harmful and beneficial things through it. However, since all properties are primary qualities or come from them and since touch is a power located in a nerve in accordance with a temperate complexion, the organ of touch should equally receive an impression from every quality coming from the primary qualities. Therefore, the perception of taste should also happen according to touch, since taste is either beneficial or harmful and it comes from the primary qualities.[21]

[19] John Blund, *Tractatus de anima*, XV, *De gustu*, § 211 (ed. Callus, Hunt, 1970, 56): "Sed contra. Appelletur A tactus in quantum ipse est apprehensivus calidi et frigidi, et appelletur tactus B in quantum ipse est apprehensivus asperi et levis; secundum hoc tam A quam B habet esse in actu per tactum sicut gustus. Qua ratione ergo dicis B esse tactum particularem, eadem ratione debes dicere gustum esse tactum particularem; vel quare non. Sicut enim sapor non est una de primis proprietatibus, similiter nec est asperitas, nec levitas, vel viscositas est de primis proprietatibus, immo de secundis."

[20] For a fourteenth century point of view of Burley, see Mansfield, Chap. 12 (footnote 11) in this volume, pp. 229–251.

[21] John Blund, *Tractatus de anima*, XV, *De gustu*, § 211 (ed. Callus, Hunt, 1970, 56–57): "Item. Instrumentum tactus est habens se in temperata complexione; unde cum nullum animal sit adeo temperatae complexionis ut homo, in nullo animali viget tactus tantum ut in homine. Unde cum tactus sit secundum complexionem receptam a primis elementis non deserit tactus animal; et est tactus apprehensivus qualitatum quae sunt ab elementis primis, scilicet calidum, frigidum, siccum, humidum; et contactus est secundum complexionem temperatam. Et tactus est in homine ad conservandum subiectum in esse, quia secundum eum percipimus nociva et apprehendimus expedientia. Sed cum omnis proprietas sit vel prima vel fluens a prima, cum tactus sit vis ordinata in nervo secundum temperatam complexionem, debet illud instrumentum tactus aeque recipere immutationem ab omni qualitate fluente a primis qualitatibus; debet ergo secundum tactum fieri apprehensio saporis, cum sapor sit expediens vel nocivus et fluens a primis qualitatibus."

The argument is not unlike that put forth by Avicenna, who considers touch to be the most important one among the five senses, since it distinguishes what is beneficial from what is harmful and it is thus indispensable to survival and, therefore, present in all animals.[22]

Following Aristotle, Avicenna seems on the one hand to assimilate taste to a specific, mediated form of touch, operating through the medium of saliva; on the other hand, he stresses how taste, while certainly useful in distinguishing edible from non-edible foods, is not necessary to survival. For, even when it is lacking (for example, when the animal is sick), the other senses can cooperate to replace its function.[23] Let us reformulate the argument as follows: a flavour is either something beneficial (namely, it belongs to a food that is nourishing insofar as it is suited to the bodily nature of a certain individual) or something harmful (namely, it belongs to something that does not satisfy the body's needs); therefore, the perception of a flavour must happen according to touch, for touch, like taste, discerns beneficial things from harmful things. Again, given that each quality is either primary or derived from it, it follows that taste, too, indirectly derives from primary qualities. Moreover, hot and cold, dry and wet, are objects of touch, which is in accordance (*secundum*) with a temperate complexion. Since man's constitution is very temperate, his sense of touch is extremely well developed.

2.5 How Flavour is Manifested to Taste

Having clarified the nature of the relationship between touch and taste, Blund asks then what the medium through which a flavour is manifested to the sense of taste is. Just like air is the medium through which sounds and smells reach the senses of hearing and of smell respectively and just like the presence of light is the necessary condition for seeing colours, through which medium does the flavour of a food or a

[22] Avicenna, *Liber de anima seu sextus de naturalibus*, pars II, cap. 3 (ed. Van Riet, 1972, 130–131): "Primus sensuum propter quos animal est animal est tactus. Sicut enim omne quod habet terrenam animam habet virtutem nutritivam et possibile est ut non habeat aliquam aliarum, sed non convertitur, sic omne quod habet animam sensibilem habet sensum tangendi et possibile est ut non habeat aliquem aliorum, sed non convertitur. Dispositio etenim nutritivae comparatione aliarum virtutum animae terrenae est qualis dispositio tactus comparatione aliarum virtutum animae sensibilis: hoc est quod prima compositio animalis est ex qualitatibus tactibilibus."

[23] Avicenna, *Liber de anima seu sextus de naturalibus*, pars II, cap. 3 (ed. Van Riet, 1972, 131–132): "Gustus autem quamvis inter gustata indicet id propter quod permanet vita, tamen possibile est ut, destructo gustu, animal remaneat animal. Alii enim sensus fortassis cooperantur ad inquirendum nutrimentum conveniens et ad discedendum a nocivo. Alii etiam sensus non iuvant ad sciendum quod aer circumdans corpus, exempli gratia, est adurens vel congelans; et quod omnino fames est desiderium sicci et calidi, et sitis est desiderium frigidi et humidi, nutrimentum enim certum est quod est huiusmodi qualitatum quas apprehendit tactus; gustus etiam delectatio est; unde saepe destruitur sensus gustus cum infirmatur et remanet animal. Ergo tactus est primus sensuum qui est necessarius omni animali terreno."

drink reach the sense of taste? And again, just like sound is the result of two sounding objects hitting one another, what is it that makes flavour manifest to taste?[24] Having outlined the problems regarding the definition of taste, which mostly concern the difficulty in differentiating it from touch and the search for an explanation of the mechanism that makes a flavour manifest to taste, Blund now responds in detail to each point raised. We thus reach the solution to the problem:

> Solution: It should be stated regarding the first objection that the definition of the sense is: taste is a power in a nerve which is spread out, so that, when an impression in the organ of taste is received by means of flavour, it is the will which turns the eye of the mind towards that modification so it apprehends the taste coming from the outside joining the passion of the sense of taste with the flavour from the outside. Nor is it required according to the meaning of the definition that either the nerve or the power of the nerve perceives, but rather the eye of the mind.[25]

Blund picks up here the first definition of taste as an ordered power (*virtus ordinata*) of the spread out (*expansus*) nerve, adding that this is a very particular nerve, able to receive the changes produced by flavours in the tasting sensory apparatus.

2.6 Voluntary Attention

It is then up to the soul, or better, to its attention, to decide whether to longer and reflect upon that change, thus apprehending the qualities of foods such as flavours. Without the soul's attention, therefore, no true perceptual content, such as perception, is possible at all. In Blund's view, voluntary attention truly connects mind and body through a perceptual (and not physiological) intentionality directed from the inner subject to the objects' qualities (which we could read as "affordances"). In fact, Blund repeats the idea that the soul is the true perceptive subject in the last lines of this passage, where he reaffirms that, although capable of receiving the changes flavour produces in its organ, the sensory nerve is not that which apprehends. It is the soul, sustained by will, which actualizes the tasting faculty, which, until the very moment it gets into contact with its proper object, is only potentially

[24] John Blund, *Tractatus de anima*, XV, *De gustu*, § 213 (ed. Callus, Hunt, 1970, 57): "Item. Ubi est auditus in effectu? Est ad ipsum aer sonum deferens. Similiter est aer medium deferens odorem ad odoratum. Quaeritur ergo quid sit medium deferens saporem ad gustum. Similiter quare ad hoc quod sit visus in effectu oportet quod ibi sit lux detegens et manifestans colorem visui. Similiter ad hoc quod sit auditus praeexigitur collisio corporum quod sonora sint, [et] per illam collisionem detegitur et manifestatur eorum sonoritas auditui. Quaeritur ergo quid sit consimili modo detegens et manifestans saporem gustui?"

[25] John Blund, *Tractatus de anima*, XV, *De gustu*, § 214 (ed. Callus, Hunt, 1970, 57): "Solutio. Dicendum est ad primum quod sensus descriptionis est: gustus est vis ordinata in nervo expanso, ut recepta immutatione in instrumentum gustus per saporem, convertens sit voluntas aciem animi ad illam immutationem, et apprehendat anima saporem extra copulando passionem quae est in sensu gustus cum sapore extra. Nec exigitur secundum sententiam descriptionis quod vel nervus vel vis nervi apprehendat, sed acies animi."

suited to receive flavours. In the *Solutio*, Blund turns once again to Avicenna's point, when he argues that the function of saliva is actually what prevents a straightforward identification between touch and taste:

> Regarding the next objection, it should be said that taste is not a kind of touch, because taste is a power located in an organ which is of a watery nature. Thus, since taste has its existence properly in a watery substance, it is thanks to the disposition of this organ of taste that it receives impressions from flavour. For it is of the nature of flavour to be in such a subject, since this subject is watery, namely, when a watery nature is predominant. Thus, it happens that the tongue itself is spongy because of a watery nature, as is clear in sponges and foam that have fine parts, because they are of a watery nature and are mixed together with parts of air. Therefore, it happens that the tongue is an organ that is flexible and mobile because it is of a watery nature, which is fluid and has airy parts mixed with it. Touch, however, is a power that derives from a temperate complexion and, thus, it is of a different species than taste, since taste is a power arising mostly according to a watery nature. Moreover, it is stated by many writers that touch is a power placed in a nerve which is of an earthy nature.[26]

This passage is of huge importance for what concerns not only the boundaries between senses, but also those between mind and body.[27] In the first part of his exposition, we recall, Blund observed that taste is generally considered as a sort of touch, given that it is exercised by contact: we do not have the experience of tasting something, unless a body endowed with taste comes into contact with our mouth, just like we cannot have the sensation of touch unless an object comes into contact with our body. By contrast, the passage under consideration presents a radically different perspective, as the starting sentence makes clear: taste, it is asserted, is *not* a kind of touch. In support of this thesis, Blund offers some accurate observations, based on experience and, broadly, on the specific nature of the two organs: the tasting organ has a watery nature (even if *esse in subiecto*), while most people agree that touch is earthy. For flavour is generated by watery substances and the tasting organ is structured so as to be wet, in such a way that even a food whose nature is dry may release flavour particles once it is introduced in the oral cavity and wetted. The tongue, here defined as spongy, i.e., able to attract water, is a flexible and mobile organ whose function is to mix or, more precisely, 'knead' (*intermiscere*) foods so that they reveal a flavour.[28]

---

[26] John Blund, *Tractatus de anima*, XV, *De gustu*, § 215 (ed. Callus, Hunt, 1970, 57–58): "Ad aliud. Dicendum quod gustus non est species tactus. Quoniam gustus est vis ordinata in instrumento quod est aquosae naturae. Unde, cum sapor habeat esse proprie in aquosa substantia, de aptitudine ipsius instrumenti gustus est ut ipsum recipiat impressiones a sapore. Saporis enim est esse in subiecto tali cum ipsum subiectum sit aquosum, ubi, scilicet aquea natura sit praedominans; et ideo contingit quod ipsa lingua est spongiosa propter aquosam naturam, ut patens est in spongia et spuma aquae quae raras habent partes, quia aquosae sunt naturae et intermiscentur partes aereae. Unde contingit quod lingua est membrum flexibile et mobile, quia ipsa est aquosae naturae quae fluida est et habet partes aereas sibi admixtas. Tactus autem est vis proveniens a temperata complexione; unde alterius speciei est quam gustus: cum sit gustus vis surgens maxime secundum aquosam naturam. Dicitur tamen a pluribus auctoribus quod tactus est vis ordinata in nervo terreae naturae."

[27] For a fourteenth century master of Arts like Buridan, see Beneduce, Chap. 13 (footnote 56) in this volume, pp. 253–275.

[28] John Blund, *Tractatus de anima*, XV, *De gustu*, § 216 (ed. Callus, Hunt, 1970, 58).

While touch functions by direct contact with an external body, taste needs a medium like saliva to carry flavour to the expanded nerve; therefore, touch and taste are two different senses. According to the physiology expounded here, the tongue is a sort of watery (and therefore mobile and flexible) sponge, which absorbs a liquid, i.e., the salivary humour. Thanks to this humour, the sensory organ perceives the flavour that is "dissolved" in the salivary liquid. Moreover, touch functions in a dry and earthy environment, through a temperate complexion, while taste is activated only in a watery environment.

Aristotle	Avicenna	Blund
Taste has nothing then, to act as a medium; although as the object of sight is colour, so the object of taste is flavour. But nothing can produce a perception of flavour without liquid; it must possess wetness actually or potentially, like salt, which is both soluble itself and solvent of the tongue.[29]	Touch itself does not transmit flavour in the same way as touch does transmit heat, when it comes into contact with something hot; but taste requires a sort of medium that receives flavour and exists without having flavour itself: this medium is the salivary humour that comes from the salivary organs.[30]	It should be said that the medium which carries flavour is saliva; and for that reason Aristotle says in the book *On the Soul* that without saliva there is no taste in act. The detection of flavour is through the contact of a flavour-bearing body with humidity.[31]

Just as air is the medium through which sounds and smells reach their respective sense organs, so saliva is the medium that conveys flavours to taste. Saliva, assisted by the tongue, without which foods may not be kneaded, has the function of releasing watery particles. The latter, Blund asserts, melt flavoursome bodies, allowing the qualities proper to foods and drinks, such as flavours, to manifest themselves.

2.7 Perception, Qualities and Sense Data

Knowledge of a sensible object is obtained through the five senses and an elaborate nervous system. In truth, each nerve ramification conveys only partial external perceptions: hearing conveys the sound of an object, sight conveys vision, taste conveys flavour, and so on.[32]

[29] Aristoteles, *De anima*, II.10 (ed. Henderson, 422a17–19, trans. Hett, 1936, 125).

[30] Avicenna, *Liber de anima seu sextus de naturalibus*, pars II, cap. 4 (ed. Van Riet, 1972, 143—my translation): "[…] ipse tactus non reddit saporem sicut tactus calidi reddit calorem, sed quasi eget medio quod recipiat saporem et sit in se non habens saporem: quod est humor salivae quod provenit ex instrumentis salivaribus."

[31] John Blund, *Tractatus de anima*, XV, *De gustu*, § 216 (ed. Callus, Hunt, 1970, 58): "Dicendum quod medium deferens saporem gustui est saliva; et ideo dicit Aristoteles in libro *De anima* quod sine saliva non est sapor actu. Detectio saporis est resolutio corporis saporiferi per subintrationem humiditatis."

[32] John Blund, *Tractatus de anima*, XVII, *De sensu communi*, § 233 (ed. Callus, Hunt, 1970, 62).

Yet we know sensible objects globally, in their perceptible wholeness. This is the work of the *sensus communis*, which oversees the completion and composition (*compositio*) of external perceptions.[33] Blund's account of *sensus communis*, expounded in chapter 17 of his *Tractatus de anima*, mediates between Aristotle's and Avicenna's. Aristotle maintains that the so-called common sense unifies the five senses, even though it is not a further sense; Avicenna, on the other hand, envisages common sense as a further internal sense, to which all sensible data are conveyed by the other five senses. Blund makes a distinction at both lexical and conceptual level, thus managing to distinguish between *sensus communis* and *imaginatio*. Every time we perceive something, after the act of the apprehensive faculty, *sensus communis* intervenes, by unifying all the partial and peripheral perceptions in common sensibles. The common sense, however, is capable of *receptio*, but not of *retentio* of the sensible, exactly as a pool of water, Blund observes, can receive a figure drawn on its surface but cannot retain it.[34] Therefore, if there are two kinds of senses (i.e., external and internal) and if the external ones only make an imperfect abstraction, tied to sensible accidents, while the internal ones abstract a whole and perfect image thanks to the soul's activity, it follows that *sensus communis* increases the degree of abstraction in the cognitive process.

Having framed the problem, Blund provides a definition of the imaginative faculty, specifying the relationship between the imagination and the common sense.

> The imagination is a power located in the far end of the anterior parts of the brain, that is, in the first concavity, since it retains the impressions received from the common sense. Therefore, since the imaginative faculty is able to retain the sensations received from the common sense, which are images of individual things, the apprehension of absent things can be carried out through the imagination; by contrast, since the common sense cannot, by its own propensity, retain images of the individual objects received from the five external senses, the apprehension of absent objects cannot happen through the common sense, but only the apprehension of present ones.[35]

---

[33] John Blund, *Tractatus de anima*, XVII, *De sensu communi*, § 237 (ed. Callus, Hunt, 1970, 63): "Ab Avicenna habetur in commento *De anima* quod sensus communis est vis ordinata in anteriori parte cerebri recipiens impressiones fluentes in ipsam a quinque sensibus exterioribus." See also § 239 (ed. Callus, Hunt, 1970, 63): "Ab Aristotele habetur in libro *De anima*, quod motus, et numerus, et magnitudo non sunt sensata alicuius sensus exterioris, immo sunt sensata sensus interioris, qui est sensus communis."

[34] John Blund, *Tractatus de anima*, XVII, *De sensu communi*, § 246 (ed. Callus, Hunt, 1970, 66): "Ut habetur ab Avicenna commento, alia est vis secundum quam est species impressionis receptio, et alia est vis secundum quam est impressionis receptae retentio, ut patet in aqua est vis receptiva impressionis, sed non vis impressionis receptae retentiva." As Blund himself recognizes, the reference to Avicenna is explicit (cf. Avicenna, *Liber de anima seu sextus de naturalibus*, pars I, cap. 5).

[35] John Blund, *Tractatus de anima*, XVIII, *Quid sit imaginatio, et quid sit ipsius differentia a memoria*, § 252 (ed. Callus, Hunt, 1970, 67): "Imaginatio est vis ordinata in anterioris partis cerebri extremitate, scilicet in prima concavitate, retinens impressiones receptas a sensu communi. Unde quia vis imaginativa retentiva est passionum receptarum a sensu communi, quae sunt similitudines rerum singularium, per imaginationem rerum absentium potest fieri apprehensio; sensus autem communis, cum de sua aptitudine non possit retinere imagines singularium receptas a quinque sensibus exterioribus, non potest per eum fieri apprehensio rerum absentium, immo praesentium tantum." Cf. Avicenna, *Liber de anima seu sextus de naturalibus*, pars I, cap. 5.

Blund thus demonstrates that, at this level, *apprehensio* is only possible thanks to *imaginatio*, which is capable of retaining images that are not immediately present to the senses. To the objection that, under this interpretation, the imagination and the memory are the same, since they perform the same functions, Blund responds as follows:

> It must be said that, among the apprehensive powers, some apprehend and operate at the same time, as Avicenna says in his commentary on the *De anima*, some apprehend and do not operate at the same time. The powers that apprehend and operate at the same time are those that apprehend things and mutually compose them or analyse them, for example the estimation and the memory. Through the memory, indeed, both apprehension and recollection of things take place, since these same things were previously apprehended. So, in memorization itself there is composition. Through the imagination, on the other hand, we only apprehend the thing whose image is inscribed in the object of that faculty, but apprehension through the imagination does not take place by combining or dividing some perception on some object. Consequently, the imagination only apprehends an object, without operating on it at the same time.[36]

If, then, someone were to further object that the memory brings about some kind of composition, the imagination would not be able to understand whether the image of a certain thing was already known through the senses, anyway; for only the memory allows us to connect a certain image to sensible perception:

> But if someone observes that imagination sees a thing that is not there and that was previously perceived, and in this way apprehends a thing seemingly by perceiving what it previously was in the sense; and therefore it operates simultaneously with its own apprehension, since the fact that the imagined thing was previously in the sense is actually a form of combination; if someone asserts so, it must be replied to this that although the imagination sees a thing because this thing was previously sensed, still through the imagination there is no perception of the fact that that thing was previously in the sense. For this process does not first occur in the imagination, but in the memory.[37]

[36] John Blund, *Tractatus de anima*, XVIII, *Quid sit imaginatio*, § 252 (ed. Callus, Hunt, 1970, 68): "Solutio. Dicendum est quod virium apprehendentium quaedam apprehendunt et operantur simul, ut habetur in commento *De anima* [Avicennae], quaedam apprehendunt et non operantur simul. Apprehendunt et operantur simul, ut vis aestimativa et vis memorialis, quae apprehendunt res et componunt eas ad invicem vel dividunt. Per memoriam enim fit apprehensio rerum et recordatio quod prius eaedem res fuerunt apprehensae; unde in ipsa recordatione est compositio; per vim autem imaginativam apprehenditur solummodo res cuius imago describitur in subiecto illius virtutis, sed non componendo vel dividendo aliquam passionem in aliquo subiecto fit per eam apprehensio. Unde vis imaginativa apprehendit simpliciter rem non simul operando."

[37] John Blund, *Tractatus de anima*, XVIII, *Quid sit imaginatio*, § 253 (ed. Callus, Hunt, 1970, 68): "Si quis autem obiciat de hoc quod imaginatio intuetur rem absentem et sensatam prius, et ita apprehendit rem ut videtur percipiendo quod ipsa fuit prius in sensu; ergo simul operatur cum sua apprehensione, quoniam rem imaginatam prius fuisse in sensu quaedam compositio est. Dicendum est ad hoc quod licet imaginatio intueatur rem secundum quod ipsa fuit prius sensata, non tamen est per ipsam perceptio quod res illa prius fuerit in sensu. Illud enim prius in imaginationem non cadit, sed in memoriam."

Since the memory lacks the active function of composition and division of images that the imagination possesses, the memory can only apprehend the common sensibles that *sensus communis* provides it with. Blund makes no reference to Avicenna's *vis imaginativa*. The Arabic philosopher had drawn a sharp distinction between *vis imaginativa* and *imaginatio*: the former actively combines and separates sensible data, the latter is an essentially passive faculty whose function is to store sensible data coming from the *sensus communis*. The imagination, that is, acquires non-material images, which, however, can also depend on material things as their model.

The imaginative faculty knows an existing object, therefore, not as matter (or as present in matter), but as the image of the object insofar as it is perceived. Even though it preserves its crucial role as a bridge between the sensible and the intelligible sphere, the imagination becomes functional to sensitivity, since the active function that the *vis imaginativa* plays in Avicenna is played by the memory and the estimation in Blund.

> Avicenna maintains that the estimation is a power located in the lobe of the brain, aimed at apprehending the non-perceptible inputs that exist in individual things and in the senses; it judges if a thing must be avoided due to its *intentio*, in case this *intentio* is harmful, or if it must be pursued due to its *intentio*, in case this *intentio* is advantageous: for example, the power that is in the sheep, judging by this power that the wolf is to be avoided, and the lamb, which is the offspring of the sheep itself, must be cared for.[38]

Avicenna had been the first to theorize on the faculty of estimation (in Arabic: *wahm* or *zann*). This is the power with which the sheep grasps that the wolf, as object of its fear, has gone away. The estimation is basically a judgement, an assessment of opportunity or danger, a calculation of advantage and disadvantage. It perceives elements such as pleasure and pain, and grasps positive or negative values in the single objects that have previously been perceived and then imagined. In the same way, the estimation understands the meaning and function of objects, and in this way allows abstraction to take a step further. In relation to the estimation, Blund provides a clever definition of *intentio*, basing himself on Avicenna.

> The Commentator [*scil.* Avicenna] calls *intentio* that individual quality that does not fall into the senses, and which is capable of either harming or benefitting a thing. Harming, like the property that belongs to the wolf, by virtue of which the sheep avoids the wolf; benefitting, like the property that belongs to the sheep, by virtue of which the lamb seeks her. [...] As it was said earlier, through the estimation sometimes composition occurs, some other times division. [...] Since through the estimation it is possible to decide what is true and

[38] John Blund, *Tractatus de anima*, XIX, *Quid sit aestimatio*, § 254 (ed. Callus, Hunt, 1970, 68–69): "Ab Avicenna habetur quod aestimatio est vis ordinata in media concavitate cerebri ad apprehendendum intentiones non sensatas quae sunt in rebus singularibus et sensibus, diiudicans utrum res sit fugienda propter intentionem si ipsa intentio sit nociva, vel appetenda propter intentionem si ipsa sit expediens: ut vis quae est in ove diiudicans quod ab hoc lupo est fugiendum, et quod huius agni, qui est agnus istius ovis, est miserendum."

> what is false, animals can discern true and false; they are therefore capable of disagreeing with one other.[39]

And he concludes:

> Animals, therefore, are not capable of apprehending universals. For the estimation is still directly connected to the apprehension of individual existing objects. Thanks to the estimation one does not apprehend that wolves in general are to be avoided, but one apprehends that *this* individual wolf, which is perceived or was perceived before, must be avoided.[40]

2.8 Perception and the Source of True Knowledge

Blund offers an Avicennian interpretation of the cognitive process; in accordance with this, we may try to draw some conclusions for what concerns taste. Once the sensory nerve, whose nature is bodily, receives the modification produced by the flavour of an external body, the first sensible image of that flavoursome body is created in the man. However, the *perception* of that image, which Blund considers to be the soul's specific activity, must be clearly distinguished from its physiological *reception*. For only the soul's will and the voluntary attention it turns to the object allow a man to grasp the truth of the water he is really drinking. In other words, it is through intentionality—understood both as a direction towards an object (*voluntas*) and as voluntary attention (*intentio*)—that the subject is aware of true knowledge. In this perspective, the sensible world with its objects and the sense organs are not sources of true knowledge, for they are deprived of intentionality. Strange as it may seem, we are not dealing here with two realities, but with one reality and two ontologies: an internal one that allows for recognising the activity of the self and is based on the activity of a bodiless subject; and an external one, according to which sensory organs are instruments of the soul. It may seem odd that a philosopher like Blund, whose anthropology is clearly based on the soul-body dualism, should hold that in the end reality is just one. But we have to consider that in a context where the

---

[39] John Blund, *Tractatus de anima*, XIX, *Quid sit aestimatio*, § 254 (ed. Callus, Hunt, 1970, 69–70): "Intentionem appellat Commentator qualitatem singularem non cadentem in sensum, que est vel rei nocitiva vel expediens. Nocitiva, ut illa proprietas que est in lupo propter quam ovis fugit lupum; expediens, ut illa proprietas quae est in ove propter quam eam appetit agnus. [...] Ut dictum est superius, per aestimationem tum fit compositio tum fit divisio. [...] Sic ergo cum per vim aestimativam possit perpendi quid sit verum et quid sit falsum, bruta animalia possunt discernere verum a falso; possunt ergo bruta animalia uti mutuo disputationibus."

[40] John Blund, *Tractatus de anima*, XIX, *Quid sit aestimatio*, § 253 (ed. Callus, Hunt, 1970, 70): "Dicendum est quod universalia non possunt apprehendi a brutis animalibus. Aestimatio autem non apprehendit nisi singularia; unde secundum vim aestimativam non apprehenditur quod a lupo sit fugiendum, sed apprehenditur quod ab *hoc* lupo sit fugiendum, qui est in sensu, vel prius fuit in sensu."

mind is not considered as a subject agent the fact that mind and soul are two separate substances does not imply that reality itself is split in two.[41] On the contrary, for Blund the ultimate unity of reality is the strongest epistemic guarantee of the accurateness, and indeed of the very possibility of every kind of knowledge. In conceiving of the bond between the soul and the body, Blund distinguishes between the physiological reception of the sensible image (proper to the body) and the moment in which that bodily modification is perceived. The latter happens only if a specific activity of the soul, the only actual perceptive subject, takes place. In Blund's philosophical and natural interpretation, the term "soul" means properly only an accidental relationship to the body, insofar as the body is enlivened and animated by it.[42] In a non-religious commentary such as the *Tractatus*, this distinction is not only technical but also assumes the special meaning of a methodological claim. In fact, theologians reject the thirteenth century's lexical and methodological changes brought about by the spread of Aristotle's philosophy and appeal to the well-known themes of the Augustinian tradition to counter the purely philosophical and natural consideration of the soul, from both the physiological and the empirical psychological standpoints.[43] But, as we have seen in analysing the "case study" of taste, this is a methodological trespassing of the field. And for a Master of the early thirteenth

---

[41] Cf. De Libera (2007, 39): "[…] le "sujet' aristotélicien" est devenu le sujet agent des modernes en devenant "suppôt" d'actes et d'opérations. Descartes n'ayant pas joué de rôle décisif en l'affaire, on ne s'etonnera pas non plus que le "sujet cartésien" soit, si non humilié, du moins subordonné à la figure leibnizienne du "suppôt d'actions", herité du Moyen Ages et de la Seconde Scolastique et qu'une attention particulière soit donnée aux principes scolastiques qui, fondant la transformation ci-devant sujet d'inhérence en sujet agissant, valent a nos yeux comme autant de règles de passage d'une formation discursive à une autre." It is an interpretation that is associated with that of Jonathan Barnes when he speaks of soul as "the capacity or ability something has for engaging in a certain activity in certain circumstances." Cf. Barnes (1971–1972, 113). For a general survey about a medieval notion of Subject of Thought, see Trifogli, Chap. 10 in this volume, pp. 191–211.

[42] Cf. here § 2.3 and footnote 8.

[43] See, for instance, the changes brought about by the Neoplatonic influence of Avicenna's thought (and not necessarily Augustinians, as E. Gilson assumes) and/or the medical tradition of Aristotelian commentators of the beginning of the thirteenth century, whose theories were supplied by the Latin translations of some philosophical and medical works: as remarked by Dunne (2012, xii) and Knuuttila (2018), the Arabic medical encyclopaedia of 'Alī ibn al-'Abbās al-Maǧūsī (in Latin Haly Abbas), which contained various remarks on emotions based on Galen's medical philosophy; the first partial translation by Constantine the African (c. 1080), called the *Pantegni*, and the complete translation by Stephen of Antioch (1127), the *Regalis* disposition; and mostly Avicenna's treatise on psychology called *Kitāb al-najāt*. See also Hasse (2000, 225): "Avicenna and his psychology dominated the structure and much of the content of psychological writings in the West for over half a century, form John Blund to Albert the Great and Peter Hispanus. Avicenna had developed a theory of the soul which combined Peripatetic philosophical argumentation with an elaborate system of faculties, based on a great deal of physiological material. Avicenna's *De Anima* offered the latest and best in philosophical subtlety and comprehensiveness in terms of scientific discoveries. It was for these reasons that he was preferred even to Aristotle by many writers of the period examined. A factor which contributed greatly to Avicenna's success was the compatibility of his theory with the teachings of the medical tradition, which had begun to influence even theological discussions of the soul. Aristotle's treatise could not compete with Avicenna's on the level of physiology."

century, who claims to be investigating scientifically the mind-body problem, this cannot but sound as an unfair encroachment and can only lead to one conclusion: that in their teaching, theologians should not deal with problems that are the responsibility of Masters of Arts.

References

Primary Literature

Aristoteles. 1936. *De anima*, ed. Henderson, with an English translation by Walter Stanley Hett. Cambridge, MA.: Harvard University Press.

Avicenna. 1952. *Avicenna's Psychology: An English Translation of Kitāb al-Najāt, Book II, Chapter VI, with Historical-Philosophical Notes ant Textual Improvements on the Cairo Editions*. Trans. Fazlur Rahman. London: Oxford University Press.

———. 1972. *Avicenna Latinus. Liber de anima. I, II, III*, ed. Simone Van Riet. Louvain/ Leiden: Peeters/Brill.

Boethius. 1906. *In Isagogen Porphyrii Commenta*, ed. Samuel Brandt. Vidonaa/Lipsiae: Tempsky/ Freytag.

Blund, John. 1970. In *Tractatus De Anima*, ed. Daniel Angelo Callus and Richard W. Hunt. London: Oxford University Press.

Secondary Literature

Barnes, Johnathan. 1971–1972. Aristotle Concept of Mind. *Proceedings of the Aristotelian Society* 72: 101–114 (rep. in *Articles on Aristotle*, vol. 4: *Psychology and Aesthetics*, ed. Johnathan Barnes, Malcolm Schofield, and Richard Sorabji, 32–41. London: Duckworth, 1979).

Bernardini, Paola. 2002. Nota su alcune tematiche dei Commenti al "De Anima" della facoltà delle arti (ca. 1250–1260). In *Il commento filosofico nell'occidente latino (secoli XIII–XV) / The Philosophical Commentary in Latin West (13–15th centuries)*, ed. Gianfranco Fioravanti, Claudio Leonardi, and Stefano Perfetti, 311–325. Turnhout: Brepols.

Burnett, Charles. 1994. The Introduction of Arabic Learning into British Schools. In *The Introduction of Arabic Philosophy into Europe*, ed. Charles E. Butterworth and Blake A. Kessel, 40–57. Leiden: Brill.

Callus, Daniel Angelo. 1943. Aristotelian Learning to Oxford. *Proceedings of British Academy* 29: 229–281.

Colker, Marvin L. 1972. *Ioannes Blund, Tractatus de Anima (edited by D.A. Callus and R.W. Hunt), Book Review*. Speculum 47 (1): 108–110.

Corrado, Lucia. 2013. *Psicologia e scienza dell'animo tra i secoli XII e XIII. Giovanni Blund e il Tractatus de anima*. Dissertation. Fisciano: Università di Salerno.

De Libera, Alain 2007. *Archéologie du Sujet*, vol. 1: *Naissance du Sujet*. Paris: Vrin.

Dunne, Michael W. 2013. *John Blund Treatise on The Soul (edited by D.A. Callus and R.W. Hunt), with a new Introduction and English Translation by Michael W. Dunne*. Oxford: Oxford University Press.

Gardinali, Mariella. 1984. Da Avicenna ad Averroé: Quaestiones super librum "De Anima", Oxford, 1250 c.a. *Rivista di Storia della Filosofia* 47 (2): 375–388.

Gutas, Dimitri. 1999. *Avicenna and the Aristotelian tradition*. Leiden: Brill.

Hasse, Dag Nikolaus. 2000. *Avicenna De Anima in the West: The Formation of a Peripatetica Philosophy of the Soul*, vol. 1. London/Turin: Warburg Institute Studies and Textes/Nino Aragno Editore.

Knuuttila, Simo. 2018. Medieval Theories of the Emotions. In *The Stanford Encyclopedia of Philosophy*, ed. Edward N. Zalta. https://plato.stanford.edu/archives/sum2018/entries/medieval-emotions/.

McEvoy, James. 1998. Liberal Arts, Science, Philosophy, Theology and Wisdom at Oxford 1200–1250. In *Was ist Philosophie im Mittelalter? Akten der X. Internationale Kongresses für mittelalterliche Philosophie der Société Internationale pour l'Étude de la philosophie médiévale, 25 bis 30 August 1997 in Erfurt*, ed. Jan A. Aerstsen and Andreas Speer, 560–570. Berlin/New York: de Gruyter.

Minio-Paluello, Lorenzo. 1969. Note sull'Aristotele Latino medievale. *Rivista di Filosofia Neoscolastica* 52 (1): 455–468.

Saccenti, Riccardo. 2016. *Un nuovo lessico morale medievale: il contributo di Burgundio da Pisa*. Roma: Aracne.

Van Steenberghen, Fernand. 1991. *La Philosophie au XIIIe siècle*. Paris: Vrin.

Weijers, Olga. 1995. *La "disputatio" à la Faculté des Arts de Paris (1200–1350 environ): esquisse d'une typologie*. Turnhout: Brepols.

Werner, Dorothée. 2005. *John Blund. Tractatus de Anima. Traktat über die Seele. Lateinisch-Deutsch*. Freiburg: Herder.

Chapter 3
Anthropology of Gilbertus Anglicus' *Compendium medicinae*

Agnieszka Kijewska

Abstract In 1950, at the Faculty of Christian Philosophy of the Catholic University of Lublin, rev. Marian Kurdziałek defended his doctoral dissertation entitled *Gilbert the Englishman and the psychological digressions in his "Compendium medicinae"*. Kurdziałek's goal was to discuss the problem of the date of the composition of this medical encyclopaedia taking into account the philosophical sources used by Gilbert. The focus of Kurdziałek's research was on Gilbert's anthropology as found in his *Compendium medicinae*, the work that follows Galen's *dictum* that the best doctor is also a philosopher. In my paper, I want to present Kurdziałek's achievements and the methods he adopted in the field of study concerning the person and work of Gilbert the Englishman against the background of the present state of research. In my opinion, the most recent research, such as that pursued by Michael McVaugh, is in full agreement with Kurdziałek's approach as far as the method of analysing Gilbert's sources is concerned, although the more recent research concentrates on a different aspect of Gilbert's work, namely that of medieval medicine *sensu stricto*. However, medieval medicine, especially when practiced in the vein of Galen's precept, contained a theory of man as its important and indispensable part. In this paper, I try to reconstruct Gilbert's anthropology in reading his *Compendium medicinae* following the principal lines traced by Kurdziałek's interpretation. One of the most important results of this reading is the conclusion that Gilbert's anthropology assumed a theory of universal hylomorphism that was typical of the eclectic Aristotelianism dominant in the philosophical thinking of the earlier parts of the thirteenth century. Thus, I can confirm Michael McVaugh's thesis that Gilbert's *Compendium medicinae* was composed at the latest early into the second half of the

The paper was financed from the resources granted to the research project for the years 2016–2020 forming part of the program "Monuments of Polish Philosophical, Theological and Social Thought of the Twentieth and Twenty-First Centuries" under the auspices of the Minister of Science and Higher Education, Project no. 0021/FIL/2016/20.

A. Kijewska (✉)
Katolicki Uniwersytet Lubelski Jana Pawła II, Lublin, Poland
e-mail: agnieszka.kijewska@kul.pl

M. Gensler et al. (eds.), *The Embodied Soul*, Historical-Analytical Studies on Nature, Mind and Action 11, https://doi.org/10.1007/978-3-030-99453-2_3

thirteenth century. Additionally, I point out that both Kurdziałek's and McVaugh's findings highlight the fact, that in the thirteenth century Aristotelian philosophy was influential not only among philosophers and theologians, but among other intellectuals as well, in particular in the milieu of physicians.

Keywords Gilbertus Anglicus · Medieval medicine · Medieval anthropology · Aristotelianism

3.1 The Shadowy Figure of Gilbertus Anglicus

With my paper I would like to pay homage to my master, Rev. Prof. Marian Kurdziałek (1920–1997), an outstanding Polish medievalist. He began his academic career in 1950 with his doctoral dissertation entitled *Gilbert the Englishman and the psychological digressions in his "Compendium medicinae"* and in this text I intend to present the main results of that work against the background of the achievements of the contemporary research into this subject. The idea for the dissertation came from Kurdziałek's mentor and supervisor, Prof. Alexander Birkenmajer, and one of his tasks was to determine the probable time of the composition of Gilbert's work (known to have originated at some time in the thirteenth century) on the basis of the evidence provided by the philosophical sources used by Gilbert in his *Compendium*.

Gilbert the Englishman, called *doctor desideratissimus*, is an extremely important figure in the history of medieval medicine;[1] attempts to get a clear picture of him, both as a person and as a thinker, stumble, however, upon the most intractable difficulties. One of the basic disagreements among Gilbert scholars concerns the question of whether Gilbert the Englishman is to be identified with Gilbert de Aquila (Gilbert de l'Egle). In his dissertation, Kurdziałek indicates the source of this controversy, which goes back to Wickersheimer's findings of 1936. Wickersheimer found the author's name written as "Gilbertus de Aquila Anglicus" in a manuscript of the *Compendium medicinae* from the last quarter of the thirteenth century preserved in Bruges.[2] In recent literature, most authors no longer engage in this debate and assume the identity of these two persons, as do, for example, Piers Mitchell,[3] A. Gonzalez-Hernandez and M.V. Dominguez-Rodrigez.[4] These authors probably follow the findings of Faye Getz, who in her entry on Gilbert for the Oxford Dictionary of National Bibliography of 2004 identifies Gilbertus de Aquila with Glibertus Anglicus. Earlier, in her edition of the Middle English translation of Gilbert's pharmaceutical writings, she notes:

[1] Cf. Esteban-Segura (2012, 18–19): "The works of Gilbertus were also popular both in Latin and in translation, and were widely copied and adapted."

[2] Cf. Kurdziałek (1950, 2–3); Kurdziałek (1963, 106–107); Riha (1994, 61).

[3] Cf. Mitchell (2004, 21–22).

[4] Cf. Gonzalez-Hernandez and Dominguez-Rodriguez (2008, 147).

> The background and career of Gilbertus Anglicus are obscure. A manuscript dated 1271 (Bruges, Bib. Pub. MS 469), about twenty years after he was dead, refers to him as Gilbertus de Aquila, Anglicus, and from this may be drawn, first, that he had a reputation beyond England and, second, that he could have come from one of the prominent English families of that time named Aquila. From other documents it may be gathered that he was a cleric in major orders and physician to the king by 1207. It may be further inferred that he was educated abroad, because English universities were unable at that time fully to educate a medical doctor. Some have offered that he must have studied at Salerno, but Paris, Bologna, or Montpellier are also possible.[5]

When reconstructing Gilbert's biography and work, it was therefore necessary to try to harmonize biographical data coming from diverse sources and, apparently, not entirely coherent. The author of the first monograph on Gilbert (1918), Henry E. Handerson, summarizes the research of his predecessors:

> Bale, Pits and Leland, the earliest English biographers, tell us that Gilbert, after the completion of his studies in England, proceeded to the Continent to enlarge his education, and finally became physician to the great Justiciar, Hubert Walter, archbishop of Canterbury, who died in the year 1205. This would place him under the reign of King John, in the early part of the thirteenth century.[6]

The reference to the fact that Gilbert was at the court of the Archbishop of Canterbury, Hubert Walter, had to be brought into agreement with another report, suggesting that Gilbert probably practised medical art in the Kingdom of Jerusalem. The Bruges copy of the *Compendium* contains the information that Gilbert cured Bertram, a son of Hugh of Jubail in the Kingdom of Jerusalem, with some eye balm (*collyrium*).[7] In an attempt to reconcile these two reports, Marian Kurdziałek suggests that Gilbert moved to Paris or Montpellier after the death of the Archbishop of Canterbury in 1205.[8]

Scholars have mainly studied Gilbert's *Compendium medicinae*, a popular medieval encyclopaedia, although it is confirmed that Gilbert was also the author of a commentary on the poem *De urinis* by Giles of Corbeil.[9] In order to determine the time of the composition of the *Compendium medicinae* more exactly, Kurdziałek verified and supplemented the biographical information by referring to additional evidence—provided by Gilbert's readings and references to newly translated philosophical works found in the *Compendium*. These references he examined in the light of the historical knowledge about the transmission and penetration of philosophical texts into the Latin West in the thirteenth century, especially the texts of Aristotle and his Arab commentators.[10]

---

[5] Getz (1991, liv–lv).

[6] Handerson (1918, 14).

[7] Cf. McVaugh (2010, 314–315); Handerson (1918, 18); Kurdziałek (1950, 5).

[8] Cf. Kurdziałek (1950, 4–5).

[9] Cf. McVaugh (2010, 301); Kurdziałek (1950, 7).

[10] Kurdziałek in fact begins his examination of the *Compendium* with a discussion of Gilbert's references to authors belonging to the School of Salerno, especially to Richard of Salerno. Cf. Kurdziałek (1950, 16–18).

Kurdziałek begins his investigation with an examination of the quotes of and references to Aristotle's texts found in the *Compendium medicinae*. The most frequently quoted Aristotelian text in Gilbert's encyclopaedia turns out to be the *Meteorology* in Henry Aristippus's translation (which was completed before 1162) from Greek, but there is also a quotation from Gerard of Cremona's version (from Arabic, finished before 1187).[11] There are no traces in the *Compendium* of William of Moerbeke's translations of Aristotle's works from the years 1260–1270.[12] Other texts by the Stagirite referred to in Gilbert's handbook include: *On the Soul*, *On Animals*, *On Generation and Corruption*, there are also references to *The Book of Causes* (*Liber de causis*), once attributed to Aristotle. The author of the *Compendium* was also familiar with Averroes's commentary on the *Metaphysics* and the *Physics*, as well as with the *Quaestiones Nicolai Peripatetici*, usually ascribed to Averroes.[13] These findings encouraged Kurdziałek to put forward the thesis that the *Compendium medicinae* was probably composed between the years 1230 and 1240, although the date can be moved to the 1250s.[14]

The doubts concerning Gilbert's life were investigated again by Faye Getz and Michael McVaugh.[15] Both scholars were acquainted with Kurdziałek's 1963 article published in German, which contained a summary of the most important theses of his doctoral work.[16] McVaugh seems to have masterfully employed a method similar to the one adopted by Kurdziałek (analysis of references to cited texts), but, in contrast, he turned his attention to Gilbert's reception and use of strictly medical sources. To reconcile the apparently incoherent reports concerning Gilbert's biography, Michael McVaugh put forward a bold and yet quite plausible hypothesis that the author of the *Compendium* was born in England after 1210 as a son of a doctor, also named Gilbert, who belonged to the court of the then Archbishop of Canterbury. Gilbert junior studied the arts in Paris or—which is also probable—in Oxford. McVaugh states:

> Charles Burnett has pointed out that in the late 1230s and early 1240s Adam of Buckfield was lecturing there (namely in Oxford) on the whole of Aristotelian "corpus vetustius," the collection of Arabic-Latin translations including among others the *Physics*, *De caelo*, *De generatione*, *Meteorologica*, and *De anima*. [...] These are largely the Aristotelian works that Gilbert knows and cites (although he also knows the *De animalibus*, on which Adam did not lecture), so that he might possibly have studied at Oxford ca. 1240, perhaps even with Adam—especially since Gilbert demonstrates a knowledge of Averroes' commentary on the *Metaphysics*.[17]

[11] Cf. Trizio (2010, 794); Riha (1994, 69).

[12] Cf. Kurdziałek (1963, 109–110).

[13] Cf. Wielgus (1973, 53 f.).

[14] Cf. Kurdziałek (1950, 71); Kurdziałek (1963, 108, 117); Kijewska (2018).

[15] Cf. McVaugh (2010, 295).

[16] Cf. Kurdziałek (1963, 106 f.).

[17] McVaugh (2010, 299).

According to McVaugh's reconstruction of the career of Gilbert the son, after his training in the arts, he studied medicine, probably in Montpellier—for there are traces of his presence on the Continent—and published the *Compendium* there in the late 1250s.[18] Montpellier had already been granted university status in 1220 and was famous for the teaching and practice of medical art, so it is quite possible that Gilbert headed there.[19] In addition, Gilbert's textbook is an excellent testimony to the presence and reception of Galen's texts, which, especially in the thirteenth century, dominated the medieval teaching of medical art, and were well represented in the library collection of the University of Montpellier, as well as in Paris or in Oxford. Michael McVaugh writes about the presence of Galen's writings in the *Compendium*:

> Finally, Paris-trained Gilbertus Anglicus seems eventually to have left medical practice to teach at Montpellier and there, probably in 1250s, composed his well-known *Compendium medicinae*. By now a still larger number of Galenic works was in use at the school: the *Compendium* cites *De crisi* and *De interioribus* [...], *De simplici medicina* and *De complexionibus* [...], but also *De morbo et accidenti*, *De criticis diebus*, and *De iuvamentis membrorum*. The sketchy picture that these scattered sources give us is of a steadily increasing accessibility of more specialized and technical Galenic works to this medical faculty during the middle third of the thirteen century.[20]

Michael McVaugh notes also the appearance of other important medical texts in the *Compendium*, such as Al-Kindi's *De gradibus*, Avicenna's *The Canon of Medicine*, as well as the writings of an Arab physician and surgeon, Razes. According to McVaugh, the references to "Verba Rasis" indicate that Gilbert's treatise was not written before 1250.[21]

3.2 From Ancient to Scholastic Medicine

One may wonder why a classical textbook of medieval medicine should include quotes from philosophical works. The Middle Ages inherited the close relationship between medicine and philosophy from ancient times. In her book *Medieval and Early Renaissance Medicine* Nancy Siraisi states:

> Interaction between philosophy and medicine occurred from the earliest history of Greek science. For example, concepts shared with or derived from pre-Socratic natural philosophy (such as that of the four elements—earth, air, fire, and water) are present in some of the Hippocratic treatises, and Plato propounded physiological theories in his cosmological *Timaeus*. Aristotle's influence was especially important. In the first place, Aristotle, who

[18] Cf. McVaugh (2010, 323). In this context the *terminus post quem* for the *Compendium*'s composition moves to about 1250.

[19] Cf. Dumas (2015, 28–29).

[20] Cf. McVaugh (2019, 384).

[21] Cf. McVaugh (2010, 309 footnote 29, 312–313); cf. Gilbertus Anglicus, *Compendium medicinae*, f. 310ra: "Verba Rasis".

> was himself interested in medicine, strongly asserted a relation between medicine and natural philosophy with the remark, "But it behoves the natural scientist to obtain also a clear view of the first principles of *health* and *disease*... Indeed we may say of most physical inquirers, and of those physicians who study their art more philosophically, that while the former complete their work with a disquisition on medicine, the latter start from a consideration of nature."[22]

Aristotle's precept to search for the first causes and principles of health and disease gave rise to the medical sect of rationalists (or dogmatists), who believed that the most important task of medicine is rationally to investigate the causes of various physiological phenomena (health/disease) and construct a relevant theory by deduction. The empiricist sect appeared as a reaction to this overly theoretical approach. Supporters of this sect pointed out that the goal of medicine is to heal, not to create a theory that has value only when it serves the practice. The methodist sect, founded at the beginning of the first century AD, proposed a treatment based on a few simple principles that are easy and quick to master.[23]

Galen, reporting the positions of various medical schools, tried to reconcile the positions of the empiricists and the rationalists in such a way that "while his works are mainly rationalist in their approach, they also contain empiricist elements."[24] R. James Hankinson states: "Galen cherished a vision of medicine as an Aristotelian axiomatic science; but in a modern jargon, reason only supplies the context of justification, while the phenomena provide that of discovery."[25] Philosophy played a central role in the combination of these two positions. Peter N. Singer believes that in the theory of Galen, an author of fundamental importance for medieval medicine, philosophy played a significant role primarily in three areas: logic, natural philosophy and ethics.[26] The theory of proof helped Galen rationally to explain material reality and distinguish what is true from what is false, certain knowledge from probable knowledge. By adopting certain philosophical concepts and theories Galen engaged in ongoing philosophical debates (e.g., between teleologism and mechanicism). When taking part in discussions about the concept of the soul, its structure and powers he was close to the Platonic concept of the soul, because of the ethical involvement of his philosophy.[27]

Reflecting on the state of the medical art of his time, Galen notes: "Doctors will pay lip-service to Hippocrates, to be sure, and look up to him as to a man without peer; but when it comes to taking the necessary steps to reach the same rank themselves—well, they do quite the opposite."[28] According to Galen, the reason why his contemporaries do not grow to the level of Hippocrates is that the study of

---

[22] Quotation from: Aristoteles, *Sense and Sensibilia*, 1, 436a17–21 (in Siraisi, 1990, 2–3).

[23] Cf. Siraisi (1990, 3–4); Horden (2011, 41–42).

[24] Siraisi (1990, 5).

[25] Hankinson (2003, 296).

[26] Cf. Singer (2016).

[27] Cf. Singer (2016); Donini (2008, 186).

[28] Galen, *The Best Doctor is Also a Philosopher*, 1 (in Galen, 1997).

philosophy is neglected. Formulating the principle that the best doctor is also a philosopher, Galen sets the following requirements for a potential doctor:

> He must be practiced in logical theory in order to discover the nature of body, the differences between diseases, and the indications as to treatment; he must despise money and cultivate temperance in order to stay on the course. He must, therefore, know all the parts of philosophy: the logical, the physical, and the ethical [...]. If, then, philosophy is necessary to doctors with regard both to preliminary learning and to subsequent training, clearly all true doctors must also be philosophers.[29]

These recommendations harmonized perfectly with the structure of the scholastic teaching of medicine, so it is clear why the medieval reception of Galen's texts was closely related to the penetration of specific philosophical texts and doctrines. Anna Maria Urso puts it in the following way:

> As was already the case in Alexandria, also in the West in the Middle Ages the parallel spread of philosophy played a crucial role in the assimilation and exegesis of Galen's thinking, in particular Aristotelian logic and physics, which, already well documented in Salerno, dominated the Arts curriculum with which all students started their academic careers. The link between Aristotelian philosophy and Galenic medicine, moreover, is also visible in some historical circumstances: for example, in the fact that in the twelfth century physicians were 'among the pioneers in introducing Aristotelian ideas into Western Europe' and that some translators of Galen also translated Aristotle and vice versa.[30]

Gerard of Cremona (d. 1187) was such a translator of both Galen's and Aristotle's texts, including pseudo-Aristotle's *The Book of Causes* and the texts of Arabic philosophers. Brian Long asks where Gerard's interest in Galen's thought came from, as Gerard was essentially a translator of philosophical texts. His answer is in the spirit of what Anna M. Urso suggested:

> In the first half of the twelfth century, after all, medical texts proved to be an important source of philosophical knowledge, and Aristotelian natural philosophy in particular. It may even have been the case that clerics in the Latin West had become increasingly medicalized in the course of the long twelfth century: medical manuscripts circulated in extremely large numbers in monasteries and cathedrals, and medical influences appear to have spread widely.[31]

Thus, medieval medicine had a strong theoretical (philosophical) foundation, which was supported by the fact that it had become a "university" discipline. As a university discipline medicine was

> a *scientia* in the Aristotelian sense of claiming to offer certain and universally true knowledge, derived, by syllogistic reasoning, from accepted premises. Indeed its relationship to the Aristotelian physical and ethical works that became available to European scholars in Latin translation was so close that many 'disputed questions' [...] focused on the divergences between Galen and Aristotle.[32]

---

[29] Galen, *The Best Doctor is Also a Philosopher*, 3 (in Galen, 1997, 33).

[30] Urso (2019, 363–364).

[31] Long (2019, 352–353).

[32] Horden (2011, 41).

It is not surprising then that the anthropology in the *Compendium medicinae* seems to follow Galen's aforementioned principle that the best doctor should also be a philosopher. Introducing psychological considerations Gilbert states that in the matters of the soul, ignorance is the source of shame and stupidity, the source of poverty.[33]

3.3 In Search of the Definition of the Soul

Medieval psychology certainly strove to determine the essence of the soul, its main properties, based on the findings of appropriate authorities. As shown by Kurdziałek, most medieval authors wrote their treatises according to a certain template, the archetype of which was Aristotle's *On the Soul*, which he begins with historical observations (*opiniones*). Gilbert does not break away from this scheme and cites several definitions of the soul.[34]

The first definition of the soul quoted by Gilbert is the definition (*opinio*) attributed to Pythagoras, who claims that the soul is a number moving itself (*Pythagoras dicit: anima est numerus seipsum movens*).[35] In Gilbert's work the definition is explained as follows: some numbers have the character of a point, like the number one, some of lines, like the number two, some of surfaces, like the number four, some of solids, like the number eight. There are four characteristics of the soul: intellect, knowledge, reason and opinion. Intellect is linked with the number one because it understands what is simple, and therefore it is compared to a point. Reason understands the form as it exists in the body and hence it is compared to a line. Opinion cognizes form as a form and is compared to a plane number. Knowledge comprehends form as it exists in a subject and a subject in a form and therefore it is compared to a cubic number.[36] Gilbert interprets this part of the definition of the soul, which states that it is *seipsum movens*, in such a way that the soul is both the efficient and the final cause of movement in the body. He explains it on the basis of the process of seeing, showing that the movement of which the soul is the source is circular: the efficient and the final causes are identical there. The soul sends the

[33] Gilbertus Anglicus, *Compendium medicinae*, f. 243va: "Hic enim intendimus dicere aliquid de anima, quia ignorantia parit verecundiam et stoliditas paupertatem."

[34] Kurdziałek (1950, 78); cf. Aristoteles, *De anima*, I.2 (ed. Henderson, 1936, 403b).

[35] Kurdziałek (1950, 78). The source of this Pythagoreans definition is Aristoteles *De anima*, I.2 (ed. Henderson, 1936, 404b).

[36] Gilbertus Anglicus, *Compendium medicinae*, f. 243va: "Numerorum aut quidam est punctualitas, ut unitas, quidam linearis, ut binarius, quidam superficialis, ut quaternarius, quidam cubitus, ut octonarius. Quattuor sunt proprietates animae scilicet: intellectus, scientia, ratio, opinio. Intellectus comparatur unitati, quoniam comprehendit simplicia et ita comparatur puncto. Item ratio comprehendit formam prout est in corpore et ita comparatur lineae. Opinio comprehendit formam prout formam et ita comparatur numero superficiali. Scientia comprehendit formam ut est forma in subiecto et subiectum formae et ita comparatur numero cubito."

spiritus[37] to the optic nerve (*nervus opticus*), from there a ray runs through the eye, which mixes with the scattered forms and returns, notifying the soul of the fact.[38]

Another definition of the soul quoted by Gilbert is the one which he ascribes to Macrobius, namely that the soul is a receptive intellect, but such a definition is not found in the writings of Macrobius.[39] He is also familiar with Isaac Israeli's view that the soul is a shadow of the divine Intelligence and its threefold creation: sensitive, rational and vegetative.[40]

Gilbert cites Aristotle's definition of the soul in three wordings: 1) A soul is the principle of the movement of an organic natural body capable of movement (*Anima est principium motus physici corporis organici potentialiter movens*); 2) A soul is the first actualisation (*entelecheia*) of an organic natural body (*Anima est prima entelecheia corporis physici organici*); 3) A soul is the first perfection of an organic natural body which has life in actuality (*Anima est perfectio corporis organici physici entelecheiae vitam habentis*).[41] Gilbert knows the exact meaning of the term *entelecheia*, mainly thanks to the commentary of Chalcidius, who, quoting Aristotle's definition of the soul, comments on it as follows: "The soul is the first perfection of the organic body that has life in potency." And he explains a little further that "the form by which individuals are shaped, is called by Aristotle *entelecheia*, that is 'absolute perfection', and therefore 'the soul of the natural, organic body is an entelechy'."[42]

Gilbert introduces Plato's considerations about the nature of the soul claiming that Plato states the same as Aristotle but in different words.[43] According to Plato the soul is a source/cause of circular motion and in this respect the soul is similar to its

---

[37] For the concept of spirit within this volume, see Panarelli, Chap. 7 in this volume, pp. 115–125.

[38] Gilbertus Anglicus, *Compendium medicinae*, f. 243va: "Se ipsum movens dicitur quia anima est causa efficiens et finalis ipsius motus. Verbi gratia: Anima dirigit spiritum ad nervum opticum et ita transmittit radium visibilem usque ad oculum. Radius eum invenit aera transformatum et revertit et renuntiat animae. Et ita est causa efficiens et finalis et ita videtur moveri orbiculariter." To explain the process of sensory perception, reference was made to the theory of "vital spirits" (*spiritus vitales*). This theory, initiated by Erasistratus and Galen, and taken over by Nemesius of Emesa, whose texts found a reception at the school of Salerno, claimed that it is these "vital spirits"—subtle particles of matter—that move with bodily arteries and transmit information. Cf. Longrigg (1993, 210–215); Raczyńska (1997, 44f.); Bednarczyk (1995, 139 and f., 201 and f.).

[39] Cf. Kurdziałek (1950, 80); Gilbertus Anglicus, *Compendium medicinae*, f. 243va: "Macrobius dicit: anima est intellectus in passione: intellectus praedicatur active vel passive."

[40] Cf. Kurdziałek (1950, 81).

[41] Cf. Kurdziałek (1950, 81); Aristoteles, *De anima*, II.1 (ed. Henderson, 1936, 412a–412b); Gilbertus Anglicus, *Compendium medicinae*, f. 243vb: "Aristoteles inquit anima est principium motus physici corporis organici potentialiter movens. Vel anima est prima entelecheia corporis physici organici. Vel anima est perfectio corporis organici physici entelecheiae vitam habentis." Cf. Gilbertus Anglicus, *Compendium medicinae*, f. 244rb.

[42] Cf. Plato, *Timaeus a Calcidio translatus commentarioque instructus* (ed. Waszink, 1962, 235–237).

[43] Gilbertus Anglicus, *Compendium medicinae*, f. 244va: "Plato vero dicit idem sed diversis verbis." Ibidem, f. 244rb: "Aristoteles loquebatur de rebus vere et essentialiter et philosophice, Plato figurative, casualiter et mathematice."

Creator. This circular motion is a cognitive motion: God knows everything simultaneously and the soul is able to know multiple things.[44] As God is the efficient cause of all things contained in the macrocosm, so the soul is the efficient and final cause of multiple activities contained in its own microcosm.[45]

In his *Compendium* Gilbert the Englishman supports the following Aristotelian definition of the soul: a soul is the first actualisation (*entelecheia*) of an organic natural body, and bases his psychological considerations on it. But when he analyses this definition in more detail, he seems to interpret it, according to Kurdziałek, in the spirit of the doctrine of universal hylomorphism, which is typical for eclectic Aristotelianism.[46] Gilbert's answer to the question of whether the soul is simple (*simplex*) is both affirmative and negative. He admits different sorts of composition, and, consequently, different meanings of simplicity. The soul is singular because it is not composed of numerical parts and is not a mixture (*mixtum*) and cannot be mixed with anything. Yet, the soul consists of matter and form.[47] Any compound in nature is a compound of matter and form. However, the doctrine of hylomorphism, as presented in Gilbert's *Compendium*, is a special version of universal hylomorphism. Kurdziałek talks about a "version" of universal hylomorphism, because it is difficult to determine on the basis of Gilbert's text whether he limits the hylomorphic composition exclusively to the structure of the soul, or whether he merely limits his interests to anthropology. The author of the *Compendium* does not deal with the question of whether celestial bodies (*intelligentiae*) also have such a structure and what the nature and origin of this matter is. He only states that celestial bodies are corporeal but incorruptible and distinguishes various sorts of incorruptibility. Only the First Cause is absolutely incorruptible (*quantum ad totum*).[48]

The claim that the soul is composed of matter and form allows Gilbert to show the immortality of the soul on the basis of Aristotelian psychology. Thus, the fact

---

[44] This interrelation between cognition and causality is clearly a Neoplatonic element in Gilbert's doctrine. Cf. Gersh (1973, 112).

[45] Gilbertus Anglicus, *Compendium medicinae*, f. 244va: "Dicit enim animam moveri a se ipsa circulariter, hoc autem dictum est, per causam, quia anima est causa circularis motus. Incipit enim ab aliquo puncto et terminatur in idem [...]. Inde etiam patet per illud quod cum anima in multis sit creatori suo simillima in homo habet privilegium quod sicut deus est simplex, ita anima et sicut Deus omnia simul comprehendit, ita anima plura. Sicut enim Deus est causa efficiens omnium quae fuerunt et fiunt in macrocosmo, ita anima est efficiens et finalis causa multarum operationum quae fiunt in microcosmo suo."

[46] Cf. Van Steenberghen (1966, 414 and f.).

[47] Gilbertus Anglicus, *Compendium medicinae*, f. 243vb: "Anima autem in se simplex est sed est composita ex materia et forma. Ergo est composita. Et dico ad hoc quod diversa sunt genera compositionum et sic hoc nomen simplex multipliciter dicitur: quandoque eum privat commixtionem, ut in aqua, quandoque partium numerum, quandoque corruptionem. Et sic est anima simplex."

[48] Kurdziałek (1950, 83); Gilbertus Anglicus, *Compendium medicinae*, f. 243vb: "Et dico quod superiora corpora sunt corporea, non tamen corruptibilia, quemadmodum aliquid est incorporeum et illud est incorruptibile. Et dico quoddam incorruptibile esse quantum ad totum ut prima causa et quoddam corruptibile quantum ad totum ut caduca, quoddam est corruptibile quantum ad partem et incorruptibile quantum ad partem et non quantum ad totum ut elementum ignis, cuius est et toto non desinere."

that the rational soul consists of matter and form is a guarantee of its substantiality. Gilbert, following Aristotle, enumerates three parts of the soul, sc. rational, vegetative and sensitive. However, this statement appears to stand in direct opposition to his former claim that the soul is simple and has no parts. Does Gilbert really contradict himself? Gilbert seems to have been vaguely aware of the problem and tried to avoid it by means of a subtle use of terminology, that is why he spoke so much about the unity of the human soul using the terms *potentiae/potestates* instead of the term *partes animae*. He explains that Aristotle, when speaking of the "parts" of the soul, in fact means its three powers: vegetative, sensitive and rational.[49] The fact that the vegetative soul precedes both the sensitive and the rational one in time is completely clear to our physician on the basis of embryogenesis. The vegetative soul prepares matter and consequently produces an animated seed.[50] The sensitive and the rational soul appear in man at the same time, but the former one is earlier in the order of nature, the reason for this ordering being that the rational soul, which is a substance in its own right and not just a form, cannot unite with the body without the help (*sine adminiculo*) of another principle which is a pure form.[51] The *vegetabile* and what follows it, i.e., the *sensibile* and the *rationale*, combine into one complete form of man, which Gilbert calls humanity (*humanitas*) and against this background he emphasises the unity of the human soul, which is without parts.[52]

Unfortunately, Gilbert does not explain with more precision what the principle of this combination is and how it is realized, which makes it difficult to determine precisely what the relationship between these three kinds of soul consists in within the single form of humanity. Kurdziałek, however, notes that a comparison which Gilbert drew from Aristotle but independently expanded and commented on it may help to shed some light on the issue of the mutual relationships between these powers. We can represent the vegetative soul as a kind of triangle, because we can distinguish the following three powers within it: *virtus nutritiva* and *augmentativa*, each of them having three modes of operation, and *generativa* (just as there are three sides which constitute a triangle). In its turn, the sensitive soul is similar to a square of which two triangles can be made: it has six powers. These powers include the

---

[49] Gilbertus Anglicus, *Compendium medicinae*, f. 244va: "Quaeritur utrum homo habeat tres animas quod videtur velle Aristoteles in Topicis dicens homo triplicem habet animam, et dico triplicem potestatem animae."

[50] Gilbertus Anglicus, *Compendium medicinae*, f. 245vb: "Dico quod anima vegetabilis praecedit sensibilem ad hoc ut disponat materiam. Efficitur autem semen animatum in testiculis, unde omne semen genitivum animatum est."

[51] Kurdziałek (1950, 84); Gilbertus Anglicus, *Compendium medicinae*, f. 245vb: "Anima autem sensibilis non praecedit illam rationalem tempore sed natura praecedat. Eadem enim dispositio naturalis debetur sensibili et animae rationali. Sine tamen sensibili non potest advenire anima rationalis, quia anima rationalis non pure est forma, sed usia. Unde advenire non potest sine aliquo adminiculo eius quia sit pure forma. Ideo adveniente illa [sc. anima sensibilis], advenit anima rationalis et illa recedente, recedit anima rationalis."

[52] Gilbertus Anglicus, *Compendium medicinae*, f. 246ra: "Immo ex praecedente vegetabili et subsequentibus rationali et sensibili componitur forma hominis perfecta quae humanitas appellatur et ita non concedo quod in homine sint tres animae, sed unica, illa tamen habet tres partes."

already mentioned *virtutes* of the vegetative soul and, additionally, the powers of *motus voluntarius*, *virtus phantastica* and the sense of touch. The intelligent soul can be compared to a circle, because in addition to all the above-mentioned potencies it also has a mind (*intellectus*) and reason (*ratio*).[53]

According to Kurdziałek, Gilbert starts from the Stagirite's assumptions and develops them consistently in the spirit of a theory that later found the most perfect and consistent expression in Aquinas's psychology.[54] According to that theory, in a hylomorphic unity embracing different levels of perfection the lower forms or perfections lose their actual reality as independent and merge within the unity imposed by the highest form in the composition, which alone is the true formal principle of the relevant hylomorphic compound. If Gilbert failed to present this kind of solution derived from Aristotle's doctrine in a very clear and convincing way, it was because the author of the *Compendium medicinae* was too much swayed by Neoplatonism in his interpretation of the Stagirite's theses. Although he admits that the rational soul is the form of the body, he also defends its substantial character, which elevates it *supra corpus*. Furthermore, he claims that every form in itself is perfect, simple and general, and this form is multiplied and individuated by accidents. For instance, the vegetative souls in Socrates and in Plato are not different in themselves, as concerns their species, but they are different because of different accidents and this difference is called in *Compendium* material difference. Thus, according to Gilbert, a conjunction of accidents can be regarded as the principle of individuation.[55]

3.4 The Problem of the Soul's Immortality

The issue, which significantly influenced psychological discussions in the thirteenth century, was the question of the immortality of the soul. Encountering Aristotle's philosophy of nature at the turn of the thirteenth century, Christian thinkers were

---

[53] Kurdziałek (1950, 84–85); Aristoteles, *De anima*, II.3 (ed. Henderson, 1936, 414b). Gilbertus Anglicus, *Compendium medicinae*, f. 244ra: "Anima autem vegetabilis comparatur triangulae figurae propter triplicem eius virtutem secundum nutritivam quae dividitur in appetere, digesti et expulsi. Et augmentativam, quia arbor quandoque crescit, quandoque stat, quandoque declinat. Et virtute generativa. Anima sensibilis comparatur quadrato eo quod ex quadrato duo possunt fieri trianguli. Habet enim sex virtutes tres ab anima vegetabili dictas et res a se, scilicet motum voluntarium et virtutem phantasticam et tactum. Anima rationalis comparatur figurae circulari, habet enim omnes praedictas virtutes et rationem et intellectum."

[54] For Aquinas's psychology within this volume, see Trifogli, Chap. 10 in this volume, pp. 197–200.

[55] Kurdziałek (1950, 85); Gilbertus Anglicus, *Compendium medicinae*, f. 245ra: "Et dico ad hoc quod omnis rei forma perfecta; pura forma dico in se simplex est et communis, multiplicatur tamen et individuatur per accidentia, quibus omnino subtractis nulla invenitur circa formam pluralitas, nulla singularitas. Verbi gratia: anima vegetabilis in Socrate et anima vegetabilis in Platone non differunt quantum est in se, sed differunt propter accidentia supervenientia, quia alia sunt in Socrate et alia in Platone. Talis differentia materialis appellatur. In se autem non differunt quia sunt eadem species."

struck by Aristotle's rationalism and naturalism and his lack of understanding for this very issue.[56] This is why Gilbert the Englishman put special emphasis on the question of the immortality of the soul, and more specifically, the immortality of its rational part. The vegetative and the sensitive souls are destructible, since they are strictly dependent on the bodily organs from which they receive forms. The intellect, however, which is the integral part of the intelligent soul, depends only on the first cause and hence it is characterized by indestructibility.[57]

The fact that the rational soul does not depend for its existence on any corporeal organ is not the only argument for its immortality. In his *Compendium*, Gilbert also outlines other arguments, which Kurdziałek summarizes as follows: *Doctor desideratissimus* states that the vegetative and sensitive souls assume bodily forms and their "privations", therefore they are destructible. On the other hand, the rational soul receives composite forms from the Supreme Cause and for this reason it has to be regarded as indestructible. Again, since only the rational soul is capable of eternal happiness and destined for it, it is only this soul that is indestructible.[58] This argument is based on the principle that like knows like: only the rational soul can receive forms directly from the Supreme Cause, which means that the rational soul is, in some respects at least, like the Supreme Cause, and this very fact points to its indestructibility. One can detect here the influence of the Avicennian concept of the two ways of cognition: one is the way of genetic empiricism and the other that of illumination.[59] The soul is a *tabula rasa*, without any built-in mental concepts, but as it assimilates itself to God, which it does through knowledge and virtues, the soul acquires ability to participate in the divine cognitive power.[60]

Gilbert the Englishman formulates another argument for the indestructibility of the rational soul in terms of the Aristotelian concepts of act (*energeia*) and potency (*dynamis*). In Gilbert's specific terminology, the vegetative and sensitive powers are called pure forms, which means that they are principles organizing matter (which is a potency) and therefore depending on it for existence. When the matter (potency) is dissolved, the related pure forms (acts), in this case the vegetative and sensitive principles, also disappear. Meanwhile, the rational soul is not a pure form, but

[56] For Aquinas's defence of the independence of thought from the body in this volume, see Trifogli, Chap. 10 in this volume, pp. 197–200.

[57] Kurdziałek (1950, 85–86); Gilbertus Anglicus, *Compendium medicinae*, f. 243vb: "Proprietates animae rationalis sunt tres, scilicet vegetatio, ratio, sensus. Proprietates animae vegetablis et sensibilis dependentes sunt ab ipso organo, organum autem est corruptibile, quare et anima talis. Intellectus autem sive ratio quae est proprietas animae rationalis, dependens est a prima causarum causa, quare est incorruptibilis. Item anima vegetabilis et sensibilis participant formas corporales et privationes earum quare sunt corruptibiles."

[58] Kurdziałek (1950, 85–86); Gilbertus Anglicus, *Compendium medicinae*, f. 244ra: "Anima rationalis participat formas compositas a summa causa prodeuntes quare est incorruptibilis. Item anima rationalis ad beatitudinem aeternam creata est quae est incorruptibilis."

[59] Cf. Hasse (2013, 111 ff.).

[60] Gilbertus Anglicus, *Compendium medicinae*, f. 244rb: "Dicit Aristoteles illam [sc. animam] esse locus specierum, habet enim virtutem divinam inquantum est delectans in virtutibus et scientiis et speculatur suum Creatorem. […] Est enim ut tabula rasa."

consists of matter and form; therefore, it can still exist after the body's destruction.[61] According to the author of the *Compendium*, the process of disintegration of individual things is not caused by the fact that they contain matter, i.e., an element of potency. Destruction is the result of contradictory properties that different components have: under the influence of an external agent, those contradictory accidents cause suffering (*passio*), which leads to destruction. Every natural form that organizes some matter introduces with itself into the compound certain secondary accidents and perfections. For instance: the form of fire introduces with itself the qualities of dryness and warmth, the form of the air introduces the qualities of moistness and warmth, and so on. While the vegetative soul is characterized by vegetative powers, and the sensible soul is characterized by nutrition, growth and generation, the rational soul's specific properties are intellection and reasoning. But the qualities of the rational soul, shared with supracelestial essences, are not intrinsically related to contrary elements such as *actio* et *passio*, i.e., they do not involve an internal strife of mutually opposed forces that may bring about the destruction of the compound and this ensures that, in contrast to the vegetative and sensitive principles, the rational soul is wholly immune to destruction.[62]

The *corollarium* of the thesis of the rational soul's indestructibility is the affirmation of its immutability. Gilbert's position on this matter is expressed by Kurdziałek in the form of a syllogism:

Esse autem alterabile, corruptibile, passibile—paria sunt;
Sed est incorruptibilis
Ergo impassibilis et inalterabilis.

The soul, as it exists in itself, is not subject to motion (*motus*) because all motion is essentially connected with the body, nor is it subject to change (*alteratio*). The soul may accidentally undergo changes, such as a passage from joy to sadness or vice versa, but changes of this sort only concern the soul insofar as it is united with the body in one compound. That is why we say that a man may be joyful, but not a soul.[63]

[61] Kurdziałek (1950, 87); Gilbertus Anglicus, *Compendium medicinae*, f. 245va: "Anima autem rationalis cum non sit pure forma sed potius composita ex materia et forma. Neque enim potentiale neque actuale contrahit ex corporali dispensatione, immo potius in se habet potentiam et actum et ita destructo corpore non destruitur."

[62] Gilbertus Anglicus, *Compendium medicinae*, f. 245va: "Item aliter dico quod corruptio in rebus non contigit ex materia, nec ex corporeitate sed ex diversitate accidentium sequentium corporeitatem. Ubi notandum quod omnis prima perfectio adveniens in materia secum introducit quasdam contrarias perfectiones et etiam quaedam accidentia. Quae accidentia illi formae sunt debita: ut igneitati debetur caliditas et siccitas, aeritati caliditas et humiditas, animae naturali vegetabilitas, sensibili—nutritio, augmentatio et generatio; rationali intellectus et ratio. Contingit genera naturalia esse corruptibilia propter talium accidentium contrarietatem, quia nec sit passio nisi ratione contrarietatis, neque sit corruptio naturalis nisi per actionem et passionem. Accidentia autem quae sequuntur formas corporum supracaelestium et animae rationalis non habent contrarietatem et ideo non est in eis actio neque passio, quare non corrumpunt. Contrarietates autem sequuntur formas sensibiles et vegetabiles."

[63] Kurdziałek (1950, 88); Gilbertus Anglicus, *Compendium medicinae*, f. 244ra: "Item quaeritur utrum anima rationalis sit alterabilis quod videtur. Iusta enim potuit fieri non iusta. Esse autem

3.5 The Problem of Intellect

The intellect and its place and role in the structure of the soul was an extremely important issue for the psychology of the thirteenth century, resulting from the reception of Aristotle's *On the Soul*.[64] Gilbert the Englishman presents "a strongly Avicennian interpretation of Aristotle's teaching about the intellect"[65] in his *Compendium*, although he does so in a somewhat confused way. He claims that the intellect is twofold: practical and agent. Practical intellect depends on the senses and its function is the same as that ascribed by Aristotle to potential intellect. Practical intellect is the one that distinguishes between good and evil, and it is practical intellect on which the perfection of the rational knowledge of nature depends.[66] Gilbert refers, somewhat idiosyncratically, to the other kind of intellect, the agent intellect, with the term *intellectus regens*. Strange as this may appear, Gilbert does not specify the nature or function of agent intellect in any place of his work. The mere juxtaposition of practical and agent intellects as parts of the primary division of the intellect into parts is by itself a very peculiar device, because both Aristotle and Avicenna, and almost all thinkers of the peripatetic tradition, have divided intellect, according to the fundamental functions it performs, into practical and theoretical (*speculative*).[67]

Gilbert also introduces another division of the human intellect according to that intellect's relation to the human soul. The intellect in man is twofold—Gilbert remarks—because there is an intellect that belongs to the soul itself, which does not contain images (*phantasmata*) and is connected with the eternal, divine intellect, and thanks to this intellect the soul has intellectual knowledge. The other form of intellect is the intellect connected (with the body), which uses images, and all the other lower powers of the soul are subordinated to it. It is destructible and is made capable of knowing and understanding thanks to enlightenment coming from that higher form of intellect which belongs to the soul itself. Moreover, over and above the connected intellect there is a joined intelligence, which theologians call an angelic or divine substance, which is separated from the soul. This intelligence becomes intellect for the soul when it comes to know the forms that are in this intelligence. Gilbert calls this kind of intellect formal intellect, and the intellect which is

---

alterabile, corruptibile, passibile, paria sunt. Sed est incorruptibilis, ergo impassibilis et inalterabilis."

[64] Cf. Black (2010, 320 ff.).

[65] Kurdziałek (1950, 91). Deborah Black (2010, 329) states: "One reason for this appears to have been the perceived compatibility between Avicenna's emanationist account of knowledge acquisition and the Augustinian doctrine of divine illumination, which led some thinkers to identify Avicenna's agent intellect with God himself."

[66] Gilbertus Anglicus, *Compendium medicinae*, f. 186rb: "Intellectus autem duplex est: est enim intellectus practicus et activus. Et intellectus hic est qui boni et mali est discretivus, huic debetur perfectio scientiae rationalis et naturalis."

[67] Cf. Kurdziałek (1950, 92).

connected, material intellect,[68] because in its essence it is passive and its function is to receive forms. Two ways of intellectually knowing come from these two forms of intellect: *inventio* (discovery) and *resolutio* (solving). An important feature of this theory is its fairly novel concept of an intelligence which is separate from individual intellects, identified by theologians with an angelic intelligence or even the divine substance. This intelligence is "full of forms", which is why Gilbert calls it "formal intellect", contrasting it with the "material or combined intellect", which receives cognitive forms. *Resolutio* is related to intelligence, while *inventio* is referred to the material intellect. In that way *inventio* indicates the reception of a cognitive form, whose source and foundation is in the separated intellect.[69] Gilbert is not always clear in his discussions, for the takes fragmentary theories and concepts form different sources and combines them together, to form a somewhat eclectic whole. Thus, the theory of separate intelligence identified as an angelic intelligence, he attributes to theologians. Gilbert's main interests lies in the functioning of the human cognitive process and in its physiological foundations, not in the theoretical subtilities. As a result, he comes up with a theory which is a peculiar combination of genetic empiricism and illumination theory, as in Avicenna's philosophy. Such an Avicennian theory of knowledge was typical of the eclectic Aristotelianism of the thirteenth century.

3.6 Conclusion

In this paper, I made an attempt to reconstruct the anthropology of Gilbert the Englishman as contained in his *Compendium medicinae*, starting from its interpretation proposed by Marian Kurdziałek in his doctoral dissertation of 1950. Marian Kurdziałek's master was Aleksander Birkenmajer, who already in the interwar period highlighted the role played by physicians and naturalists in opening the way for the reception and appropriation of Aristotelian and Arab philosophy and science

[68] Kurdziałek (1950, 97): "Because *intellectus solius animae* is full of forms, while *intellectus coniuncti* is a passive cognitive power, our author calls the first of them *formalis* and the other one *materialis*. The last term comes via Avicenna from Alexander of Aphrodisias. The term is also used in more or less the same meaning by a contemporary to Gilbert William of Auvergne (†1249)."

[69] Gilbertus Anglicus, *Compendium medicinae*, f. 186rb: "Et est duplex: est enim intellectus qui est solius animae et est sine phantasmate et est in perpetuis et hoc intellectu intelligens est anima; et est intellectus coniuncti, et non est sine phantasmate, eo quod ad ipsum aliae virtutes ordinantur, et hic est corruptibilis, et intellectus fit illuminante intellectu, qui est solius animae; et est intelligentia supra intellectum coniunctum licet ponant theologi intellectivam hanc angelum esse vel aliquam divinam substantiam ab anima separatam et in hac est intellectus sine phantasmate, et apprehendit eas quae in anima sunt formae a creatione quibus intelligit. Omnis enim intelligentia est plena formis. Unde dicitur talis intellectus formalis, et alius qui est coniuncti et est materialis. Secundum hoc intellectu distinctae sunt duae viae in cognitione, una per inventionem, alia per resolutionem. Inventio enim debetur intellectui materiali, resolutio—formali. Inventio vero subiecta resolutioni, sicut materiale—formali."

in the Latin West.[70] Kurdziałek adopted a method of studying the text of the *Compendium* that proved useful in determining the probable time of the *Compendium*'s composition. Central to that method was examination of the textual, especially philosophical, sources used by the author in his work; this provided important indications in the epoch of translation and reception of a great number of philosophical and scientific texts. The results obtained by Kurdziałek in his investigations proved fairly substantial, which was quite an achievement given the difficult conditions in which he was forced to work: due to the political circumstances at that time he had no access to major libraries containing Gilbert's manuscripts and had to rely only on the printed edition of 1510; he had also very limited access to literature on Gilbert. Kurdziałek's findings have been largely confirmed and in part supplemented by the most recent research. When presenting Gilbert's anthropology, Kurdziałek placed it in the milieu of the then newly discovered Aristotelianism as seen in the light of Avicenna's reinterpretation (Gilson's *aristotèlisme avicennisant*), the current of thought which later came to be called eclectic Aristotelianism. The two most salient features of this type of Aristotelianism were: the theory of universal hylomorphism and the tendency to combine genetic empiricism with the theory of illumination in the theory of knowledge. We can safely affirm that these two features characterized the philosophical anthropology professed by Gilbert the Englishman, although he was sometimes vague in his treatment of philosophical matters and his formulations are occasionally ambiguous. The theory of universal hylomorphism is not discussed for its own sake in his work, but mostly applied to the discussion of the structure of the human soul. Likewise, his involvement with the theory of illumination is limited to the statement that human intellect functions by receiving intelligible forms from a separate intelligence. Still, the application of these abstruse philosophical theories to purely medical context, as found in Gilbert's work, is of interest and certainly indicative of the dynamism of the intellectual life in the thirteenth century.

In recent years, Michael McVaugh conducted research on Gilbert the Englishman, analysing the *Compendium medicinae* from the point of view of the evidence for the reception of Greek and Arab medical texts it contained, including Avicenna's *Canon of Medicine*. This work had been translated into Latin by Gerard of Cremona before 1187 but did not receive immediate recognition, probably due to its difficulty and novelty. It was only after 1230 that physicians started to appreciate the value of the *Canon* and gradually, fragmentarily, began to include it in their considerations, first by referring to the name of Avicenna (e.g., Henry Harpenstraeng, d. 1244), then by quoting his statements (e.g., *Summa medicinalis* by Walter of Agilon), and finally by incorporating the *Canon* into the academic curriculum. According to Michael McVaugh, the *Compendium medicinae* belongs to the later reception phase of the *Canon* because Gilbert consciously and selectively uses Avicenna, often reaching for Aristotle to explain some of his ideas. McVaugh gives an interesting example of such use of Avicenna (Canon I. 1.4) in Gilbert, namely in the paragraph devoted to

[70] Cf. Birkenmajer (1930).

the illness called hectic fever (*febris hectica*). Gilbert uses the term *ros* to describe the type of moisture (*humiditas*) present in the body that has not yet become nourishment. A second type of moisture is that which has already been transformed into nourishment, but not yet into the substance of the body. The list is made complete with a third type of moisture that is present in the members of the body from its birth and ensures the continuous existence of the body; this is called *humiditas radicalis*. Fever occurs when unnatural heat consumes particular types of moisture. Using Galen's image of the lamp, Gilbert explains that hectic fever occurs when different types of moisture are consumed, just like a lamp's flame burns oil first, then the wick itself, which means that the lamp gets destroyed by its own flame. When hectic fever burns up the last type of moisture, the *humiditas radicalis*, what follows is the death of the individual.[71]

In Michael McVaugh's view, the approach, the range of scientific and philosophical references and the language of the *Compendium* are the same as appear in the commentaries from Montpellier from the mid-thirteenth century and place this work alongside the productions of the Montpellier milieu. The fact that Gilbert used the term *ros* and used the metaphor of the lamp is a clear proof that he was involved in discussions related to the reception of the Avicennian medicine.[72] Altogether, all these medical works are a testimony to the importance of Avicenna's *Canon* for medieval scientifically minded physicians. Kurdziałek, for his part, showed how important Avicenna was to Gilbert in the forming of his conception of man, especially his conceptions of human intellection and cognition. This concurrence of the findings of two scholars of different generations, belonging to different milieus, is a beautiful example of a fruitful dialogue of scholars concerned about the message of a text from a distant past, conducted above the divisions of time, space and social geography.

In conclusion, it can be securely affirmed that Gilbert's anthropology belongs to the philosophical eclecticism, in which Aristotelian philosophy and science was interpreted through the prism of Avicenna's reinterpretation and often in a Neoplatonic spirit.[73] It seems, therefore, that it is plausible from this point of view to date the composition of the *Compendium* to the middle of the thirteenth century, because it is only in the second half of that century that eclectic Aristotelianism gave way to more elaborate and sophisticated reinterpretations of Aristotle, namely to orthodox and heterodox Aristotelianism.

---

[71] Cf. McVaugh (2010, 310–311); McVaugh (1974, 266 ff.); Kijewska (2018, 35 ff.).

[72] Cf. McVaugh (2010, 311).

[73] Kurdziałek (1950, 132).

References

Primary Literature

Aristoteles. 1936. *De anima*, ed. Henderson, with an English translation by Walter Stanley Hett. Cambridge, MA: Harvard University Press.

Galen. 1997. *Selected Works. Trans. P.N. Singer*. Oxford: Oxford University Press.

Gilbertus Anglicus. 1510. *Compendium medicinae*. Lugdunum.

Plato. 1962. *Timaeus a Calcidio translatus commentarioque instructus*, ed. J.H. Waszink. London/ Leiden: Brill.

Secondary Literature

Bednarczyk, Andrzej. 1995. *Galen, główne kategorie systemu filozoficzno-lekarskiego*. Warszawa: Uniwersytet Warszawski.

Birkenmajer, Alexander. 1930. *Le* rôle *joué par les médecins et les naturalistes dans la réception d'Aristote au XII et XIIIe siècles*. Varsovie.

Black, Deborah L. 2010. The Nature of Intellect. In *The Cambridge History of Medieval Philosophy*, ed. Robert Pasnau and Christina Van Dyke, 320–333. Cambridge: Cambridge University Press.

Donini, Pierluigi. 2008. Psychology. In *The Cambridge Companion to Galen*, ed. J.R. Hankinson, 184–209. Cambridge: Cambridge University Press.

Dumas, Geneviève. 2015. *Santé et société à Montpellier à la fin du Moyen Age*. Leiden/ Boston: Brill.

Esteban-Segura, Laura. 2012. *System of Physics (GUL MS Hunter 509, ff. 1r–167v). A Compendium of Mediaeval Medicine including the Middle English Gilbertus Anglicus*. Bern/ Berlin/Bruxelles/Frankfurt/New York/Oxford/Wien: Peter Lang.

Gersh, Stephen. 1973. *From Iamblichus to Eriugena. An Investigation of the Prehistory and Evolution of the Pseudo-Dionysian Tradition*. Leiden: Brill.

Getz, Faye Marie. 1991. *Healing and Society in Medieval England. A Middle English Translation of the Pharmaceutical Writings of Gilbertus Anglicus*. Madison: The University of Wisconsin Press.

Gonzalez-Hernandez, Ayoze, and M. Victoria Dominguez-Rodriguez. 2008. Migraine in Gilbertus Anglicus "Compendium medicinae". The Case of MS Sloane 3486 and Welcome MS 537. *Journal of the History of Neuroscience* 17: 147–159.

Handerson, Henry Ebenezer. 1918. *Gilbertus Anglicus. Medicine of the Thirteenth Century*. Cleveland: Hard Press.

Hankinson, R. James. 2003. Philosophy and Science. In *The Cambridge Companion to Greek and Roman Philosophy*, ed. David Sedley, 271–299. Cambridge: Cambridge University Press.

Hasse, Dag Nikolaus. 2013. Avicenna's Epistemological Optimism. In *Interpreting Avicenna. Critical Essays*, ed. Peter Adamson, 109–119. Cambridge: Cambridge University Press.

Horden, Peregrine. 2011. Medieval Medicine. In *The Oxford Handbook of the History of Medicine*, ed. Mark Jackson, 40–59. Oxford: Oxford University Press.

Kijewska, Agnieszka. 2018. Mariana Kurdziałka droga do mediewistyki. In *Marian Kurdziałek – mediewista*, ed. Monika A. Komsta, 11–55. Lublin: Wydawnictwo Uniwersyteckie KUL.

Kurdziałek, Marian. 1950. *Gilbert Anglik i dygresje psychologiczne w jego „Compendium medicinae"*. Dissertation. Lublin: Katolicki Uniwersytet Lubelski.

———. 1963. Gilbertus Anglicus und die psychologischen Erörterungen in seinem "Compendium medicinae". *Sudhoffs Archiv für Geschichte der Medizin* 47: 106–112.

Long, Brian. 2019. Arabic-Latin Translations: Transmission and Transformation. In *Brill's Companion to the Reception of Galen*, ed. Petros Bouras-Villianatos and Barbara Zipser, 343–358. Leiden/Boston: Brill.

Longrigg, James. 1993. *Greek Rational Medicine. Philosophy and Medicine from Alcmaeon to the Alexandrians*. London/New York: Routledge.

McVaugh, Michael. 1974. The "Humidum Radicale" in Thirteenth-Century Medicine. *Traditio* 30: 259–283.

———. 2010. Who was Gilbert the Englishman? In *The Study of Medieval Manuscripts of England: Festschrift in Honor of Richard W. Pfaff*, ed. George H. Brown and Linda E. Voigts, 295–324. Tempe: Brepols.

———. 2019. Galen in the Medieval Universities, 1200–1400. In *Brill's Companion to the Reception of Galen*, ed. Petros Bouras-Villianatos and Barbara Zipser, 381–392. Leiden/Boston: Brill.

Mitchell, Piers D. 2004. *Medicine in the Crusades: Warfare, Wounds and the Medieval Surgeon*. Cambridge: Cambridge University Press.

Raczyńska, Barbara. 1997. Filozoficzna teoria człowieka w głównych traktatach salernitańskich (Aforismi cum glosulis Ursonis Salernitani; Quaestiones Nicolai Peripatetici). *Acta Mediaevalia* 9.

Riha, Ortrun. 1994. Gilbertus Anglicus und sein "Compendium medicinae". Arbeitstechnik und Wissensorganisation. *Sudhoffs Archiv für Geschichte der Medizin* 78: 59–79.

Singer, Peter N. 2016. Galen. In *The Stanford Encyclopedia of Philosophy*. https://plato.stanford.edu/entries/galen. Accessed 25 Jan 2020.

Siraisi, Nancy G. 1990. *Medieval and Early Renaissance Medicine. An Introduction to Knowledge and Practice*. Chicago: The University of Chicago Press.

Trizio, Michele. 2010. Appendix B: Medieval Translation. In *The Cambridge History of Medieval Philosophy*, ed. Robert Pasnau and Christina Van Dyke, 793–802. Cambridge: Cambridge University Press.

Urso, Anna Maria. 2019. Translating Galen in the Medieval West: The Greek-Latin Translations. In *Brill's Companion to the Reception of Galen*, ed. Petros Bouras-Villianatos and Barbara Zipser, 359–380. Leiden/Boston: Brill.

Van Steenberghen, Fernand. 1966. *La philosophie au XIIIe siècle*. Louvain/Paris: Publications Universitaires.

Wielgus, Stanisław. 1973. Quaestiones Nicolai Peripatetici. Note d'introduction. *Mediaevalia Philosophica Polonorum* 17: 57–155.

Chapter 4
A Stain on the Bronze: Some Medieval Latin Commentators on *De insomniis* 2, 459b23–460a32

Christina Thomsen Thörnqvist

Abstract In a curious and much-discussed passage in *De insomniis*, Aristotle claims that if a woman looks into a mirror during her menstrual period, her gaze will cause a blood-red stain to appear on the mirror's surface. Modern commentators have struggled with the passage and its authenticity has been both questioned and defended. In the medieval Latin reception of Aristotle's treatises on sleep and dreams, many of the commentators devote considerable attention to Aristotle's account of the mirror phenomenon and try to harmonise it with his overall theory of perception as well as with his theory of motion. The paper analyses the development in a number of Latin commentaries on *De insomniis* from the twelfth and thirteenth centuries.

Keywords Aristotle · Albert the Great · Scholasticism · *Parva naturalia* · Perception · Vision · Menstruation

4.1 Introduction

Aristotle's *De insomniis* 2, 459b23–460a32 has long puzzled scholars. Having spent the beginning of the treatise on defining *enýpnion* as belonging to the imaginative capacity of the sensitive soul (458a33–459a22), Aristotle turns to an account of the process of dream formation: Sense-impressions received when we are awake leave remnants in our sense-organs. As long as we are awake, the continuous flow of sense-impressions from the external world to our senses claims all our attention, but

The present study is a result of the research programme *Representation and Reality: Historical and Contemporary Perspectives on the Aristotelian Tradition*, funded by Riksbankens jubileumsfond, Sweden.

C. Thomsen Thörnqvist (✉)
University of Gothenburg, Gothenburg, Sweden
e-mail: christina.thornqvist@gu.se

M. Gensler et al. (eds.), *The Embodied Soul*, Historical-Analytical Studies on Nature, Mind and Action 11, https://doi.org/10.1007/978-3-030-99453-2_4

when we fall asleep and the common sense together with all the particular senses are deactivated, the more discrete remnants of previous sense-impressions become observable to us in the form of the *phantásmata* that constitute our dreams (459a23ff.).

As empirical support for his conclusion that the sense-organs have the capacity of storing the perceptual remnants that generate our dreams, Aristotle provides some examples of how sensory stimuli can continue to affect the sense-organs also after the object perceived is gone, such as, for instance, the optical phenomenon that occurs when we look into the sun and then shut our eyes (459b7–459b23). So far so good. But in 459b23, the trouble starts. It first seems that Aristotle's intention is to elaborate on the sense-organs' sensitivity to stimuli, because the passage is introduced by the following claim:

> That the sense-organs are rapidly sensitive even to slight differences is shown by what happens with mirrors.[1]

However, the example of the mirror as evidence of the high sensitivity of the sense-organs seems to be left hanging in the air, because Aristotle then makes a sharp turn to the claim that the particular example of mirrors demonstrates that not only are the sense-organs affected by the sensible object, the sense-organ itself also acts upon the object. This is proved, Aristotle claims, surprisingly enough, by the fact that when a menstruating woman looks into a mirror, her vision will act upon the mirror and generate a red cloud on its surface:

> At the same time, it is clear from this case that the organ of sight is not only affected by, but also acts upon, its object. For in extremely clean mirrors, when women look into them during their menstrual period, the mirror surface takes on a sort of blood-red cloud. In fact, if the mirror is a new one, it is not easy to get the stain out, although it is easier with an old one. The reason is, as we have said, that the organ of sight is affected not only by the air, but is also active and imparts movement, just as shining objects do. In fact the organ of sight is just such an object and one that possesses colour.[2]

Aristotle offers the following explanation to this alleged phenomenon:

> One may reasonably suppose, then, that during menstrual periods the eyes are in the same state as any other part of the body. Furthermore, they are full of blood-vessels by nature. Hence, when menstruation occurs, owing to disorder and turbulence of the blood, the difference in the eyes is invisible to us, and yet it is present (for the nature of semen and of the menses is the same). The air is moved by the eyes, and makes the air extending over the mirror's surface to be of a certain quality, i.e., that by which it is affected itself. And this air in turn affects the surface of the mirror.[3]

This description of how the organ of sight acts upon the sensible object is followed by a lengthy explanation of why the stain is more difficult to remove from new mirrors than from old ones (460a12–23).

---

[1] Aristoteles, *De insomniis*, 2, 453b23–25. I quote from David Gallop's translation (1996) throughout this chapter; here, 93.

[2] Aristoteles, *De insomniis*, 2, 459b27–30 (Gallop, 1996, 93).

[3] Gallop (1996, 93).

The transition in 459b23 is awkward enough and the subsequent return to the account of dream formation in 460a32 no less so; at the end of the passage, Aristotle picks up the thread which he seems to have dropped in 459b23–24 and repeats his claim that the phenomenon described proves that the organ of sight is not only affected by the sensible object but also acts upon it (460a25–26).[4] He then provides, as he claims, further evidence in the form of the observation that oil and wine take on smells of odorous objects placed in or close to them (460a26–32). But what is all this supposed to prove? In 460a32–460b3, the account of the affection of vision on the sensible object closes with an explicit exhortation that we should now return to the original inquiry from which we took off, that is, with accepting the conclusion that external sense-impressions continue to affect the sense-organs also after the object perceived is gone.

De insomniis 2, 459b23–460a32 has become notorious for several reasons. The abrupt transitions in both ends, the extent and content of the passage in relation to the main topic, the fact that it is difficult to imagine that Aristotle would really have believed in the phenomenon described without empirical evidence—these observations have all been adduced as arguments for the conclusion that the passage is spurious. In more recent days, however, scholars tend to defend the passage's authenticity, even if David Gallop stands out somewhat in joining several of his predecessors in questioning its authenticity.[5] Already David Ross, building on Drossaart Lulofs,[6] did not find the passage overly suspicious[7] and, more recently, Philip van der Eijk has defended it in a knowledgeable and convincing way.[8]

Authentic or not, the passage was part of the Aristotelian text transmitted to the Latin West, and for the medievals, just as for any modern reader who expects Aristotle to provide a philosophical account in the *Parva naturalia* that is at least reasonably consistent with his theoretical framework as a whole, the passage presented a serious challenge. In addition to all the other conundrums generated by it, Aristotle's claim that the organ of vision acts upon the sensible object is seemingly in direct conflict with his theory of perception, where vision is not a reciprocal activity, but an affection of the visual power and its organ that occurs when the quality of the sensible object is transmitted from the object to the eye via the medium.[9] So how did the medievals deal with this awkward passage? The present chapter will offer an account of how medieval commentators on Aristotle's treatises on sleep and dreams approached the challenge of reconciling Aristotle's repeated claim in *De*

[4] Aristoteles, *De insomniis*, 2, 460a23–26: "From the above, then, it is plain that movement is generated even by slight differences, that perception is rapid; and that the sense-organ that perceives colours is not only affected by them but also acts reciprocally" (Gallop, 1996, 95).

[5] Gallop (1996, 145).

[6] Aristoteles, *De insomniis et de divinatione per somnum* (ed. Drossaart Lulofs, 1947, xxx–xxxiii).

[7] Aristoteles, *Parva naturalia* (ed. Ross, 1955, 272).

[8] Aristoteles, *Parva naturalia III, De insomniis, De divinatione per somnum* (trans. Van der Eijk, 1994, 167–193). Note that a helpful summary is found in Van der Eijk (2005, 180, n. 24).

[9] See, for instance, Aristoteles, *De anima* II.7.

insomniis that the organ of sight acts upon the sensible object with their understanding of Aristotle's overall theory of perception.[10]

4.2 Adam of Buckfield and Albert the Great

The earliest known Latin exposition of *De insomniis* 459b23–460a23 is found in Adam of Buckfield's (ca. 1220–before 1294) literal commentary on Aristotle's treatises on sleep and dreams dated to between the later 1230s and the early 1240s.[11] The major part of Adam's exposition of the passage on the woman and the mirror is little more than a paraphrase of Aristotle, but Adam does make a brief attempt to offer an explanation to the phenomenon: It is not *qua* seeing that the woman's vision acts upon the mirror but insofar that it is "polluted" and in a "badly mixed state", just as the gaze of the basilisk. At the same time, Adam surprisingly enough concludes that it is insofar as it perceives that vision in this case is active, and that the colouring of the mirror is caused by an emission of "beams" (*radii*) from the sense-organ affected by the menstrual blood.[12] Adam's exposition provides no explanation of how the beams radiating from the woman's eyes transfer the affection of the menstrual blood on the eye to the mirror; the actual nature of this action of the organ of vision upon the external object without the body and the external object being in contact with each other remains an open question.

If Adam's commentary was finished around 1240, it must have appeared more or less at the same time as Albert the Great's (ca. 1200–1280) exposition of the *Parva naturalia* in his *De homine*, which he completed around 1242.[13] Albert here discusses the phenomenon of the menstruating woman and the mirror under the question whether the sense-organs are related to the sensible species exclusively as patients or also as agents.[14] He affirms the former, but also states that the sense organs act upon the sensible objects *per accidens*. Albert's account of how the sense-organ can act upon an external object at a distance is to provide the later tradition with its stock explanation of *De insomniis* 459b23–460a23. It is an efficient

[10] Aristoteles, *De sensu*, 2, 438a25–27 is often quoted as evidence of Aristotle's adherence to the intromission theory: "In general, it is unreasonable to hold that the organ of sight sees because something goes out from it and that this reaches as far as the stars, or that it goes out a certain distance and naturally unites with something, as some say" (trans. Miller, 2018, 73).

[11] On the chronology, see Burnett (1996). Adam's commentary is edited in the Vivès edition of Aquinas: *Doctoris angelici divi Thomae Aquinatis opera omnia XXIV*, ed. Stanislas E. Fretté (1875, 293–310); the text is available online: http://www.corpusthomisticum.org/xsm.html.

[12] Adam of Buckfield, *In De insomniis* (ed. Fretté, l. 3): "Et ita non inficit speculum inquantum videns, sed inquantum inquinatum vel incomplexionatum, ut in basilisco; et ideo sequitur quod visus inquantum sentit, aliquid agat. Et, quia videns, radios emittit, hujusmodi infectio est naturalis inquinatio, ut patet."

[13] On the date, see Weisheipl (1980, 565–77) and Albertus Magnus, *De homine* (ed. Anzulewicz and Söder, 2008, xiv–xv).

[14] Albertus Magnus, *De homine* (ed. Anzulewicz and Söder, 2008, 368: 41–369: 8).

solution, because Albert provides us with the transmitter that we so sorely need for the passage to fit with Aristotle's general principles of motion. Albert concludes that the visual sense-organ does indeed act accidentally upon objects in cases where hot vapour is released from it and affects objects in its surroundings. In the case of the menstruating woman, according to Albert, it is vapour and *spiritus* emanating from the eyes that transfer the qualities of the menstrual blood to the air which, in turn, transmits it to objects nearby. Albert carefully points out that even though it only shows on the surface of the mirror, also other objects close to the menstruating woman are affected. He ties his explanation nicely to Aristotle's comparison to wine and oil in 460a26–32; this phenomenon too, Albert claims, is due to the fact that vapours, in this case *vapores odorabiles*, pass from an object to the liquid and change it.[15]

Albert's second exposition of *De insomniis* in his commentary on the *Parva naturalia* (dated to the late 1250s[16]) contains a more extensive exposition of 459b23–460a32. The explanation that the menstrual blood affects the mirror via vapour emanating from the eyes is basically the same as in *De homine*, but Albert here ties his theory much closer to Aristotle's text; the vapour, Albert points out, is generated by the menstrual blood which is "boiling" in the eye—quite clearly Albert's interpretation of Aristotle's *dià tarachḕn kaì phlegmasían haimatikḕn* (διὰ ταραχὴν καὶ φλεγμασίαν αἱματικὴν) in 460a6–7—and, hence, evaporates.[17] In this context, Albert also devotes considerable effort to describing the chain reaction in the air, expounding more systematically on 460a9–11 than in the corresponding exposition of the passage in *De homine*. He finishes off by elaborating on 453b23–25 and stating that two conclusions may be drawn from Aristotle's example of the menstruating woman and the mirror: (1) Just as the mirror is "easily, rapidly and

[15] Albertus Magnus, *De homine* (ed. Anzulewicz and Söder, 2008, 368: 67–369: 8): "Concedendo ultimas rationes dicimus organa sensuum tantum pati in sentiendo et nihil agere nisi per accidens, scilicet in quantum vapores calidi et spiritus subtiles resolvuntur ab organis et inficiunt quaedam vicina sibi. In distillatione enim sanguinis menstrui decurrit sanguis ab omnibus venis, et cum venae sanguinis sint in oculo, spiritus et vapores, qui egrediuntur ab oculo, inficiuntur illo sanguine et inficiunt speculum, ut superius est determinatum, et alia corpora vicina inficiunt, licet infectio non ita appareat, sicut etiam vinum vel oleum inficitur ex odoribus vicinis propter evaporationes odorabiles transeuntes in vinum vel oleum."

[16] On the date, see Weisheipl (1980, 570), and Donati (2018, 170–71). Donati is currently preparing a critical edition of Albert's commentary on Aristotle's treatises on sleep and dreams for the Cologne edition and has generously given me permission to quote Albert's text here from her forthcoming edition.

[17] Albertus Magnus, *De somno et vigilia*, II, tr. 1, cap. 6 (ed. Donati, in progress = ed. Borgnet, 1890, 165b): "In mulieribus autem menstruatis rationabile est *oculos* affici sanguine, *sicut* et omnes reliquae *partes* corporis earum afficiuntur; naturaliter *enim* secundum suam compositionem *existunt pleni venarum*; venae autem sunt viae sanguinis; *ideoque, cum fiunt menstrua, propter turbationem et fervorem sanguineum*, licet *nobis incerta sit differentia* sanguinis illius qui discurrit ex *oculis*, scimus tamen quod *inest* ipsis aliqua differentia et portio *menstrui* sanguinis, sicut et in ceteris membris [...]. Et cum fervet et resolvitur vaporabiliter, *movetur aer ab ipso* et serpit alteratio ista et motus usque ad aerem *qui in speculis* est tangens speculi superficiem propter continuitatem qua continuatur totus aer ab oculo usque ad speculum."

profoundly" affected by very slight external changes, so the sense-organs are highly sensitive to small "particles" (*particulae*) inside the body; the sense-organs react even more rapidly to these, because unlike external stimuli, they have a direct connection to the sense-organs via the nerves. (2) If the eye can transfer the affection of the menstrual blood to an external object as demonstrated by Aristotle's example, then the body's internal organs, when affected by sensible forms, must be able to transfer this affection to the particular senses and to do so with even greater force.[18]

In his excellent article "*Der Sturz des Kamels und die Befleckung des Spiegels: Fernwirkungstheorien in arabischen und lateinischen Kommentaren zu Aristoteles* De insomniis", Dag Hasse demonstrates how it became a standard ingredient in the medieval explanations of *De insomniis* 459b23–460a32 to associate the alleged phenomenon of the menstruating woman and the mirror with theories on *fascinatio*, the imaginative power's alleged capacity to act upon an external object at a distance. Hasse relies on a limited selection of sources—Adam's and Albert's commentaries on *De insomniis*, the slightly later commentaries by Radulphus Brito (†1320/21) and John of Jandun (1280/9–1328) on the *De anima*, and the much later commentary by Agustino Niso (†1538) on *De insomniis*. Hasse himself laments the fact that little work so far has been done on the medieval Latin tradition on the *Parva naturalia*.[19] In the course of the last five years a number of thirteenth- and fourteenth-century commentaries on Aristotle's treatises on sleep and dreams have been edited for the first time: The question commentaries by Geoffrey of Aspall (†1287), Simon of Faversham (†1306), James of Douai (fl. ca. 1270) and Walter Burley (ca. 1275–1344/5) have all been made accessible in modern editions, and so has the commentary on *De insomniis* by Brito.[20] I believe that an analysis of these recently

[18] Albertus Magnus, *De somno et vigilia*, II, tr. 1, cap. 6 (ed. Donati, in progress = Borgnet, 1890, 166a–b). "Ex omnibus his quae dicta sunt arguimus duas conclusiones, quarum una est: si facile et celeriter et usque in profundum sui [...], sicut speculum a parvis differentiis patitur ab extra, quod similiter facile et celeriter potest moveri organum sensus a particulis intra, cum illae sint organis sensuum coniunctae in eodem subiecto et continuus sit nervus ab interioribus ad organa veniens, exteriora autem sint diversa ab eis subiecto et non habentia adeo naturalem continuitatem usque ad ipsa. Secunda est quod, cum organum sensus agat secundum passionem innaturaliter et accidentaliter acceptam in aliud corpus quod non est secum in eodem subiecto, quod multo fortius organa interiora et instrumenta informata sensibilibus formis agere possunt in organa sensuum propriorum secundum formam quae ut perfectio in perfectibili et naturaliter est in eis sicut actus in potentia."

[19] Hasse (2016, 532).

[20] Geoffrey of Aspall = ed. Ebbesen (2014), Simon of Faversham = ed. Ebbesen (2013), James of Douai = ed. Ebbesen (2015), Walter Burley = ed. Thomsen Thörnqvist (2014), Radulphus Brito = ed. Ebbesen (2016). In addition, I am currently preparing critical editions of two anonymous commentaries on the treatises on sleep and dreams preserved in the MS Rome, Biblioteca Angelica, 549, of which one has been ascribed to Siger of Brabant (see Pinborg 1969); both commentaries have been consulted here as well. For an overview of the content of these and a number of other medieval *quaestiones* on the *De sensu et sensato*, *De memoria et reminiscentia* and the treatises on sleep and dreams, see Ebbesen et al. (2015).

As mentioned above, this chapter concentrates on the medieval reception of *De insomniis*, 2, 459b23–460a32 in the commentary tradition on the *Parva naturalia*; Adam of Buckfield's and Albert the Great's works are studied with a particular focus on their influence on the later tradition.

edited commentaries makes it possible to supplement and modify the conclusions drawn by Hasse in several ways. As I will show, there is, in addition, some further information of relevance to be found in a wider range of sources by Albert and Adam.

Hasse draws our attention to some notable differences between Adam's and Albert's expositions in their respective commentaries on *De insomniis*. He points out that both Adam's claim that the eye acts on the mirror from a distance not by a chain reaction in the air but by an emission of beams, and his comparison of the menstruating woman's gaze to that of the basilisk's echo medieval discussions of *fascinatio*. In particular, Adam's account seems close to al-Kindi's *De radiis*, where al-Kindi, contrary to both Avicenna and al-Ghazali, claims that *fascinatio* reaches the object through an emission of beams from the eyes.[21] Albert's explanation, on the other hand, Hasse remarks, contains none of these features.[22] Furthermore, Hasse observes that Brito ascribes the affection of the mirror to vapour emanating from the eyes, describing the phenomenon as a stream of vapour rather than as a chain reaction in the air, which seems to position Brito in between, as it were, Adam and Albert.[23]

While Hasse is quite right in pointing out that the reference to *fascinatio* is present in the Latin reception of *De insomniis* from Adam of Buckfield onwards, the differences he identifies in Albert's and Adam's interpretations seem less clear-cut when a broader range of sources is taken into account. Some examples:

(i) When introducing the passage in his *De insomniis* commentary, Adam explicitly adheres to Aristotle's theory of the process as a chain reaction in the medium: The menstrual blood colours the eye, which, in turn, colours the medium *successive* until it reaches the mirror and colours its surface.[24]

(ii) In his commentary on *De anima*, Adam expounds *De anima* II.7, 419a15ff, where Aristotle argues against Democritus on the vital role of the medium in the perceptual process. Adam here emphasises the chain reaction while again comparing the phenomenon to the lethal power of the basilisk: Neither phenomenon, Adam claims explicitly, is a case of emission; it is the organ of vision that affects the medium and causes a chain reaction in it which finally reaches the object.[25]

For studies that focus on Albert's theories of action at a distance from a broader perspective, see the studies by Francis J. Kovach (1979) and Alessandro Palazzo (2009).

[21] Hasse (2016, 527, 534).

[22] Hasse (2016, 533): "Bemerkenswert an diese Kommentar is auch das, was fehlt. Albertus bietet keine Theorie der Strahlen-Emission, keine Verweise auf Fernwirkungslehren bei Avicenna und Gazali und keine weiteren Beispiele für Fernwirkungen wie den Bösen Blick oder den Basilisk."

[23] Hasse (2016, 531).

[24] Adam of Buckfield, *In De insomniis* (ed. Fretté, l. 3): "Unde sicut semen generatur ex superfluitate ultimi digesti, sic et menstruum; et ita est in mulieris oculo, et oculus infectus movet aerem sibi contiguum, et ille alium successive, usque ad aerem contiguum speculo: qui aer est infectus cum simili passione, qua inficitur oculus: et sic inficit superficiem speculi."

[25] A transcription of the MS Bologna, Bibl. univ. 2344 by J. Ottman (2018) is available online (http://rrp.stanford.edu/BuckfieldDAn2.shtml): Adam of Buckfield, *De anima*, II.8, 53–55: "Ex

(iii) Hasse notes that Brito and Jandun, just as Adam, argue in defence of Aristotle's general position that perception is a passive process that it is not insofar as the eye is seeing that it acts on the external object.[26] This is not found in Albert's commentary on the *Parva naturalia* that Hasse has consulted, nor, as mentioned above, does Albert make any references to *fascinatio* there. On the other hand, Albert's commentary on the *De anima* contains both. Commenting on the *De anima* II.5, Albert refutes a number of alleged proofs that sensation is an active power. Among these, we find the phenomenon of the menstruating woman and the mirror adduced as such an invalid proof. Albert here, just as Adam and the later commentators, compares the phenomenon to the power of the basilisk's poisonous gaze. But all this is utterly stupid, Albert claims, because "this action does not belong to the eye insofar as it has sensation, but rather insofar that it is a part of the body that generates vapour; because it is the vapour resolved from the eye that causes the contamination, not the visual capacity that is in the eye."[27]

To sum up, Hasse's study suggests a split in the tradition between Albert on the one hand and Adam and the later commentators on the other, but the examples above demonstrate, I believe, that there is good reason to modify this view, and, not least, to note that Adam's position is even more ambivalent than it appears when his commentary on *De insomniis* is considered on its own.

quo videtur aperte velle quod videmus intussuscipientes et non extramittentes. [...] De muliere autem patiente menstruum et de animali solo visu interficiente potest dici quod non extramittatur aliquid a corpore cum inficitur speculum a muliere aut interficitur aliquod animal a basilisco, sed aer contiguus instrumento videndi inficitur, qui infectus inficit aera sibi proximum, et sic quousque inficiatur sibi ultimum."

[26] Hasse (2016, 532; 534).

[27] Albertus Magnus, *De anima* (ed. Stroick, 97: 81–90; 98: 24–28): "Adhuc autem inducunt quarto loco experimenta; quia vident oculos menstruatarum inficere aërem et specula et aliquando oculos etiam intuentium eas et oculos basilisci spargere venenum, eo quod, ut dicunt, visu interficit serpens, qui basiliscus vocatur; hoc autem, ut dicunt, non fit nisi agendo in aërem et in obiectum, et ita dicunt sensum esse potentiam activam. Sed haec omnia et similia frivola sunt, quoniam demonstrative probatum est in praehabitis, sensum esse potentiam passivam et non activam. [...] Quod autem ultimo inducunt, omnino stultum est, quia talis actio non est oculi, inquantum est sensus in ipso, sed potius, prout est pars corporis evaporantis; vapor enim resolutus ab oculo est ille qui inficit, et non visus, qui est in oculo."

4.3 Some Thirteenth- and Fourteenth-Century *Quaestiones* on *De insomniis* 2, 459b23–460a32

Turning to the newly edited question commentaries on *De insomniis*,[28] the following may be observed. Geoffrey of Aspall, whose work is the one closest chronologically to Adam and Albert, and more precisely dated to the early 1260s, is of particular interest here, because Geoffrey is the only commentator of the ones here studied who displays no obvious dependence on Albert. In his edition of Aspall's *quaestiones* on the treatises on sleep and dreams, Sten Ebbesen concludes that there is no indication in Aspall's commentary that he used Albert's works.[29] Subsequent studies of the reception of various topics in the Latin tradition on *De somno et vigilia* and *De insomniis* have corroborated Ebbesen's observation,[30] and the case of the woman and the mirror turns out to be no exception. Aspall evidently represents a kind of intermediary stage in the further development in that he still discusses the phenomenon under the question whether sensation is only a passive power or also an active one, whereas in the later commentaries, the phenomenon constitutes a *quaestio* of its own, usually under the title *utrum oculus mulieris menstruosae habeat inficere speculum* or something similar. Aspall's independence of Albert in this case shows itself through the absence of the theory of vapour as transmitter of the qualities of the menstrual blood to the external object. Aspall, as Adam, paraphrases Aristotle's description of the chain reaction in the air without further explanation of its nature or efficient cause, and then proceeds along the following lines: There is a twofold meaning of *visus*, which can refer not only to the capacity of seeing but also to the organ of vision, and it is with respect to the latter that the woman's vision infects the air. The former can be further divided; with respect to the reception of the sensible species seeing is a purely passive capacity, but seeing *qua* apprehending or judging is an action.[31] Aspall's rather clever remark that some capacities closely related to

---

[28] Geoffrey of Aspall, *Quaestiones super librum "De somno et vigilia"* (ed. Ebbesen, 2014, 331–332); Simon of Faversham, *Quaestiones super librum "De somno et vigilia"* (ed. Ebbesen, 2013, 139–141); Radulphus Brito (ed. Ebbesen, 2016, 61–64); James of Douai (ed. Ebbesen, 2015, 70: 19–73: 15); Walter Burley (ed. Thomsen Thörnqvist, 2014, 495: 1–497: 5); Siger of Brabant (?), *Quaestiones super librum "De somno et vigilia"* (ed. Thomsen Thörnqvist, MS München, BSB, Clm. 9559, ff. 49vb–50ra); Anonymus Angelicani, *Quaestiones super librum "De somno et vigilia"* (ed. Thomsen Thörnqvist, MS Roma, Bibl. Angelica, 549, f. 110rb).

[29] Ebbesen (2014, 261).

[30] See, for instance, Thomsen Thörnqvist (2016, 301–302) and Thomsen Thörnqvist (2022).

[31] Geoffrey of Aspall, *Quaestiones super librum "De somno et vigilia"* (ed. Ebbesen, 2014, 332): "Ad hoc potest dici ad praesens quod visus potest dupliciter considerari: aut ratione suae virtutis aut ratione organi. Ratione organi agit visus aliquo modo, sicut patet per experimentum de muliere patiente menstruum, quia, sicut dicit Aristoteles, oculus infectus sanguine menstrui inficit aerem sibi proximum, et ille aer inficit alterum aerem sibi proximum, et sic usque ad speculum; et sicut ipse dicit, organum visivum [[et]] est aliquo modo agens, est enim lucidum. Si ratione virtutis visivae, hoc potest esse dupliciter: aut quantum ad receptionem tantum, et sic est tantum passivum; aut ratione apprehensionis et iudicii, et sic aliquo modo agit."

the visual capacity are arguably active and not passive has to my knowledge no counterpart elsewhere.

After Aspall, the Latin commentaries here studied offer a more or less unified interpretation of *De insomniis* 459b23–460a32. To the question whether a menstruating woman can act upon the mirror the way described by Aristotle the standard arguments *quod non* can be summarised as follows.

(1) Relying on Aristotle's statement in *Physics* 7 that the movements of the mover and the moved must occur simultaneously,[32] it is claimed that the mirror is not affected by the eye.
(2) If the woman's vision *qua* visual capacity were affecting the mirror, the phenomenon could occur also at a large distance, but it is assumed that the mirror is not affected at a great distance, etc.
(3) If it were vapour emanating from the eyes that causes the discoloration of the mirror, vapour from other parts of the body would have the same effect, but this is assumed not to be the case.

These standard arguments are met with the following standard responses:

Re 1: The movement of the eyes and the movement of the mirror do not occur simultaneously but the movement of the eye and of the medium do.
Re 2: It is not *qua* visual capacity that the eye affects the mirror but insofar as the eye is *evaporativus*.
Re 3: Other parts of the body do, in fact, also emit vapour, but due to its nature as particularly porous the eye emits larger quantities than other body parts.

In their solutions to the *quaestio*, the commentators typically connect Aristotle's account of the menstruating woman and the mirror to the phenomenon of *fascinatio*, but in a new and quite different way. Whereas Albert never describes the phenomenon of the woman and the mirror as an instance of *fascinatio*, but only occasionally compares the two phenomena as examples of action at a distance,[33] the commentators here studied turn the tables and use Albert's theory of vapour as the transmitter as the basis for an alternative explanation of fascination; in their view, fascination does not occur, as Avicenna claims in his commentary on the *De anima*,[34] by the agent's imaginative power taking control of an external body without any intervening medium. Instead, when the woman is agitated and imagines that she injures an external object, her imaginative power generates heat in her body, which in turn generates vapour, which in due course reaches the eyes, bringing harmful substances

[32] Aristoteles, *Physica*, VII.1, 242a23–26.

[33] On the relation between *fascinatio* in Albert and his interpretation of *De insomniis* 459b23–460a32, see also Palazzo (2009, 147; 153; 206).

[34] For Avicenna's account on action at a distance, see, in particular, *Liber de anima seu sextus de naturalibus*, pars IV, cap. 4 (ed. Van Riet, 54: 80–67: 75).

with it, and finally, via the air, reaches the external object and affects it in a harmful way.[35]

The accounts are far from identical. For example, the nature of the harmful substance varies: Brito claims that, when the woman is no more fertile, the menstrual blood is stored in her uterus where it decays and generates poisonous substances that are transmitted with the vapour to the external object and harm it.[36] Simon of Faversham is less specific but refers to certain *vapores mali*, which are clearly harmful because of the poisonous nature of the menstrual blood from which they are generated;[37] whereas, for instance, James of Douai refers much more vaguely to certain bodily *humores*, which are not explicitly associated with the menstrual blood.[38] The efficient cause of the phenomenon also varies somewhat. James and Simon both claim that fascination requires that the woman imagines that the object is harmed,[39] whereas Brito is less specific and seems to regard it as sufficient that the woman is agitated while she looks at the object.[40] Burley and Simon both claim that the process starts at the location of imagination wherefrom it gradually spreads to other parts of the body, whereas the other commentators are less specific about the starting-point.[41] These details aside, the commentators all convey the view that the

---

[35] As an example, see Radulphus Brito (ed. Ebbesen, 2016, 63): "Et per illum modum fit fascinatio, et non ut dicit Avicenna, quia post quadraginta communiter mulieres non concipiunt secundum Philosophum, et tunc desinunt menstrua fluere, et tunc conservantur in matrice sicut ibi putrefacta et sunt principium malorum venenorum, et cum tales vetulae commoventur ex aliquo, tunc fit magna agitatio istorum menstruorum per corpus, et tunc pars subtilis illorum evaporat<ur> per oculos, et illa pars quae sic exit per oculos inficit aerem contiguum oculis, et ille aer infectus inficit alium, et ideo, si sint aliqui multum passibiles sicut pueri, statim interficerent ipsos aspectu."

[36] See Radulphus Brito (ed. Ebbesen, 2016, 63), quoted in the previous footnote.

[37] Simon of Faversham (ed. Ebbesen, 2013, 140): "Menstruum est cuius contactu fruges non germinant, herbae arescunt, muscae moriuntur, et si canes illud gustent in rabiem efferentur; glutinum factum est quod nec acu nec ferro dissolvitur, sed corpore polluto ipso menstruo sponte dispergitur." In Burley, the connection between the *humores* generated in the process of *fascinatio* and the menstrual blood is even vaguer; see Walter Burley (ed. Thomsen Thörnqvist, 2014, 496: 5–12).

Simon's examples of the poisonous effect of the menstrual blood clearly originate from ancient sources; see Pliny, *Naturalis historia*, VII, 15 (ed. Mayhoff, 1909, 22–23): "Sed nihil facile reperiatur mulierum profluvio magis monstrificum. acescunt superventu musta, sterilescunt contactae fruges, moriuntur insita, exuruntur hortorum germina, fructus arborum, quibus insidere, decidunt, speculorum fulgor aspectu ipso hebetatur, acies ferri praestringitur, eboris nitor, alvi apium moriuntur, aes etiam ac ferrum robigo protinus corripit odorque dirus aera, et in rabiem aguntur gustato eo canes atque insanabili veneno morsus inficitur. quin et bituminum sequax alioquin ac lenta natura in lacu Iudaeae qui vocatur Asphaltites, certo tempore anni supernatans non quit sibi avelli, ad omnem contactum adhaerens praeterquam filo <***> quem tale virus infecerit. etiam formicis, animali minimo, inesse sensum eius ferunt, abicique gustatas fruges nec postea repeti."

[38] See James of Douai (ed. Ebbesen, 2015, 72: 18–22).

[39] James of Douai (ed. Ebbesen, 2015, 72: 17–18); Simon of Faversham (ed. Ebbesen, 2013, 140); cf. Walter Burley (ed. Thomsen Thörnqvist, 2014, 496: 5–6).

[40] See Radulphus Brito (ed. Ebbesen, 2016, 63), quoted in n. 33 above.

[41] Simon of Faversham (ed. Ebbesen, 2013, 140): "Si vetula imaginetur ad maleficium et corruptionem alicuius pueri, elevantur vapores mali circa imaginativam"; cf. Walter Burley (ed. Thomsen Thörnqvist, 2014, 496: 7).

phenomenon of the menstruating woman acting on a mirror at a distance is essentially identical to the phenomenon of *fascinatio* where a woman can kill another living being merely by looking at it and wishing it harm. This feature in the Latin commentaries on *De insomniis* does not appear out of the blue. Albert and Thomas Aquinas seem to have provided the basis for it jointly.

Albert's criticism of Avicenna's theory on *fascinatio*, his own solution to the passage in *De insomniis*, where the mover and the moved are brought into contact with each other,[42] together with the fact that he compares the menstruating woman to the basilisk as the conventional example of *fascinatio* lay the foundation of the later development. However, as mentioned, Albert never identifies the phenomenon of the woman and the mirror as a case of *fascinatio*. Thomas Aquinas, on the other hand, is able to help out with that when he, in several instances, sets up his own version of Albert's "Aristotelian" theory against Avicenna's explanation of fascination, where no physical contact is needed, and presents the former as the correct explanation. But to my knowledge, Thomas never ascribes the harmful effect of the woman's eyes to the menstrual blood; instead, it is the *spiritus* in Thomas' exposition that is infected by the bodily process generated by the woman's emotional change. For instance, Thomas claims in his commentary on the Galatians that when her *appetitus sensitivus* is affected by the perception or imagination of the object, this affection in turn changes the body, more precisely its *spiritus*, the *spiritus* in turn affects the air via the eyes, and finally the object.[43]

4.4 Conclusion

From the observations above, I believe we may conclude that Aristotle's passage on the menstruating woman and the mirror in 459b23–460a32 was associated to the theories of fascination at a very early stage of the Latin commentary tradition. The connection must have been made in sources pre-dating both Adam and Albert, because neither of them seems to be drawing on the other, and the Arabic adaptation

[42] In accordance with Aristotle's fundamental principle in *Physics* VII.2.

[43] See, for instance, Thomas Aquinas, *Super Epistolam ad Galatas lectura* (ed. Cai, 591–592): "Unde non est verum quod Avicenna dicit de huiusmodi fascinatione. Et ideo dicendum, quod ad imaginationem seu apprehensionem hominis, quando fortis est, immutatur sensus, seu appetitus sensitivus: quae quidem immutatio non est sine alteratione corporis et spirituum corporis, sicut nos videmus quod ad apprehensionem delectabilis movetur appetitus sensitivus ad concupiscentiam, et ex inde corpus calefit. Similiter ex apprehensione timendi, frigescit. Immutatio autem spirituum maxime inficit oculos, qui infecti rem per aspectum inficiunt, sicut patet in speculo mundo, quod ex aspectu menstruatae inficitur. Sic ergo quia vetulae obstinatae in malitia et durae sunt, ex forti apprehensione immutatur appetitus sensitivus, et ex hoc, sicut dictum est, infectio maxime fit a venis ad oculos, et ex oculis ad rem perspectam. Unde quia caro pueri mollis est, ad earum invidum aspectum inficitur et fascinatur. Et quandoque quidem ad hunc effectum Daemones operantur." For a more elaborate account of Aquinas' view, see Palazzo (2009, 145–146).

of the *Parva naturalia*, the peculiar *Kitāb al-Ḥiss wa-al-maḥsūs*, did not include *De insomniis* 459b23–460a32.[44] Albert then clearly provided an explanation of the phenomenon that fitted it into the general Aristotelian framework by presenting a theory which requires material contact between the agent and the patient and wards off the problem that Aristotle suddenly appears to be advocating an extramission theory of vision. With Geoffrey of Aspall, Albert's explanation had not yet become the standard solution, but in all the other medieval Latin commentaries on *De insomniis* here analysed, Albert's theory seems to have developed into an alternative "Aristotelian" explanation of *fascinatio*, where the alleged phenomenon of the menstruating woman and the mirror in *De insomniis* is presented as a process that is either essentially the same as *fascinatio* or an actual instance of the same, where the menstrual blood not only colours the object, but harms it because of its poisonous effect.

It should be noted that the medieval tradition on *De insomniis* 459b23–460a32 efficiently excludes the possibility that any corresponding phenomenon can be caused by men's semen. The commentators carefully connect the dots, referring to that fact that even though, according to Aristotle, both semen and menstrual blood are by-products of the digestive process,[45] the woman's digestion is less efficient than the man's because her body is colder.[46] Since it is the excessive amount of blood caused by the woman's deficient digestion that starts the whole process, the most well-known example of *fascinatio* in the medieval tradition, the phenomenon of the "evil eye", is inevitably and with solid support in Aristotle's other biological works, confined to the female sex.

[44] An edition of the work is being prepared by Rotraud Hansberger; see, for instance, Hansberger (2017).

[45] For semen as a residue of the digestive process, see Aristoteles, *De generatione animalium*, I.18, 724b24–725a4.

[46] See Aristoteles, *De generatione animalium*, I.19, 726b31–727a2 (trans. Peck): "Now (1) the weaker creature too must of necessity produce a residue, greater in amount and less thoroughly concocted; and (2) this, if such is its character, must of necessity be a volume of bloodlike fluid. (3) That which by nature has a smaller share of heat is weaker; and (4) the female answers to this description, as we have said already. From which we conclude that the bloodlike secretion which occurs in the female must of necessity be a residue just as much <as the secretion in the male>. Of such a character is the discharge of what is called the menstrual fluid." Cf., for example, Radulphus Brito (ed. Ebbesen, 2016, 61); James of Douai (ed. Ebbesen, 2015, 71.19–22); Simon of Faversham (ed. Ebbesen, 2013, 139–149); Walter Burley (ed. Thomsen Thörnqvist, 2014, 495.17–20). For Albert's theories on the role and consequences of indigested semen in the reproductive process, see Loconsole, Chap. 6 in this volume, pp. 91–113.

References

Primary Literature

Adam of Buckfield. *In De anima*. Transcription of MS Bologna, Bibl. univ. 2344, by Jennifer Ottman. The Richard Rufus of Cornwall Project: http://rrp.stanford.edu/BuckfieldDAn2.shtml.
Albertus Magnus. 1890. *Parvorum naturalium pars prima. Alberti Magni Opera omnia* 9, ed. Auguste Borgnet. Paris: Apud Ludovicum Vivès.
———. 1968. *De anima. Alberti Magni opera omnia* 7/1, ed. Clemens Stroick. Münster: Aschendorff.
———. 2008. *De homine. Alberti Magni opera omnia* 27/2, ed. Henryk Anzulewicz and Joachim Roland Söder. Münster: Aschendorff.
———. Forthcoming. *De somno et vigilia*, ed. Silvia Donati. Münster: Aschendorff.
Aristoteles. 1947. *De insomniis et De divinatione per somnum*, ed. Hendrik Joan Drossaart Lulofs. Leiden: Brill.
———. 1943. *Generation of Animals*, trans. Arthur Leslie Peck. Cambridge, MA: Harvard University Press.
———. 1955. *Parva naturalia. A Revised Text with Introduction and Commentary*, ed. and trans. David Ross. Oxford: Clarendon Press.
———. 1994. *Parva naturalia III: De insomniis, De divinatione per somnum. Werke in deutscher Übersetzung* 14.3, trans. Philip. J. van der Eijk. Berlin: Akademie Verlag.
———. 1996. *On Sleep and Dreams. A Text and Translation with Introduction, Notes and Glossary*, trans. David Gallop. Warminster: Aris and Phillips.
———. 2018. *On the Soul and Other Psychological Works. A New Translation*, trans. Fred D. Miller Jr. Oxford: Oxford University Press.
Avicenna. 1968. *Liber de anima seu Sextus de naturalibus IV–VI*, ed. Simone Van Riet. Louvain: Peeters; Leiden: Brill.
Geoffrey of Aspall. 2014. *Quaestiones super librum* De somno et vigilia: *An Edition*, ed. Sten Ebbesen. *Cahiers de l'Institut de Moyen-Âge Grec et Latin* 83: 257–341.
James of Douai. 2015. *James of Douai on Dreams*, ed. Sten Ebbesen. *Cahiers de l'Institut du Moyen-Âge Grec et Latin* 84: 22–92.
Pliny the Elder. 1909. *C. Plinii Secundi Naturalis historiae libri XXXVII*, vol. II: *Libri VII–XV*, ed. Karl F. Mayhoff. Leipzig: Teubner.
Radulphus Brito. 2016. *Radulphus Brito on Memory and Dreams: An Edition*, ed. Sten Ebbesen. *Cahiers de l'Institut de Moyen-Âge Grec et Latin* 85: 11–86.
Simon of Faversham. 2013. *Quaestiones super librum* De somno et vigilia*: An Edition*, ed. Sten Ebbesen. *Cahiers de l'Institut du Moyen-Âge Grec et Latin* 82: 90–145.
Thomas Aquinas. 1953. *S. Thomas Aquinatis Super Epistolas S. Pauli lectura*, ed. Raffaele Cai. Turin: Marietti.
Walter Burley. 2014. *Walter Burley's Expositio on Aristotle's Treatises on Sleep and Dreaming: An Edition*, ed. Christina Thomsen Thörnqvist. *Cahiers de l'Institut de Moyen-Âge Grec et Latin* 83: 379–515.

Secondary Literature

Burnett, Charles. 1996. The Introduction of Aristotle's Natural Philosophy into Great Britain: A Preliminary Survey of the Manuscript Evidence. In *Aristotle in Britain during the Middle Ages*, ed. John Marenbon, 21–50. Turnhout: Brepols.

Donati, Silvia. 2018. Albert the Great as a Commentator of Aristotle's De somno et vigilia: The Influence of the Arabic Tradition. In *The Parva naturalia in Greek, Arabic, and Latin Aristotelianism*, ed. Börje Bydén and Filip Radovic, 169–209. Dordrecht: Springer.

Ebbesen, Sten, Christina Thomsen Thörnqvist, and Véronique Decaix. 2015. Questions on *De sensu et sensato, De memoria et reminiscentia* and *De somno et vigilia*. A Catalogue. *Bulletin de philosophie médiévale* 57: 59–115.

Hansberger, Rotraud. 2017. Representation of Which Reality? "Spiritual Forms" and "ma'ni" in the Arabic Adaptation of Aristotle's *Parva naturalia*. In *The Parva naturalia in Greek, Arabic and Latin Aristotelianism: Supplementing the Science of the Soul*, ed. Börje Bydén and Filip Radovic, 99–121. Cham: Springer.

Hasse, Dag Nikolaus. 2016. Der Sturz des Kamels und die Befleckung des Spiegels: Fernwirkungstheorien in arabischen und lateinischen Kommentaren zu Aristoteles' De insomniis. In *ΣΩMA: Körperkonzepte und körperliche Existenz in der antiken Philosophie und Literatur*, ed. Thomas Buchheim, David Meißner, and Nora Wachsmann, 525–541. Hamburg: Felix Meiner.

Kovach, Francis Joseph. 1979. The Enduring Question of Action at a Distance in Saint Albert the Great. *The Southwestern Journal of Philosophy* 10 (3): 161–235.

Palazzo, Alessandro. 2009. Albert the Great's Doctrine of Fascination in the Context of his Philosophical System. In *Via Alberti: Texte–Quellen–Interpretationen*, ed. Ludger Honnefelder, Hannes Möhle, and Susana Bullido del Barrio, 135–215. Münster: Aschendorff.

Pinborg, Jan. 1969. Die Handschrift Roma Bibliotheca Angelica 549 und Boethius de Dacia. *Classica et Mediaevalia* 28: 373–393.

Thomsen Thörnqvist, Christina. 2022. How Dreams are Made: Some Latin Medieval Commentators on Dream Formation in Aristotle's *De insomniis*. In *Forms of Representation in the Aristotelian Tradition*, vol. 2: *Dreaming*, ed. Christina Thomsen Thörnqvist and Juhana Toivanen, 150–177. Leiden: Brill.

———. 2016. Sleepwalking Through the Thirteenth Century: Some Medieval Commentaries on Aristotle's De somno et vigilia 2.456a24–27. *Vivarium* 54 (4): 286–310.

Weisheipl, James Athanasius. 1980. Albert's Works on Natural Science (libri naturales) in Probable Chronological Order. In *Albertus Magnus and the Sciences. Commemorative Essays*, ed. James Athanasius Weisheipl, 565–577. Toronto: Pontifical Institute of Mediaeval Studies.

Chapter 5
Pygmies, Twins, Monsters: Human Nature and Its Borderlines in Albert the Great

Evelina Miteva

Abstract Albert the Great put the human being at the centre of his philosophical interest, praising the human being as "the most perfect of all living beings." In this paper, I explore two cases in which human nature is stretched to the extreme in order to obtain a more complex view of what Albert thought of human nature. First, I discuss pygmies, who seem to him to be an intermediary stage between the animal and human realms. By examining their intellectual capacities, as well as their consequent ethical and social behaviour, Albert identifies what is specific to the human being. Pygmies possess a "shadow of reason" (*umbra rationis*). He draws clear line between animals and humans, based on the division of the soul's potencies. In the second case study, I focus on twins. In *De XV problematibus* and in *De fato*, Albert discusses why twins have different fates. Twins share the same hour of birth and hence are influenced by the same stars and constellations in their lives, but they may nevertheless lead entirely different lives and have different skills and inclinations. Albert offers a naturalistic explanation of the diverging fates of twins. In the conclusion, I look into Albert's concept of *monstra*, i.e. those creatures that fail to fulfil the intention of their nature (*praeter intentionem naturae*), in order to illumine the way he saw human nature as a norm and his perception of deviations from this norm.

Keywords Albert the Great · Human nature · Twins · Pygmies · Intention of nature

---

This article has been written within the project "Human Being As a Cross-point: the Beginnings of the Modern Concept of Man in the Middle Ages" (PN-III-P1-1.1-TE-2016-2351), Babeş-Bolyai University, Cluj-Napoca.

E. Miteva (✉)
BBU Cluj Napoca, Cluj-Napoca, Romania
e-mail: evelinamiteva@yahoo.co.uk

M. Gensler et al. (eds.), *The Embodied Soul*, Historical-Analytical Studies on Nature, Mind and Action 11, https://doi.org/10.1007/978-3-030-99453-2_5

5.1 Introduction

Homo animal nobilissimum. Homo inquantum homo, solus est intellectus. Homo nexus dei et mundi.

These statements are some of Albert the Great's most famous slogans and, at the same time, some of the most enthusiastic definitions of the human being. Yet, what remains of these shiny definitions once we step out of the theory of intellect and turn to the multifaceted world of particular and individual manifestations of human nature? What can be said of the human intellect and free will, when we consider the cases in which human nature was believed to take a less-than-perfect shape, such as women, children, Ethiopians, twins, and foolish or monstrous people? In this chapter, I will focus on three cases in which human nature presses against its limits.

Each of the case studies from the field of natural philosophy I discuss, i.e., the pygmies, twins, and monstrosity, pertains to a broader philosophical problem. I will first ponder pygmies. The African tribe was known to medieval scholars from ancient sources and exhibited, according to them, various human-like behaviours, thus raising the question of what specifically constituted human nature.[1] If reason is what distinguishes humans from animals, and pygmies exhibit some traits of reason, such as speech and poetry, how should the distinction be upheld? My second case study focuses on twins. Twins presented several philosophical challenges to the scholars, such as the possibility for a single human nature to generate two individuals at once, or the difference in the fates of these individuals despite their common origin and hour of birth. Driven by his fascination with the unusual, Albert often revisited the case of twins, and below I will focus on the problems associated with the causes of the generation of twins and with their fates. To finish with, I will address the notion of the monster, which, technically speaking, refers to various deviations from human nature, drawing attention to the question of the "intention of nature" and the very idea of human nature. In doing so, I will highlight the tension between particular cases of human nature and its general definition as *animal rationale*, which has not received proper attention yet. What is the "norm" that the definition of "human nature" imposes on individuals, and what ramifications do they face if they deviate from a presumed ideal status?

5.2 Pygmies and Human Reason

In *De animalibus*, in the chapter dedicated to migrating animals (*de mutatione animalium de loco in loco*), Albert relates[2] that cranes, migrating from Greece to Egypt, where the river Nile flows, are said to arrive at the land of brute animals, such as

[1] Characteristically, the *quaestio* of Peter of Auvergne dedicated to pygmies is called *Utrum pygmei sint homines*. On the topic of pygmies in Peter of Auvergne and in Albert the Great, cf. Köhler (2008, 419–444); Köhler (1992, 718–735); Koch (1931, 194–213).

[2] Albertus Magnus, *De animalibus*, VII, tr. 1, cap. 6 (ed. Stadler, 1916, 521: 25 ff.).

monkeys (*simia*) and pygmies (*pygmei*). Pygmies "are called human beings" (*homines vocantur*), because they are very similar to people and have an erect posture. They are said to be one cubit tall (*cubitales*—etymologically, this is basically the meaning of the Greek word *pygmeos*), to reach reproductive maturity at the age of three, and to die when they are eight or even younger. It is reported that they live near the springs of the Nile and fight against the cranes migrating to their lands. "Et est rei veritas," affirms Albert, and goes on to say that many travellers have seen this kind of "animal" (*talia enim animalia, quae pigmei dicuntur, multi viderunt*), and that pygmies also have tiny horses, which they ride. This description of pygmies and their customs is followed by an account of their social characteristics. Even if pygmies are able to perform typically human operations, such as horse-riding, they lack the "usage of reason, the sense of shame and honesty, do not honour justice or practise any organization of state."[3] They do bear some similarity to humans; for example, they have language and speech, yet these are imperfect.

Albert reiterates this general depiction on various occasions, primarily in *De animalibus*, and adds various details to it. He apparently has no doubt that pygmies exist. However, despite the lengthy descriptions of human traits in pygmies, he is convinced that they are not human beings in the proper sense of the term. Nonetheless, he is also mostly reluctant to call them "animals." Albert usually calls pygmies *similitudines hominis*, for instance, when describing certain bodily features that only humans and some humanlike species have, such as breasts on the chest: this is something only humans, pygmies and monkeys have.[4] Only the human and pygmy females have identically structured vulvae, and, accordingly, the two species copulate in the same manner.[5] With scientific zeal and without any moral judgment at this point, Albert devotes equal attention to all of the features involved in the comparison. The list of similarities goes on to include several physiological characteristics: Humans and pygmies cannot sleep while standing. Since the heat of the body cools down in sleep and retracts downwards and inwards in the body, erect animals with heavy heads, such as humans and pygmies, need to lie down to sleep.[6] Both species have an erect posture, which makes pygmies a species that resembles human beings

[3] Albertus Magnus, *De animalibus*, VII, tr. 1, cap. 6 (ed. Stadler, 1916, 521: 35–36): "Sed usum rationis non habent nec verecundiam neque honestatem, neque iustitiam colunt, neque indicium reipublicae exercent."

[4] Albertus Magnus, *De animalibus*, II, tr. 1, cap. 2 (ed. Stadler, 1916, 228: 3–6): "Unum autem pectoris membrum manifestum est mamilla, quam solus homo habet ante pectus suum, et hominis quaedam similitudines, sicut est simia et pygmeus et alia quaedam animalia homini similia, de quibus inferius loquemur."

[5] Albertus Magnus, *De animalibus*, II, tr. 1. cap. 3 (ed. Stadler, 1916, 228: 3–9); ibidem, XXI.1.3 (ed. Stadler, 1920, 1332: 26–30): "[…] et feminae eorum mamillas habent pendentes in pectore sicut femina hominis, et habent vulvas magnas respectu suae quantitatis et sunt positae in fine ventris, ita quod fissura vulvae ascendit versus umbilicum sicut disponitur vulva mulieris: et coeunt haec animalia femina in dorso posita et masculo incumbente ei sicut coeunt homines."

[6] Albertus Magnus, *De animalibus*, XII, tr. 2, cap. 3 (ed. Stadler, 1916, 844: 37–40): "Et quia calor agitatur ad inferius et interius, ideo erit somnus. Et ideo animal erecti corporis in somnis non potest elevare suum caput, et praecipue ponderosi capitis, sicut homo et pigmaeus."

the closest of all animals.[7] Besides, only humans and pygmies have the frontal part of their heads formed into a face.[8] Only humans, pygmies and monkeys are unable to move their ears (unless humans have exercised to do so since a young age, through the muscles linked to the jaw).[9] Albert also points out to the roundness of the head, resembling the sphere, as another similarity, and notices that it consists of three different compartments (*cellulae*), which is particularly interesting in the light of the brain anatomy.[10]

The shape of the hands, the tongue and the lips is a special case among the physiological signs of similitude. The former is commonly associated with reason, while the latter two are linked to articulate speech. At this point, the dissimilarities between humans and pygmies begin to prevail in Albert's description, suggesting why he thought they were not humans. Humans and pygmies share the capacity for speech. The tongues of pygmies are formed so that they can speak like humans.[11] Albert is careful to point out that theirs is not some "animal language," but articulate and meaningful speech. Humans and pygmies alike boast *sermo*, that is, an "articulated and written utterance, intimating the concept of the human mind."[12] The idea that pygmies form mental concepts and voice these in structured speech suggests that they are endowed with rationality, and that would be a strong point for defining them as human beings. Albert reflects on this apparent conundrum and repeats, on several occasions, that pygmies make use of language and, hence, of concepts.[13] The use of mental concepts and of language is a typically human feature,[14] so while

[7] Monkeys, on the other hand, which are also often compared to humans due to their resemblances, stand on their back feet only occasionally. Albertus Magnus, *De animalibus*, XIV, tr. 2, cap. 1 (ed. Stadler, 1920, 963: 6–9); cf. ibidem, XII, tr. 2, cap. 3 (ed. Stadler, 1916, 844: 38).

[8] Albertus Magnus, *De animalibus*, XII, tr. 3, cap. 6 (ed. Stadler, 1916, 888: 26–32). Monkeys, once again, do not have a "face" (*facies sive vultus*).

[9] Albertus Magnus, *De animalibus*, I, tr. 2, cap. 4 (ed. Stadler, 1916, 63: 12–17).

[10] Albertus Magnus, *De animalibus*, XXI, tr. 1, cap. 3 (ed. Stadler, 1920, 1332: 19–21).

[11] Albertus Magnus, *De animalibus*, XII, tr. 3, cap. 5 (ed. Stadler, 1916, 879: 25–29): "Creatio enim linguae in homine et pigmaeo et forte in aliis quibusdam loquentibus ad duplicem est usum, ad gustandum scilicet sapores et ad sermocinandum."

[12] Albertus Magnus, *De animalibus*, IV, tr. 2, cap. 2 (ed. Stadler, 1916, 398: 2–7): "Est enim sermo vox articulata et litterata, conceptum mentis hominis denuntians, et his non competit nisi homini et paucis quae ad hominem habent affinitatem sicut pigmeus et id quod supra nominavimus marytonmorion et huiusmodi."

[13] Albertus Magnus, *De animalibus*, I, tr. 1, cap. 3 (ed. Stadler, 1916, 18: 14–15): "Adhuc autem formantium voces quaedam formant ad diversos conceptus quos habent, sicut homo et pigmeus." Cf. ibidem, IV, tr. 2, cap. 2 (ed. Stadler, 1916, 398: 2–7, footnote 12).

[14] Albertus Magnus, *De animalibus*, VIII, tr. 6, cap. 2 (ed. Stadler, 1916, 672: 24–31). In animals, vocalizations are rather expressions of a certain desire or they are caused by fantasy rather than concepts (*affectus delectationem magis quam conceptus aliquos significantes*). Humans, on the contrary, use and combine concepts freely in their language. In the same vein, animals cannot make plans for the future, but rather act out of instinct. This is why even when it seems that birds plan for their future when building nests, they do so out of instinct. Nature cannot act freely and all swallows build their nest in the same way, Albert's example states. Cf. Albertus Magnus, *De animalibus*, VIII, tr. 2, cap. 3 (ed. Stadler, 1916, 592: 1–23); Albertus Magnus, *De XV problematibus*,

Albert seems fascinated by the idea that pygmies have a meaningful language (*sermo*), he is also invariably careful to make it clear that their concepts are based on the imagination alone and not on the intellect.[15] Albert takes time to explain how language is possible for a non-human tribe. He examines the faculties of the soul that are involved in the formation of this—articulate, but non-abstract—language. The imagination, the memory and the estimation (*imaginatio, memoria, aestimatio*) contribute to this process, and, for each of them, Albert insists on the particular and individual character of the information collected by the senses and communicated through words.[16] The language of pygmies is limited to sense perceptions; it is a reaction to stimuli and particular situations. For Albert, the distinction between the sensitive and the rational soul is the demarcating line between humans and non-humans. No matter how developed the sensitive soul of pygmies may be, seemingly fostering human capacities such as language, it still remains merely a bodily capacity. The immaterial and immortal intellectual soul is a characteristic of the human being alone.

The same reasoning holds true for the hands, which in the Aristotelian line of thought are considered the supreme physiological expression of the human being's intellect. The hand is *organum organorum*,[17] because it is a "clever" tool capable of human-specific actions, such as writing, craftsmanship and art. Pygmies are indeed capable of using their hands in many ways like humans (*utuntur manibus in multis usibus sicut homo*), but are unable to use them to produce artworks.[18] The distinction that Albert makes here is one between the intellect as able to act freely, that is, to create, and the sensitive soul as being guided by nature and inborn instinct.[19] Alber draws

1 (ed. Geyer, 1975, 33, 49–53); Albertus Magnus, *De natura et origine animae*, tr. 2, cap. 8 (ed. Geyer, 1955, 31: 86–32: 1); Albertus Magnus, *Physica*, II, tr. 1, cap. 2 (ed. Hossfeld, 1987, 79: 5–16); ibidem, II, tr. 3, cap. 2 (ed. Hossfeld, 135: 58–72).

[15] Albertus Magnus, *De animalibus*, IV, tr. 2, cap. 2 (ed. Stadler, 1916, 400: 33–38): "Locutio enim cum intellectu appropriatur homini soli: et omne quidem loquens habet vocem, sed non convertitur. Et si aliquod genus simiarum simile homini loquitur, hoc est ex imaginatione simplici et non ex intellectu, sicut pigmeus et forte id quod vocavimus superius maritonmorion, et forte quaedam huiusmodi alia inveniuntur."

[16] Albertus Magnus, *De animalibus*, XXI, tr. 1, cap. 3 (ed. Stadler, 1920, 1329: 12–32). Cf. ibidem (ed. Stadler, 1920, 1331: 22–42), where Albert discusses if some animals could have imperfect syllogisms.

[17] Albertus Magnus, *De animalibus*, XXI, tr. 1, cap. 2 (ed. Stadler, 1920, 1328: 36–37): "Manus est instrumentum intellectus." Cf. ibidem, XXI, tr. 1, cap. 1 (ed. Stadler, 1920, 1324: 21–22): "[...] Organum organorum, et organum intellectus operativi." Cf. Aristoteles, *De partibus animalium*, IV.10, 687a2 ff.

[18] Albertus Magnus, *De animalibus*, XXI, tr. 1, cap. 3 (ed. Stadler, 1920, 1328: 40–41); ibidem, XXI.1.3 (ed. Stadler, 1920, 1332: 23–24).

[19] Albertus Magnus, *De animalibus*, XIV, tr. 2, cap. 2 (ed. Stadler, 1920, 964: 35–965: 6): "Cum enim manus sit organum non unius sed multarum speciei operationum et genere (universaliter enim omnis generis fiunt manu) erit signum eius quod universale principium est omnis generis operationum. Hoc autem non est natura aut virtus animae in modum naturae operans: haec enim non omnis generis operationes, sed unius formae et speciei operationem perficiunt sicut ignis ignire et omnis hyrundo omni hyrundini similiter facit nidum: manu autem et textura et sutoria et aedificatoria et quaelibet artium non uno modo, sed omni modo perficitur. Erit igitur manus sig-

an even more curious line between the animal language of pygmies and the language of humans. He admits that as the sensitive soul of pygmies is furnished with highly developed faculties, they rely on their memory, imagination and estimative powers in order to form and pronounce sensible intentions, i.e., to speak. But this process only accounts for poetical syllogisms, the worst syllogisms of all, according to Albert. Even though one could argue that bad reasoning is (human) reasoning anyway, Albert takes this is as a clear sign that pygmies do not use reason.[20]

Certain social practices are also a consequence and a sign of pygmies' lack of reason and their incapacity to build abstract syllogisms. They do not follow laws and do not have a "perfect society." They are like savages (*homines silvestres*) or like "morons" (*moriones*), incapable of correct and universally valid discernment and, therefore, incapable of forming a proper society based on universal rules and laws. The difference is that "morons" are endowed by nature with a human intellect, which has however been damaged or impeded by an accident or "by melancholy."[21]

Albert permits some sensible concepts and partial practical syllogisms even in animals and discusses, among other topics, whether animals could make free decisions based on these practical syllogisms.[22] In his view, pygmies are a special case of a highly developed animal species that may deliberate, has speech like humans and even exhibits similar facial traits (unlike monkeys), but lacks the "form" of humans, that is, the intellect.[23] Appearance is not a form, says Albert: *figura non est forma*;[24] and even if he calls pigmies the people of the woods (*homines silvestres*) at

---

num principii quod conceptu generaliter omnium formarum omnis generis operationis et contemplationis est principium. Natura enim sicut diximus et etiam forma in modum naturae operans semper in operando manet secundum unam dispositionem." Cf. ibidem, I, tr. 1, cap. 3, 18–21: "Sicut pigmeus, qui non sequitur rationem loquelae, sed naturae instinctum." Cf. footnote 13.

[20] Albertus Magnus, *De animalibus*, XXI, tr. 1, cap. 3 (ed. Stadler, 1920, 1329: 1–4): "Et hoc signum est quod nihil habet de iudicio rationis: propter quod etiam rhetoricis persuasionibus in loquendo non utitur neque poeticis quae tamen imperfectiores sunt omnium rationum: et ideo semper silvestris manet nullam prorsus civilitatem custodiens."

[21] Albertus Magnus, *De animalibus*, XXI, tr. 1, cap. 3 (ed. Stadler, 1920, 1328: 26–32): "Et ideo pigmeus nihil omnino percipit de rerum quidditatibus nec umquam percipit habitudines argumentorum: et sua locutio et sicut locutio morionum qui naturaliter stulti sunt eo quod non perceptibiles sunt rationum. Sed in hoc est differentia quod pigmeus habet privationem rationis ex natura, morio autem habet per accidens ex melancolia vel alio accidente non privationem rationis, sed potius privationem usus rationis."

[22] Oelze (2018, 150–155); Miteva (2018b, 11–28); Tellkamp (2016, 109–123); Tellkamp (2013, 205–222).

[23] On the role of intellect as a substantial form, see Trifogli, Chap. 10 in this volume, pp. 191–211; in the genetic transmission of the intellectual capacity, see Loconsole, Chap. 6 in this volume, pp. 91–113.

[24] Albertus Magnus, *De animalibus*, XI, tr. 2, cap. 4 (ed. Stadler, 1916, 796: 24–31): "Multa autem physicorum in figura convenientium diversas habent formas suarum quidditatum: et si talibus propter figurae convenientiam eadem attribuantur, peccatum erit sicut in divisione aequivoci: et ideo cavendum est ab huiusmodi aequivocatione. Licet enim homo et simia et pigmeus in figura plurium membrorum conveniant, non tamen consequens est, quod conveniant in forma et accidentibus quae formam communiter consequuntur." Cf. ibidem, I, tr. 1, cap. 3 (ed. Stadler, 1916, 17: 33–35): "Si qui autem homines sunt silvestres sicut pigmeus, non secundum unam et eandem rationem nobiscum sunt dicti homines, sed habent aliquid hominis in quadam deliberatione et loquela."

times, this is only a figurative expression, and he takes care to avoid all ambiguity and to clarify that they are not real human beings. They have "something human,"[25] are "similar to humans"[26] and have a "shadow of reason,"[27] but they are nonetheless clearly "animals," even if they are the most perfect of animals and closest to humans in the hierarchy of creatures.[28]

Albert did not see any pygmies; nor is he likely to have known people who saw pygmies. He relied on the accounts that he found, such as the passages about the tiny horses of pygmies and their war with cranes in Aristotle's *Historia animalium*. A chapter in Augustine's *De civitate Dei* which describes certain races that descend from Noah contains some other details that appear in Albert, for example, that "women conceive in their fifth year and do not live beyond their eighth."[29] To these sources Albert added some further theoretical considerations stemming from the idea that pygmies only possess the sensitive soul. Albert's full-fledged descriptions of pygmies and his conclusions about their non-human nature were based on theoretical arguments from natural philosophy. When Europeans started travelling, conquering and colonizing in the sixteenth and seventeenth centuries, they came in touch with short people (no more than 150 cm tall) as well and had then at their disposal a science and a lexicon immediately to identify them as pygmies, i.e., a funny and inferior race of people lacking reason.[30]

5.3 Twins and Human Fate

Twins have been an attractive case study for centuries for all kinds of reasons, including the study of the nature of individuality, the status of extraordinary events in nature, or the medical process that leads to this rare occurrence.[31] Twins are an exception from the "norm" of human children being born one at a time. This excep-

[25] Albertus Magnus, *De animalibus*, I, tr. 1, cap. 3 (ed. Stadler, 1916, 17, 35, footnote 22).

[26] Albertus Magnus, *De animalibus*, I, tr. 2, cap. 4 (ed. Stadler, 1916, 63: 13–14); ibidem, XXI, tr. 1, cap. 3 (ed. Stadler, 1920, 1332: 10); ibidem, IV, tr. 2, cap. 2 (ed. Stadler, 1916, 400: 35). Occasionally this slogan *hominis similitudo* refers to monkeys as well, cf. ibidem, XIV, tr. 2, cap. 3 (ed. Stadler, 1920, 969: 24–25). In a synonymous manner *similitudo* is rendered also by *affinitas*, in *De animalibus*, IV, tr. 2, cap. 2 (ed. Stadler, 1916, 398: 5–7): "Ad hominem habent affinitatem sicut pigmeus et id quod supra nominavimus marytonmorion et huiusmodi."

[27] Albertus Magnus, *De animalibus*, XXI, tr. 1, cap. 3 (ed. Stadler, 1920, 1328: 18); ibidem (ed. Stadler, 1920, 1328: 23); ibidem (ed. Stadler, 1920, 1329: 5); ibidem (ed. Stadler, 1916, 1329: 28).

[28] Albertus Magnus, *De animalibus*, XXI, tr. 1, cap. 3 (ed. Stadler, 1920, 1328: 4–6): "Pigmeus qui loquitur, cum tamen sit irrationabile animal: et ideo quantum ad animales virtutes, post hominem videtur pigmeus esse perfectius animal." Cf. ibidem (ed. Stadler, 1920, 1329: 29–30): "Magis tamen secundum naturam vicinus est bruto quam homini ut patet per antedicta."

[29] Aristoteles, *Historia animalium*, VIII.12, 597a4; Augustinus, *De civitate Dei*, XVI, cap. 8 (ed. Dombart and Kalb, 1955, 508: 1 ff.).

[30] Lüthy (2013, 11–28).

[31] Cf. Zuccolin (2017, 2019); Resnick (2013); Thijssen (1987).

tion was treated either as a sign of particular vigour and strength in the process of generation, as in the Hippocratic tradition, or as a deviation from a principle. The latter position was embraced by Aristotle, and it was largely adopted by Albert. In this framework, twins are, as the Aristotelian position claims, a special case in reproduction, caused by a defect or an abnormality. The philosophically underpinned notion of the "abnormality" of twins resulted from a deficit of knowledge. In this part, I will look into Albert's attempt to explain the generation of twins from the perspective of natural philosophy, and I will then examine how natural philosophy served to explain their different fates.

It is in *De animalibus* again that Albert systematically discusses the causes of twin births.[32] At the very beginning of the chapter on twin births, he admits that he does not know why exactly such events take place. In fact, the entire reproductive process was a great unknown in the Middle Ages, and physicians and philosophers were equally at a loss as to the process of conception and gestation. Spermatozoa and egg cells were obviously a much later discovery, which was only made possible by the discovery of the microscope; but even the anatomical knowledge of the reproductive organs was still uncertain at the time.[33] Thus, Albert starts from the basic idea that if two human beings are produced, this must be due to a division of the original reproductive material. Thus, the question arises what this reproductive material involved in the conception is, and what the possible causes of its division can be. The question of the reproductive material was actually one of the major and longstanding debates between medics and philosophers from Antiquity on, since it posed the problem of the female contribution to conception.[34] Albert sides with the Aristotelian position that only the male sperm has formative power in conception, but he permits some limited contribution from the "female sperm" as well. This means that twins must be formed either by the division of the sperm or by the division of the female semen. Albert systematically inspects these divisions. The cause of the division of the sperm lies either with the father or with the mother. On the part of the father, the sperm can be divided either because it comes from two testicles, or because it is ejaculated in several "impulses," usually no more than three or four, where every subsequent impulse is of lesser power and pleasure to the man than the preceding one(s). When the interval between these impulses is too long, the sperm "remains divided" (*remanet sperma divisum a spermate*), instead of reaching the

[32] Albertus Magnus, *De animalibus*, IX, tr. 2, cap. 6 (ed. Stadler, 1916, 727: 9–10): "Et est digressio vere declarans causam geminorum et modum et tempus naturalis partus."

[33] Spermatozoids could be seen after the discovery of the microscope in the seventeenth century, but this did not lead to any essential change in the ideas about human reproduction. It was only in 1882 that the female reproductive cells were described. Even then, however, it took a while for the predominant theory of "preformation" to change. The theory claimed that the entire future embryo was "preformed" in the sperm. And while ovaries had been known to medics since Antiquity, it was not before the Renaissance that the uterine tubes were discovered by Gabrielle Falloppio in the fifteenth century.

[34] Siraisi (1980); Vinaty (1981); de Asúa (1997); Miteva (2018a).

uterus all at once.[35] On the part of the mother, Albert distinguishes three possible causes of twin conception. One of them is related to the female semen, which, analogously to the male sperm, is ejaculated in impulses, and these can cause the division of the sperm. Another possible cause is that the uterus "swallows" or "sucks up" the sperm, which happens in several sequences; thus, the sperm can be absorbed at different places in the womb, producing two foetuses.[36] The movement of the womb is the third mother-related cause of the formation of multiple foetuses. As in the case of men's ejaculation, also these three cases on the female's side (that is, female ejaculation, contractions of the matrix and movement of the womb) are regarded as causes for the formation of multiple foetuses, as well as expressions of the male and female orgasms respectively. However, Albert neither assesses orgasm in moral terms nor regards it as a condition of a better—or worse—conception rate; instead, he treats orgasm merely as a scientific explanation of the occurrence of multiples.[37] This is rather unusual, since the female orgasm in particular has traditionally been associated either with better chances for conceiving, or, on the contrary, with a risk for the formation of an embryo through the movements of the womb.[38]

All of these explain the generation of twins in one act of sexual intercourse. Albert thus seems to share the view that twins are mostly generated in one act, since the phrase *ex uno coitu* basically occurs without any further elaboration whenever

---

[35] Albertus Magnus, *De animalibus*, IX, tr. 2, cap. 6 (ed. Stadler, 1916, 727: 14–31). Cf. Albertus Magnus, *Quaestiones super De animalibus*, X, q. 4 (ed. Filthaut, 1955, 216: 29–31); ibidem, XV, q. 17 (ed. Filthaut, 1955, 270: 39–40).

[36] Albertus Magnus, *De animalibus*, IX, tr. 2, cap. 6 (ed. Stadler, 1916, 727, 32–728, 3). In *De animalibus*, X, tr. 1, cap. 2 (ed. Stadler, 1916, 739: 15–30), Albert restates in a more generic manner that the abundance of semen absorbed at different places in the womb is the cause of the generation of twins. In both passages, Albert's wording describing how the womb "swallows" (*glutit*) and "sucks up" (*sugit*) the male semen is strongly reminiscent of the so-called "upsuck" theory, which has survived until the present day. The "upsuck" theory claims that the spasms of the uterus during orgasm help the sperm get to the egg and fertilize it. This theory remained influential well into the twentieth century, but it is heavily contested today. Cf. Shackelford and Pound (2006, 189, 203).

[37] The same principle applies not only to humans but generally to all animals. Cf. Albertus Magnus, *De animalibus*, VI, tr. 3, cap. 2 (ed. Stadler, 1916, 487: 23–27): "Et aliquando pariunt gemellos, si cibum habuerint convenientem, maxime si aries aut caper talis sit naturae in habundantia spermatis, quod gemellos consueverit generare, et si mater similiter habuerit hanc consuetudinem ex habundantia humiditatis et motus matricis in delectatione concubitus."

[38] Albertus Magnus, *De animalibus*, IX, tr. 2, cap. 6 (ed. Stadler, 1916, 728: 4–11): "Mulier in coitu tres istas habet delectationes, scilicet proiciendo spermati similem humorem, et tunc delectatur totiens, quotiens proicit, sicut et vir, et glutiendo sperma virile, et tunc totiens delectatur quotiens glutit, et quando matrix movetur, et tunc totiens delectatur quotiens motum dat matrix: propter quod quaedam feminae apprehendunt collum vesicae in vulva et trahunt, ut sic mota matrice delectentur." Cf. Resnick (2013, 351) and Zuccolin (2019, 157) quote the example of John of Jandun, who warns that the woman's excitement during intercourse may produce "monster twins," i.e., conjoined twins.

the causes of twin births are discussed.[39] This stands in tacit opposition to the Aristotelian take on the causation of twins. Aristotle famously offers a pejoratively colored interpretation of the origin of twin births, ascribing them to consecutive instances of impregnation. He exclusively discusses cases of twins that are caused by superfetation, and the sole examples he provides are those generated in adulterous acts, due to which the twins have two different fathers. As these examples are cited when Aristotle describes women's alleged particular lustfulness when pregnant, they bear clearly negative moral connotations.[40] But even in the passages in which Aristotle simply refers to the biological reasons for the generation of twins, he categorizes it as abnormal. Produced by a superabundance of generative material, i.e., by an imbalance in nature, "such creatures seem rather to be monstrosities, because their formation is contrary to this general rule and to what is usual."[41] Thus, Aristotle gives an explicitly negative description of the phenomenon of twins, be it as an unusual occurrence or as an immoral adulterous superfetation.

While Albert's study of the causes of the generation of twins in *De animalibus* abides by the terms of natural philosophy and withholds moral judgment, elsewhere he endorses the Aristotelian paradigm. In his commentary on the *Physica*, the generation of twins is directly linked to the generation of monstrously deformed people, such as the one born "in our times" (*tempora nostra*) who had eleven mouths, eleven tongues and twenty-two incompletely formed lips. This malformed foetus was stillborn.[42] In both cases, the underlying cause is the division of the superabundant generative material. The initial material exceeds the amount needed for the normal process of generation. If this material becomes proportionally divided, twins are born; in other cases, malformations are engendered, such as conjoined twins or superfluous bodily organs.

The unusual case of having twins[43] stirred several more theoretical and practical doubts and confusions. Another question of lively interest was, for instance, how many multiples may be conceived at a time. In his *Historia animalium*, Aristotle claims quintuplets (i.e., five children) is the maximum that a woman can carry, "especially in certain parts of the world," and Albert adheres to this number, also repeating Aristotle's example of an Egyptian woman who had twenty children at

---

[39] Albertus Magnus, *De animalibus*, X, tr. 1, cap. 2 (ed. Stadler, 1916, 739: 14): "In uno et eodem coitu [...]."; ibidem (ed. Stadler, 1916, 739: 16–17): "[Gemini] frequenter sunt ex uno concubitu."; ibidem, X, tr. 1, cap. 4 (ed. Stadler, 1916, 744: 38); in *De animalibus*, IX, tr. 2, cap. 6 (ed. Stadler, 1916, 727: 15–16), having listed all the causes of the generation of twins in one coitus, Albert only cursorily mentions the possibility of superimpregnation.

[40] Aristoteles, *Historia animalium*, VII.4, 585a3–24.

[41] Aristoteles, *De generatione animalium*, IV.4, 772a36–772b1.

[42] Albertus Magnus, *Physica*, II, tr. 3, cap. 3 (ed. Hossfeld, 1987, 137: 63–138: 4). Cf. *De animalibus*, XVIII, tr. 2, cap. 2 (ed. Stadler, 1920, 1223: 32–34).

[43] One should bear in mind that twin births were far rarer before the invention of fertility treatments. According to statistically underpinned medical research, twin birth rates have risen by 50–70% since 1980, as studies from France (Pison, Couvert, 2004) and the U.S. (Martin et al. 2012) show. Before, pregnancies with more than two babies were high-risk pregnancies, with a low survival rate due to premature birth, which made twins a highly exceptional phenomenon.

four births, many of whom survived.[44] However, Albert also gladly cites Avicenna's story of a woman who aborted 22 foetuses and another woman who aborted 70 formed foetuses.[45] Albert has himself been told "by a trustworthy and experienced physician" about a noble woman in Germany who gave birth to 60 babies, five at a time, none of which lived for more than half a year.[46] The same trustworthy medic has also told Albert of another noble woman he treated who aborted 150 well-formed foetuses, the size of a human little finger, who were pulsating and showed various signs of life, and even had well-formed fingers and toes, as subtle as hair.[47] In the good tradition of Aristotle and Avicenna, Albert takes pleasure in telling such stories, especially as first-hand accounts, and mentions another case of a woman he knew of in Worms who had a completely flat chest, like a man, and yet gave birth to many children, once even to triplets.[48]

One of the main topics of interest—an interest that has not in the least decreased in our times—is the fate of twins. Albert addresses this question in the first article of *De fato* (1256), which ponders whether there *is* such a thing as fate and astral influence, and addresses the problem of free will and free action. Albert formulates the question concerning the fate of twins on the basis of the biblical story of Jacob and Esau, the fraternal twins of Rebekah and Isaac (Gen. 25). Jacob and Esau were so dissimilar that they argued even in their mother's womb. At birth, one of them was dark and hairy, was going to become a hunter, and was his father's favourite; the other was a smooth and calm boy, his mother's favourite. Jacob tricked Esau into giving his firstborn right over to him. In using the case of Jacob and Esau to introduce the problem of fate, Albert followed Augustine, who repeatedly referred to this

44 Aristoteles, *Historia animalium*, VII.4, 584b34–36; Albertus Magnus, *De animalibus*, IX, tr. 1, cap. 5 (ed. Stadler, 1916, 693: 14–15): "Maior autem numerus ad quem aliquotiens perveniunt nati, est quinque."

45 Avicenna actually talks of 33 fetuses. Cf. Avicenna, *De animalibus*, IX, cap. 10, 44r. Cf. Albertus Magnus, *De animalibus*, IX, tr. 1, cap. 5 (ed. Stadler, 1916, 693: 22–24).

46 Albertus Magnus, *De animalibus*, IX, tr. 1, cap. 5 (ed. Stadler, 1916, 693: 17–20): "Et narratum est mihi a veridico et experto medico, quod quaedam nobilis mulier in Germania peperit sexaginta filios qualibet vice quinque, et nullus eorum venit ad aetatem dimidii anni."

47 Albertus Magnus, *De animalibus*, IX, tr. 1, cap. 5 (ed. Stadler, 1916, 693: 24–32). The status of such eyewitness accounts in Albert is an open issue in the literature. For example, Siraisi (1980) mentions the story of the 150 aborted fetuses as a case in which Albert "showed himself perhaps too ready to believe what his medical friends told him." Demaitre and Travill (1980) interpret Albert's use of eyewitness accounts (taking the example of what a certain "experienced and trustworthy" midwife told him about pregnancy, *Dixit autem michi una matrona fide digna in talibus multum experta* [*De animalibus*, IX, tr. 2, cap. 5, ed. Stadler, 1916, 724: 28–29]) as a corrective to the theoretical discourse of Aristotle and Avicenna. For her part, Weisser (1985, 305) argues that Albert frequently refers to popular *topoi*, as in the case of the midwife account, which actually comes from the *Corpus Hippocraticum*.

48 Albertus Magnus, *De animalibus*, IX, tr. 1, cap. 1 (ed. Stadler, 1916, 679: 11–17): "Adhuc autem differens est in annis pubertatis elevatio et tumor mamillarum in puellis, quoniam quaedam earum acquirunt mamillas magnas valde et quaedam parvas, et paucae inveniuntur quae etiam nullas habent nisi sicut viri: talis enim inventa est femina apud nos in civitate Wangionum quae nunc Wormatia vocatur: quae tamen plures enixa est infantes, et aliquando enixa est tres simul gemellos."

story in his refutation of astral causality.[49] For Augustine, the fact that these twins were born almost simultaneously, the second holding the heel of the first as they were coming out of their mother's womb, a detail often depicted in medieval illustrations of the scene, but had completely contrasting destinies is a strong argument against knowledge through astrological predictions. Even if the fraction of time between the births of the two is taken into consideration, this marginal difference cannot possibly feature or matter in astrological charts, thus making astrologists' predictions useless, if not simply wrong. While Jacob and Esau were born not even a minute apart from each other, they had such different fates that no astrologist could have calculated this. For Augustine, this is a strong argument in his polemic against astral determinism.

Albert does not explicitly disagree with Augustine, but he slightly shifts the focus. He poses the question in the manner of Augustine: How could it be that the twin brothers, "born of one coitus [of their mother] with their father," had the same birth circle, i.e., the same horoscope, yet such different fates? This may prove that fate does not exist. Albert's first shift is towards natural philosophy: while Augustine's argument hinged on the moment of the twins Esau and Jacob were born, Albert focuses on the moment of conception. The conception, as described above, takes place consecutively, according to the different impulses in the movement of the semen. Even if we ignore the time difference in the conception of the twins (and not their birth, as in Augustine), it is still true that the centres of their hearts, from which the embryos start forming, are differently positioned, and hence their birth charts should be designed accordingly, with different horizons, angles and zodiac houses. The heart is the principal organ from which the entire organism develops (and Albert compares its action to that of the celestial first mover),[50] and this explains the different fates of the twins.[51] Albert thus argues for the existence of fate as an influence of celestial bodies on terrestrial matter. There is fate, and it is bound up with the movement of the heavenly spheres. But fate is a *forma media*, a middle form between the necessary and the possible. It is a phenomenon of the middle ground between the necessity of the celestial movement and the contingency

---

[49] Albertus Magnus, *De fato*, art. 1, arg. 19 (ed. Simon, 1975, 67: 52–57). Cf. Augustinus, *De civitate Dei*, lib. 5, cap. 4 (ed. Dombart and Kalb, 1955, 131, 27 ff.); Augustinus, *De doctrina christiana*, lib. 2, cap. 22, n. 33–34 (ed. Martin, 56, 1 ff.; Augustinus, *Confessiones*, lib. 7, cap. 6, §10 (ed. *Verheijen,* 1990, 99, 67–79). Cf. Zuccolin (2019, 75–124).

[50] Albertus Magnus, *De fato*, art. 1 (ed. Simon, 1975, 66: 17–23); *De animalibus*, I, tr. 1, cap. 1 (ed. Stadler, 1920, 27: 33–35): "Propter quod constat ista tria non aeque principalia esse, sed cor esse primum principium aliorum: et ideo cor in hiis est sicut in sphaeris caelestibus est primus motor motus." Cf. Aristoteles, *De partibus animalium*, l.3, c.3, 665a10–12.

[51] Albertus Magnus, *De fato*, art. 2 (ed. Simon, 1975, 71: 30–47): "Ad hoc quod obicitur de geminis, dicendum, quod in uno concubitu semen per vices proicitur et per vices a matrice glutitur, et sic non esse unam horam conceptus geminorum, tamen etiamsi demus eos in una hora concipi, centrum tamen cordis eorum, a quo incipit conceptus formari, non est unum; et mutato centro necesse est mutari totum circulum, et sic horizon eorum non est unus, neque anguli sunt iidem nec eadem domorum dispositio. Et sic tota periodus efficitur diversa, et per consequens adhaerens rebus natis dispositio fatalis necessario variatur."

of the material world, thus only inclining things toward a certain path but not binding them.[52]

For Albert, fate is in fact inaccessible to our contingent human knowledge, but this does not mean that it does not exist. Albert studies the problem of whether fate is knowable by resorting to the case of twins once again. Often, says Albert, when twins are of different genders, the boy dies at birth and the girl lives; and there is no way to know on the grounds of their respective stars and astral circles why this happens. The problem is not only that the fates of the new-borns drastically differ, but also that, theoretically speaking, the boy should survive, given the purported superior strength and constitution of the male gender. In fact, such mixed-gender twins are said to result from a deficit. The girl was not really intended by nature, but nature just "could not do better" (*melius fieri non possit*).[53] In general, for Albert the female sex is always generated by accident and because of a defect at a certain point in conception. Nature always aims to replicate itself in its best possible and universal form, which is male; but if for some reason (excessive humidity or coldness, disproportionate matter and form, etc.) it fails, its secondary goal is to be preserved in a particular form, and so a female is generated. In the specific case of twins, sex is defined by the interaction of the generative principles, and if the vigour of the sperm only partially prevails over the matter of the forming embryos, the one in which the sperm has conquered matter becomes a boy, and whenever the sperm succumbs to matter, a girl is conceived.[54]

While animals often bear twins that survive, human twins of different genders rarely survive, claims Albert, because of the general fragility of the human complexion and the tenderness of matter (*delicatio complexionis humanae et teneritudo materiae*). That is why only few, exceptionally strong women can bring such complicated pregnancies to a good end.[55] The imbalance in foetal development is the major reason why gender-different twins are particularly at risk. Being the more complex and perfect of the two, the male foetus requires greater heat and generally more time to develop. The female foetus, which is imperfect anyway, is able to survive even in less than perfect circumstances. Thus, at the hour of birth, the boy is

---

[52] Albertus Magnus, *De fato*, art. 2 (ed. Simon, 1975: 68, 50–54): "Haec autem talis forma media est inter necessarium et possibile; necessarium enim est, quidquid est in motu caelestis circuli, possibile autem et mutabile, quidquid est in materia generabilium et corruptibilium." For a more detailed study on Albert's take on astral necessity and free will, cf. Palazzo (2011, 65–95); Palazzo (2008, 55–78), this article contains an extensive list of further literature on the topic; Anzulewicz (1999, 263–277); Anzulewicz (2000, 141–152).

[53] Albertus Magnus, *De fato*, art. 4 (ed. Simon, 1975, 74: 22–24): "Sexus femineus incidit ex defectu, nec natura particularis umquam intendit facere feminam; sed cum melius fieri non possit ex natura universali, fit adiutorium generationis et non generans proprie."

[54] Albertus Magnus, *De animalibus*, XVIII, tr. 2, cap. 3 (ed. Stadler, 1920, 1224: 32–1225: 5): "Si enim sperma impraegnans inveniat habundans humidum et vincat ipsum perfecte et in toto dividet ipsum et faciet duos gemellos in sexu simili maris. Si autem aequaliter vincatur in toto et in partibus, faciet duas gemellas sorores. Si autem in una parte divisa vincatur et in alia vincat, erunt gemini in dispari sexu unus mas et alter femina."

[55] Albertus Magnus, *De animalibus*, IX, tr. 1, cap. 5 (ed. Stadler, 1916, 694: 1–4).

either born prematurely or holds back the birth of the female, thus putting her in jeopardy. Balance is a crucial factor, which is why twins of the same gender have better chances of survival, while a pregnancy with a boy and a girl is more hazardous.[56] Provided with this naturalistic explanation of the conception of a boy and a girl, as well as of the higher mortality rate of boys in twins of different genders, Albert can now freely argue that fate can in fact be known. However, as demonstrated above, different fates result from a defect of the natural principles involved in conception, and not from the disposition of the stars.[57]

While Augustine uses the case of twins as an argument for the vanity and insubstantiality of astrology and against astral determinism, Albert takes that same case and turns it into a piece of natural philosophy. He cites the example of twins to assert the existence of fate governed by stars, as well as its knowability, contrary to the spirit of Augustine's arguments. For Albert, fate is a natural phenomenon, as the example of twins shows. We can, indeed, have knowledge about the natural causes of the conception and birth probabilities of twins. While the "lower" part of astronomy, which deals with the material world and predicts changeable material things, can only yield/offer a very approximate, conjectural knowledge, its "higher" part, which investigates the movement of the stars, that is, fate, has an undisputable scientific dimension to it. Astronomy is an indubitable and syllogistic science of the stars, but studying their influence on the changeable material beings is uncertain. However, the general principles of astronomy are knowable.[58] Just like the generation of twins, their different fates are natural phenomena, and these can be explained on the grounds of natural philosophy. Yet, as shown above, to elucidate the generation of twins, it took not merely a discussion of the physiological process but also a reflection on the theoretical views about the respective roles of the father and the mother in conception.[59] The discussion pivoted on the supposed "abnormality" of the phenomenon and had implicit moral implications.

---

[56] Albertus Magnus, *De fato*, art. 4, arg. 74, 33–43; Albertus Magnus, *Quaestiones super De animalibus*, IX, q. 24–28 (ed. Filthaut, 1955, 213, 15–29); Albertus Magnus, *De animalibus*, XVIII, tr. 2, cap. 3 (ed. Stadler, 1920, 1224: 32–1225: 5).

[57] Albertus Magnus, *De fato*, art. 4, arg. 74 (ed. Simon, 1975, 16–32): "Ad hoc quod quaeritur de disparitate sexus in geminis, dicendum quod sexus femineus semper fit per occasionem defectus alicuius principiorum. [...] Unde dispar sexus in geminis provenit ex defectu principiorum naturalium in altera parte seminis et non ex periodo caelesti."

[58] Albertus Magnus, *De fato*, art. 4, *solutio* (ed. Simon, 1975, 73: 36–56).

[59] On the relation between medicine and philosophy, see Kijewska, Chap. 3 in this volume, pp. 37–56; Panarelli, Chap. 7 in this volume , pp. 115–132; Beneduce, Chap. 13 in this volume, pp. 253–275.

5.4 Conclusion: Humans or Monsters? The Intention of Nature

It is an ostensible paradox that, in Albert's terms, twins are monsters, but pygmies are not. Albert rediscovers Aristotle's teleology and puts it into action in his natural philosophy. Nature always aims at the good (*operatio naturae semper tendit ad bonum*), unless it has been hindered from achieving this good end, for instance, because its operations have been impeded (*operatio naturae ordinatur ad bonum, nisi impediatur*). This results from the disobedience of matter (*inoboedientia materiae*), as in the formation of boy and girl twins. When this happens, the goal of nature cannot be fulfilled (*frustrari debito fine*). Such children are "by-products" of a natural process (*praeter intentionem naturae*) in which some irregularity has occurred.[60] Born despite nature's good intentions, they are called monsters.[61] *Monstrum* works almost as a *terminus technicus* in this case to denote the cases in which nature has failed to act at its best because of an impediment.

In his *Physics*, Albert systematically investigates such impediments and provides a taxonomy of the natural causes that may lead to the birth of monstrous offspring. The meticulously studied impediments always originate in matter: there is either too much or too little of it, or it is imbalanced in its elemental qualities, or the containing matter, i.e., that of the womb, is defective.[62] Twins represent a liminal case, since the human species is naturally inclined to produce either one or two offspring.[63] However, they also fall into the category of monsters, and at times Albert compares the multitude of children produced by the abundant heat of either the efficient or the material cause to the multitude of bodily organs.[64] Even more monstrous (and, in a way, fascinating for medieval and modern scholars alike) are the cases of conjoined twins.[65] They are produced as a result of a rupture of the "pellicula media," the membrane dividing them. While they are brought

---

[60] Albert's explanation about monsters is concise. All the phrases are taken from *Quaestiones super De animalibus*, q. 5 called *Utrum in natura sit aliquod monstrum*, cf. *Quaestiones super De animalibus*, XVIII, q. 5 (ed. Filthaut, 1955, 299: 45–46; 67; 61; 62; 75). Cf. Köhler (2008, 387–411).

[61] Albertus Magnus, *Physica*, II, tr. 3, cap. 3 (ed. Hossfeld, 1987, 136: 54–55): "Monstra enim sunt peccatum illius quod in natura est propter aliquid." Cf. Aristoteles, *Physica* II.8, 199a35.

[62] Albertus Magnus, *Quaestiones super De animalibus*, XVIII, q. 6 (ed. Gilthaut, 1955, 300: 12–13): "Prima radix monstruositatis est ex parte materiae."; Albertus Magnus, *Physica*, II, tr. 3, cap. 3 (ed. Hossfeld, 1987, 136: 54–138: 60). Cf. Zuccolin (2019, 156) who presents a handy schematization of Albert's four main causes and their subcases.

[63] Albertus Magnus, *De animalibus*, XVIII, tr. 2, cap. 3 (ed. Stadler, 1916, 1223: 18–22): "Homo autem quasi est intermedium animal inter duo genera, quorum unum naturaliter generat multos filios, et aliud naturaliter generat filium unum: eo quod homo naturaliter generat filium unum, sed propter habundantiam caloris et humoris aliquando generat filios duos vel plures."

[64] Albertus Magnus, *Quaestiones super De animalibus*, XVIII, q. 6 (ed. Filthaut, 1955, 300: 10–20): "Sicut enim est in multitudine filiorum, sic est in multitudine membrorum."; Albertus Magnus, *Physica*, II, tr. 3, cap. 3 (ed. Hossfeld, 1987, 137: 52–59); Albertus Magnus, *De animalibus*, XVI, tr. 2, cap. 2 (ed. Stadler, 1920, 1223: 32–34).

[65] Cf. note 29.

forth by the fusion of matter after the rupture of the membrane, conjoined twins can also be formed in the opposite process, that is, the incomplete division of the sperm in the womb. This division occurs because of an excessive delectation and movement during the intercourse.[66]

Specimens such as twins and even women are "failures" of nature, which does not fulfil in these cases its intended plan to reproduce properly, in its best (male) and *usual* manner (one child). The notion of nature's best intent is relatively readily extended onto many other monstrous cases, such as conjoined twins with one head and two bodies or one body and two heads, hermaphrodites, people with eleven mouths or with six fingers, etc. The monster is, properly speaking, a term for what would be called a "malformation" in modern language. Monsters thus are natural occurrences and remain within their respective species; non-natural monsters are not possible (for instance, there cannot be hybrids such as pigs with human heads or cynocephali, or, for that matter, Cyclopes).

Pygmies, on the other hand, as close to the human species as they may be, are defined as brute animals, due to their presumed lack of reason. In pygmies, nature has perfectly fulfilled its intention, creating the most perfect of beasts that almost accurately approximates humans. While in twins and other monsters, nature fails to develop its full potential, human rationality is present in them, even if only potentially and damaged by the corporeal matter. In Albert's argumentation, idiots are humans because they are endowed with reason, albeit obstructed, for instance, by melancholy. Pygmies might even be smarter; but this does not make them human.

Although scholars in the Middle Ages did not explicitly address the question "What is a human being?" and anthropology emerged as an individual discipline much later, as Theodor Köhler rightly points out, it should be emphasized that Albert dedicated a long *summa*—*De Homine*—to this query. Albert certainly theorized human nature.[67] If his borderline cases suggest that he had a clear-cut notion of human nature, his theoretical reflections convey a more complex view. The image that emerges is one of twofold human nature: material, corporeal and corruptible on the one hand and immaterial, universal and linked to God, on the other.[68] And while for Albert it is theoretically plausible that these two dimensions of the essence are intrinsically connected, they nevertheless produce a puzzling tension in respect to the ultimate question: What is a human being?

[66] Albertus Magnus, *Physica*, II, tr. 3, cap. 3 (ed. Hossfeld, 1987, 137: 65–70).

[67] Cf. Anzulewicz (2013, 325–347); Anzulewicz (1998, 756–766); Köhler (2008); Guldentops (2000, 217–236).

[68] Albertus Magnus, *Ethica*, X, tr. 2, cap. 3 (ed. Borgnet, 627b): "Homo enim duo homines est."

References

Primary Literature

Albertus Magnus. 1891. *Ethica*, ed. Auguste Borgnet. Paris: Vivès.

———. 1916–1920. *De animalibus*. ed. Hermann Stadler (Beiträge zur Geschichte der Philosophie des Mittelalters 15–16). Münster: Aschendorff.

———. 1975a. *De XV problematibus*, ed. Bernhard Geyer. Münster: Aschendorff.

———. 1975b. *De fato*, ed. Paul Simon. Münster: Aschendorff.

———. 1987–1993. *Physica*, ed. Paul Hossfeld. Münster: Aschendorff.

———. 1995. *De natura et origine animae*, ed. Bernhard Geyer. Münster: Aschendorff.

———. 1955. *Quaestiones super De animalibus*, ed. Ephrem Filthaut. Münster: Aschendorff.

Aristoteles. 1953. *Generation of animals*. Trans. Philip H. Wicksteed and Francis M. Cornford. Cambridge, MA: Harvard University Press.

———. 1957. *Physica*. Trans. Arthur Leslie Peck. London: Heinemann; Cambridge, MA: Harvard University Press.

———. 1991. *History of Animals*. Ed. and trans. David M. Balme. Cambridge, MA: Harvard University Press.

———. 2001. *On the Parts of Animals*. Trans. with a Commentary James G. Lennox. Oxford: Clarendon Press.

Augustinus. 1955. *De civitate Dei*, ed. Bernard Dombart and Alphonse Kalb. Turnhout: Brepols.

———. 1962. *De doctrina cristiana*, ed. Josef Martin. Turnhout: Brepols.

———. 1990. *Confessiones*, ed. Lucas Verheijen. Turnhout: Brepols.

Avicenna. 1508. *De animalibus*. Venetiis. Digitalized edition. https://babel.hathitrust.org/cgi/pt?id=uc1.31158002275575&view=1up&seq=327.

Secondary Literature

Anzulewicz, Henryk. 1998. Der Anthropologieentwurf des Albertus Magnus und die Frage nach dem Begriff und wissenschaftssystematischen Ort einer mittelalterlichen Anthropologie. In *Was ist Philosophie im Mittelalter? Akten der X. Internationale Kongresses für mittelalterliche Philosophie der Société Internationale pour l'Étude de la philosophie médiévale, 25 bis 30 August 1997 in Erfurt*, ed. Jan A. Aerstsen and Andreas Speer, 756–766. Berlin–New York: Walter de Gruyter.

———. 1999. Der Einfluß der Gestirne auf die sublunare Welt und die menschliche Willensfreiheit nach Albertus Magnus. In *Actes de la Vème Conference Annuelle de la SEAC, Gdansk 1997*, 263–277. Warszawa: Département d'Anthropologie Historique, Institut d'Archaeologie de l'Université de Varsovie; Gdańsk: Musée Maritime Central.

———. 2000. Alberts des Großen Stellungnahme zur Frage nach Notwendigkeit, Schicksal und Vorsehung. *Disputatio Philosophica. International Journal on Philosophy and Religion* 1: 141–152.

———. 2013. *Anthropology: The Concept of Man in Albert the Great*. In *A Companion to Albert the Great*, ed. Irven Resnick, 325–346. Leiden/Boston: Brill.

de Asúa, Miguel. 1997. Albert the Great and the "Controversia inter medicos et philosophos". In *Proceedings of the PMT Conference, 19–20*, ed. K.A. Gerschbach et al., 143–156. Villanova: Villanova University Press.

Demaitre, Luke, and Travill, A. A. 1980. Human Embryology and Development in the Works of Albertus Magnus. In *Albertus Magnus and the Sciences. Commemorative Essays 1980*, ed. James A. Weisheipl, 405–440. Toronto: Brill.

Guldentops, Guy. 2000. Albert the Great's zoological anthropocentrism. *Micrologus* 8: 217–236.

Koch, Joseph. 1931. Sind die Pygmäen Menschen? Ein Kapitel aus der philosophischen Anthropologie der mittelalterlichen Scholastik. *Archiv für Geschichte der Philosophie* 40: 194–213.

Köhler, Theodor W. 1992. Anthropologische Erkennungsmerkmale menschlichen Seins. Die Frage der "Pygmei" in der Hochscholastik. In *Mensch und Natur im Mittelalter*, ed. Albert Zimmermann and Andreas Speer, 718–735. Berlin/New York: Walter de Gruyter.

———. 2008. *Homo animal nobilissimum. Konturen des spezifisch Menschlichen in der naturphilosophischen Aristoteleskommentierung des dreizehnten Jahrhunderts*. Vol. 1. Leiden/ Boston: Brill.

Lüthy, Christoph H. 2013. Historical and Philosophical Reflections on Natural, Enhanced and Artificial Men and Women. In *Engineering the Human*, ed. Bert Jaap Koops, Christoph Lüthy, et al., 11–28. Berlin/Heidelberg: Springer.

Martin, Joyce A., et al. 2012. Three Decades of Twin Births in the United States (1980–2010). *NCHS Data Brief* 80: 1–8. Digital copy: https://www.cdc.gov/nchs/data/databriefs/db80.pdf.

Miteva, Evelina. 2018a. "Iam patet igitur veritas eius quae dixit Aristoteles, et causa deceptionis Galieni." Philosophers vs. Medics in Albertus Magnus' Account on Conception. In *Irrtum. Error. Erreur*, ed. Andreas Speer and Maxime Mauriège, 107–122. Berlin: Walter de Gruyter.

———. 2018b. I want to break free: Albert the Great's Naturalistic Account of Freedom of Choice and Its Limitations. *Bochumer Philosophisches Jahrbuch für Antike und Mittelalter* 21: 11–28.

Oelze, Anselm. 2018. *Animal Rationality: Later Medieval Theories (1250–1350)*. Leiden/ Boston: Brill.

Palazzo, Alessandro. 2008. The Scientific Significance of Fate and Celestial Influences in Some Mature Works by Albert the Great: De fato, De somno et vigilia, De intellectu et intelligibili, Mineralia. In *Per perscrutationem philosophicam: neue Perspektiven der mittelalterlichen Forschung. Loris Sturlese zum 60. Geburtstag gewidmet*, ed. Alessandra Beccarisi, Ruedi Imbach, and Pasquale Porro, 55–78. Hamburg: Meiner.

———. 2011. Albert the Great's Doctrine of Fate. In *Mantik, Schicksal und Freiheit im Mittelalter*, ed. Loris Sturlese, 65–95. Köln/Weimar/Wien: Böhlau.

Pison, Gilles, Nadège Couvert. 2004. The Frequency of Twin Births in France. The Triple Influence of Biology, Medicine and Family Behaviour, *Population* 59: 765–794. Digital copy: https://www.cairn-int.info/article-E_POPU_406_0877%2D%2Dthe-frequency-of-twin-births-in-france.htm.

Resnick, Irven. 2013. Conjoined Twins, Medieval Biology, and Evolving Reflection in Individual Identity. *Viator* 44 (2): 343–368.

Shackelford, Todd K., and Nicholas Pound, eds. 2006. *Sperm Competition in Humans. Classic and Contemporary Readings*. New York: Springer.

Siraisi, Nancy. 1980. The Medical Learning of Albertus Magnus. In *Albertus Magnus and the Sciences. Commemorative Essays 1980*, ed. James A. Weisheipl, 379–404. Toronto: Brill.

Tellkamp, Jörg A. 2013. Albert the Great on Perception and Non-Conceptual Content. In *Philosophical Psychology in Arabic Thought and the Latin Aristotelianism of the 13th Century*, ed. Luis Xavier López-Farjeat and Jörg A. Tellkamp, 205–222. Paris: Vrin.

———. 2016. Aping Logic? Albert the Great on Animal Mind and Action. In *Subjectivity and Selfhood in Medieval and Early Modern Philosophy*, ed. Jari Kaukua and Tomas Ekenberg, 109–123. Cham: Springer.

Thijssen, Johannes M. 1987. Twins as Monsters: Albertus Magnus' Theory of the Generation of Twins in Philosophical Context. *Bulletin of the History of Medicine* 6: 237–246.

Vinaty, Tommaso. 1981. Sant'Alberto Magno, embriologo e ginecologo. *Angelicum* 58: 151–180.

Weisser, Ursula. 1985. Die Harmonisierung antiker Zeugungstheorien im islamischen Kulturkreis und ihr Nachwirken im europäischen Mittelalter. In *Orientalische Kultur und europäisches Mittelalter*, ed. Albert Zimmermann and Ingrid Craemer-Ruegenberg, 301–326. Berlin/New York: Walter de Gruyter.

Zuccolin, Gabriella. 2017. Two Heads Two Souls? Conjoined Twins in Theological "Quodlibeta" (1270–c. 1310). *Quaestio* 17: 573–595.

———. 2019. *I gemelli nel medioevo. Questioni filosofiche, mediche e teologiche*. Como/ Pavia: Ibis.

Chapter 6
Why Philosophers Father Foolish Children: Peter of Spain, Albert the Great, and James of Viterbo on the Transmission of Intellectual Qualities

Mario Loconsole

Abstract Frequent failures in the transmission of intellectual virtues from parents to their offspring is a recurring topic in medieval medical and philosophical texts. Nonetheless, it represents a grey area within the emergent embryonic theory of the regular transfer of "hereditary traits," a liminal phenomenon of a more general theory of generation. This study exemines early approaches to the generation of foolish children by wise men. Starting with the medical account in one of the Salernitan questions discovered in the Bodleian Manuscript (Auct. F.3. 10), the argument proceeds to Peter of Spain's short digression on the topic in a commentary on Aristotle's *De animalibus* and to Albert the Great's dedicated question in the *Quaestiones super libris "De animalibus"*. The Salernitan question is the source of Albert's physiological angle in attributing the failed intergenerational transmission of intellectual qualities to the inadequate digestion of philosophers' semen. Albert's position is considered against the background of his other works that address the generation of animals, the concept of *digestio*, and the role of *melancholia adusta* in the act of thinking. Additionally, the key points of James of Viterbo's quodlibetal question are scrutinised.

Keywords Salernitan Questions · Peter of Spain · Albert the Great · James of Viterbo · Heredity · Generation · *Digestio* · Melancholy

> Ogni pianta, ogni albero e ogni radice suole produrre il frutto suo secondo la sua specie, né mai prevaricare di quanto gli ha ordinato la gran madre natura, maestra di tutte le Cose. Solo la pianta dell'uomo è quella che varia e manca, onde molte volte si vede che d'un padre di bella presenza nasce un brutto, anzi mostruoso figlio, e d'un dotto un ignorante e

M. Loconsole (✉)
Universität zu Köln, Cologne, Germany
e-mail: mlocons2@uni-koeln.de

M. Gensler et al. (eds.), *The Embodied Soul*, Historical-Analytical Studies on Nature, Mind and Action 11, https://doi.org/10.1007/978-3-030-99453-2_6

goffo. La causa di ciò lascio disputare a chi sa, poiché io non son scolastico né cattedrante, ma un uomo dozzinale e che ha poca cognizione di simil cose.[1]

Giulio Cesare Croce

Le piacevoli e ridicolose semplicità di Bertoldino, Proemio

6.1 Introduction

It was a common belief among physicians and philosophers in the thirteenth century that physical and corporeal features—and to a certain degree intellectual qualities as well—were transmissible from parents to children. Philosophy and medicine were combined to develop a refined theory of the generation of animals which aimed to grasp the causes and dynamic of regularity in nature. The theory explained how the reproductive organs of males and females provided all the material and formal principles needed to produce beings of the same species usually similar to their parents.[2]

Nevertheless, the development of the concept of heredity was not unproblematic and encountered challenges from the models of thinking that emphasised individuality—as in the case of physiognomy—and favoured the concept of innateness.[3] Moreover, as Maaike van der Lugt and Charles de Miramon have pointed out, medieval Western culture was quite ambiguous in this respect. On the one hand, medieval society was largely structured by heredity and by law, according to which offices, powers, functions, and professions were mostly inherited; on the other, a general theory of biological heredity, which was only developed in the nineteenth century by Mendel and Darwin, was still missing.[4]

However, in the thirteenth century, scholars of philosophy and medicine who studied Aristotle's works tried to clarify and develop an account of the transmission of features from parent to child. Like begets like, humans generate humans, and, by the same token, accidental features (e.g., complexion, beauty, and colours) and body-dependent psychological traits (e.g., intellectual aptitude or incapacity) were all believed to be inheritable.[5] In medical writings, problems such as hereditary

[1] "Every plant, every tree, and every root produces fruit according to its species and never oversteps what nature —the mistress of all things—has ordained for it. Only the plant of men diverges and errs, so that ugly and even monstrous children often descend from good-looking and personable fathers, while a learned man may sire an ignorant and clumsy son. Let those who are acquainted with the matter discuss its causes, for I am neither a scholar nor a schoolmaster, but an ordinary man who knows just a little about these things." (my translation)

[2] See Cadden (1993, 105–66); Van der Lugt (2004, 33–95); Fancy (2018).

[3] See Ziegler (2008).

[4] See Van der Lugt and De Miramon (2008, 6).

[5] On the topic of the resemblance between parents and children in Aristotle and Avicenna, see Aristoteles, *De generatione animalium*, I.17–18, IV.3; Aristoteles, *De historia animalium*, VII.6; Avicenna, *De animalibus*, XVIII (Venetiis 1508, 63ra–63rb).

illnesses (*morbi hereditarii*), which were conceived as dysfunctions of the parents' bodies transmissible through their *semen*, also started to be topics of interest.[6]

As there are exceptions to every rule, medieval scholars had to justify and explain aberrant phenomena, for example, spontaneous generation or conceptions by demons,[7] which—if not correctly interpreted—could subvert their attempts to understand natural principles through science. The phenomenon of wise men fathering foolish children represents an iconic case study in which the smooth transmission of specific bodily features clashes with the faltering path of intellectual qualities. In these cases, the expectations stemming from the general organic theory of generation are not met, and the regularity of natural processes is partially undermined.

6.2 The Salernitan Question and Peter of Spain's *Quaestiones super libro "De animalibus"*

The distorted procreation of wise man is the subject of a brief Salernitan question[8] edited by Brian Lawn.[9] Since Aristotle, the transfer of formal features, therein the transmission of traits and the reproduction of parental characteristics, has been regarded as a natural prerogative of the male semen.[10] Although the actual focus of the Salernitan question is on the physiological qualities of sperm, rather than on its (in)ability to transmit the paternal features, it relies on the Aristotelian principle. The corruption of the male sperm causes the hardness of the brain, an organ responsible for the act of thinking, which accounts for the new-born's inability to use reason properly. In the question, the generation of foolish children is a purely physical matter: as the excessive and prolonged use of the intellectual faculties (*ex multa cogitatione*) makes the body and the sperm of *sapientes* dry out, children with hard brains unable to perform higher intellectual activities are engendered. Curiously enough, the process may be reversed, and obtuse people may father smart sons.[11]

---

[6] See López-Beltrán (2008); Van der Lugt (2008b).

[7] See Van der Lugt (2004).

[8] The Salernitan questions were a resource for scientific and medical teaching at Salerno. Gradually accumulated and probably representing the work of several masters, the questions mostly deal with anthropology, the various branches of medicine, zoology, the everyday behaviour of man and beast, ordinary physiological activities and processes, and simple anatomy. On the genre of Salernitan questions, see Lawn (1963); Bazán et al. (1985, 285–297).

[9] Lawn (1979, 350): "Quare sapientes generant stultos? Responsio: sapientes multum cogitant, et ex multa cogitatione et membra et sperma multum desiccantur, unde durum procreatur cerebrum quod ad rationem non est habile. Ratione contraria stulti sapientes generant." Together with Albert the Great's and James of Viterbo's questions, which will both be analysed in the following section, the question is also quoted in Joseph Ziegler's article on the transmission of hereditary traits and physiognomy, cf. Ziegler (2008).

[10] On Aristotle's theory of generation, see Preus (1970); Coles (1995); Berti (2010); Lefebvre (2016).

[11] Lawn (1979, 350).

In his *Quaestiones super libro "De animalibus"*, Peter of Spain ponders why wise men generate foolish children in a broader question which inquires why all flying animals with good wings have bad paws or feet (*quare omne volatile bonae alae est mali pedis*).[12] Peter's explanation of the phenomenon radically differs from the reasoning in the Salernitan question. He cites neither the physiological causes and implications nor the corporeal and mental condition of the father; instead, he attributes failures in the transmission of certain traits to the predominance of some virtues over others. Sometimes, birds furnished with highly developed wings indeed have no efficient paws. In them, the *virtus motiva* is quantitatively split between their motor organs: the stronger the virtue of the wings, the weaker the virtue of the paws. Referring to a theory that he ascribes to Avicenna, Peter states that, likewise, those intellectually gifted (*boni in studio*) will surely have no aptitude for public affairs (*in foro*). Peter posits the virtues as being mutually exclusive, as if—in a very classical way—the pursuit of theoretical life opposed and partially excluded the pursuit of practical and public life, including the ability to sire offspring.

6.3 Albert the Great's *Quaestiones super libris "De animalibus"*

In 1258, Albert lectured on the *Quaestiones super libris "De animalibus"* in Cologne.[13] As Lawn has observed, Albert's *Quaestiones* are apparently modelled on Peter of Spain's commentary, and several of them closely follow the Salernitan questions.[14] *Quaestio* 4 in book XVIII, which ponders why philosophers as a rule have foolish children (*Quare sapientes et philosophi ut plurimum generant filios fatuos*), may substantiate Lawn's insight. Importantly, however, the philosopher as sharing the "bad luck" which afflicts the wise man (*sapiens*) is first introduced by Albert; neither the Salernitan question nor Peter of Spain's *De animalibus* makes this distinction. Moreover, Albert explicitly emphasises that the father's hereditary

---

[12] Petrus Hispanus, *Quaestiones super libro "De animalibus"*, I, q. 2 (ed. Navarro Sánchez, 2015, 128: 211–216): "Ad aliud dicimus quod sicut dicit Avicenna, una virtus quanto magis intendit uni operationi, tanto magis remittitur respectu alterius. Et ideo dicit quod qui bonus est in studio non est bonus in foro. Et per hanc viam dicit Avicenna quod sapientes generant stultos. Et hoc est quia cum omnes animae et virtutes radicentur in una radice, quanto magis intendit uni, tanto minus alteri. Cum igitur virtus motiva sit divisa in alas et pedes, intensa in alis, minor erit in pedibus, et e contrario."

[13] The *Quaestiones super libris "De animalibus" in Cologne* were published only later—certainly after 1260, but definitely before 1300. This has been established by Ephrem Filthaut, the editor of the questions in 1955, who has shown that Moerbeke's translations of the *Metaphysics* and *Meteorologica IV* was used in the *Quaestiones*; moreover, among the eight manuscripts preserving the text, the one held in the Biblioteca Ambrosiana in Milan was drafted in the thirteenth century. See Albertus Magnus, *Quaestiones super libris "De animalibus"* (ed. Filthaut, 1955, XIV). On the drafting and the sources of Albert's *Quaestiones*, see Goldstein-Préaud (1981); de Asúa (1997).

[14] Lawn (1963, 85–86). See also Demaitre, Travill (1980, 413); De Asúa (1997, 15).

features are conveyed by sperm, which is responsible for the transmission of not only the bodily virtues but also the virtues of the soul from parent to child. Expectedly, the bodily features, as well as the intellectual virtues, of the father are transferred onto the son, so that he may resemble the parent in both respects.[15] Indeed, the principle *parentes nobiliores generant filios nobiliores* is endorsed in Albert's *quaestio*: for example, one is always tempted to prefer one foal over another on the basis of its parents' features alone,[16] as experience teaches one to trust that a good horse will bear a good horse as well. Nevertheless, there is an exception to this otherwise universal principle. Specifically, the rule appears to be not always applicable to the transmission of intellectual traits in humans, and it is evident to the senses (*patet ad sensum*) that wise men usually have foolish children.[17]

This seems quite paradoxical, since, in Aristotelian terms, philosophers are the noblest of men insofar as they are the only ones capable of fulfilling human nature and mastering the highest task of exercising and perfecting the intellectual part of the soul. Yet, the best of all men prove incapable of begetting smart children, because they fail to hand down their mental or intellectual dispositions to their offspring. Moreover, like Peter, Albert quotes Avicenna to bring in the principle that one strong virtue usually restrains another; consequently, those with a talent for learning (*in studio*) are not at ease in public life (*in foro*), and their nature is not predisposed for reproduction (*in generatione vel venereo actu*), as Albert adds in more explicit terms than Peter. Nevertheless, Albert advances a purely medical explanation for this: namely, the philosopher, consumed with thoughts and meditations, debilitates his natural virtue, which reduces his suitability for procreation. Albert thus embraces also the physiological reasoning of the Salernitan question, insisting that as philosophers excessively use their faculties of reason and imagination, they produce a weak and undigested sperm, whose deficiency partially corrupts the regular process of generation.[18] Albert's composite approach includes both the predominance theory adopted from Peter of Spain (which was attributed to Avicenna) and the physiological

[15] Albertus Magnus, *Quaestiones super libris "De animalibus"*, XVIII, q. 4 (ed. Filthaut, 1955, 299: 11–16): "In spermate non solum est virtus corporis, sed animae, alioquin animatum non produceret; ergo cum filii assimilentur parentibus in dispositionibus corporalibus ut plurimum, pari ratione assimilabuntur in dispositione animae, ut sapientiae et scientiae."

[16] Albertus Magnus, *Quaestiones super libris "De animalibus"*, XVIII, q. 4 (ed. Filthaut, 1955, 299: 30–32): "Et ideo ex bonis procedit filius bonus. Unde magis desideratur pullus unius equi quam alterius."

[17] Albertus Magnus, *Quaestiones super libris "De animalibus"*, XVIII, q. 4 (ed. Filthaut, 1955, 299: 17–19): "Oppositum patet ad sensum. Dicendum, quod sapientes ut plurimum generant pueros fatuos."

[18] Albertus Magnus, *Quaestiones super libris "De animalibus"*, XVIII, q. 4 (ed. Filthaut, 1955, 299: 18–32): "Dicendum, quod sapientes ut plurimum generant pueros fatuos, quia secundum Avicennam una virtus intensa retrahit aliam a sua operatione. Nunc autem sapientes multum intendunt imaginationi et rationi et multum profundant suas cogitationes et meditationes. Unde in eis ut plurimum debilitatur virtus naturalis, et ideo frequenter sperma eorum est indigestum et malum, et propter hoc producuntur filii mali, quia qui bonus est in studio, pravus est in foro et in generatione vel venereo actu; ideo etc. Alia autem animalia magis intendunt virtuti naturali neque impediuntur propter profundationem cogitationum."

approach suggested in the Salernitan question. Nevertheless, contrary to the Salernitan question, Albert does not ascribe the cause directly to the dryness of the philosopher's sperm, but more generally to insufficient digestion in their semen.

However, in his *Quaestiones*, Albert neither explains why the sperm of philosophers is undigested nor what it means for a substance to be or not be digested. Understandably so, for Albert's questions, which are only available in a student's version, were drafted in a didactic form that does not aim to explore the nuances of the teacher's position.[19] A comprehensive explanation of the puzzle of the philosopher's offspring may be found elsewhere. Given this, in the next section, I deal with the conception of generation in Albert's works, stressing the role of sperm and the possible implications and causes of its corruption.

6.4 Sperm, Complexion, and Generation

In his *De animalibus*, a work which deals with the generation of animals in greater detail, Albert states that the role of sperm is to absorb and bear all virtues that effect generation: it receives the virtue of the stars, the virtue of the soul from the father, and the virtue of the elements. In addition, it also educes what is needed for the formation of the foetus from the female semen and menstruum. Sperm thus serves as a vehicle of the formative virtues, all of which operate through heat.[20] The digestive heat, which also enables the animal sperm to fulfil its function, is itself informed by three virtues. The first to be listed by Albert is the virtue of fire, which informs the virtue of the stars; the latter substantiates the digested matter and informs the virtue of the soul. As a result, all these virtues are simultaneously gathered in the spirit and in the heat of the semen.[21]

---

[19] Demaitre, Travill (1980, 413).

[20] Albertus Magnus, *De animalibus*, XVI, tr. 1, cap. 6 (ed. Stadler, 1920, 1080: 28–39): "Solutionis autem dictae capitulariter summa haec est quod sperma ex hoc quod quarta digestione assimilatum est membris, actu virtuali simile est virtutibus membrorum: et quia est in actu virtuali membrorum, potest educere materiam seminis et menstrui mulieris quod est in potentia ad actum formationis membrorum: et calor est instrumentum ipsius in virtute triplici. Calor enim habet virtutem elementalem, et sic digerit congregans omogenia et separans etherogenia. Est etiam in calore caeli et sic fundat et stabilit in substante esse et firmo id in quod agit. Est item in virtute animae aut virtutis animae quae movet ipsum, et sic habet vivificare et animatum facere secundum omnem partem animae quae secundum se totum est principium vitae in corpore."

[21] Albertus Magnus, *De animalibus*, XVI, tr. 1, cap. 6 (ed. Stadler, 1920, 1081: 32–40; 1082: 1–5): "Et in omnibus his sunt tres virtutes calorem digerentem informantes: virtus ignis videlicet hoc modo quo subtiliata est virtus ignis in ita ut dictum est, subtiliter commixto: et hanc virtutem informat virtus caelestis quae fundat et substantificat ea quae digeruntur eo quod est virtus corporis perpetui et substantiam suam uno modo retinentis [...]. Et istae virtutes simul collectae sunt in spiritu et calore spermatis: et quia operantur ad substantiam animatam in quantum sunt caelestis corporis et animae, ideo dicuntur hae virtutes esse sicut ars et sunt in semine sicut ars est in instrumentis artis."

Thus, the crucial role that sperm plays in the process of generation is quite evident: as Albert says, all things are made through the operation of natural heat.[22] It seems that in the sperm, being the centre of the informing activity in generation, all formal principles—i.e., the three virtues—operate by means of heat. The elements need heat to be digested, and, at the same time, the virtue of the stars and the virtue of the soul need heat to exercise their formal effect on matter. For its part, sperm performs its function through the heat resulting from digestion and combination of all the other heats. In this scheme, which is concretised in *De animalibus*, but pervades all of Albert's natural philosophy, the male semen is tasked with moulding the menstrual blood and preparing the embryonic matter for the reception of its designated form.[23] Albert envisages formal eduction in the generation of beings as predicated on the balance between matter prepared to receive a determinate form and the virtues of the stars producing specific formal impulses. Heat is the efficient cause of the process as an instrument through which the semen fulfils its function.

Although nature pursues a defined goal (*ad finem intentum et determinatum*),[24] it may err (*peccat*), just as art does.[25] While natural processes are divinely ordained and thus purposive and oriented toward the good (*ad bonum*), they unfold in matter and may as such be hindered by a range of obstacles.[26] For example, the lack or overabundance of matter, as well as an excessive weakness or strength of the formative virtue, may cause nature to generate monsters (*monstra*), which are conceived as *peccata naturae* or simply as differing from what is regularly encountered in nature.[27] Such failures of nature are mainly caused by disturbances in the transmission of the formative virtue through sperm. When the formative virtue is too powerful, it collects redundant matter, and when it is underdeveloped, it is unable to attract

[22] Albertus Magnus, *De animalibus*, XVI, tr. 1, cap. 6 (ed. Stadler, 1920, 1080: 9–13): "Nullum igitur horum membrorum erit sine taliter formante et operante, neque sine talis virtutis operatione: quamvis enim omnia sint facta ex caloris naturalis operatione, tamen quia calor est instrumentum huiusmodi virtutis in spermate, ideo necessario calor operatur et facit in virtute eiusdem formativae virtutis."

[23] Albertus Magnus, *Physica*, VIII, tr. 2, cap. 10 (ed. Hossfeld, 1987, 102: 1–5). See also Albertus Magnus, *Physica*, II, tr. 2, cap. 3 (ed. Hossfeld, 1987, 102: 21–23): "In natura non omnis materia susceptibilis est omnis formae neque omnis forma est actus cuiuslibet materiae."

[24] Albertus Magnus, *Physica*, II, tr. 3, cap. 3 (ed. Hossfeld, 1987, 138: 15).

[25] Albertus Magnus, *Physica*, II, tr. 3, cap. 3 (ed. Hossfeld, 1987, 136: 27–39): "Peccatum fit tam in his quae secundum artem sunt, quam in his quae secundum naturam sunt, et hoc apparet tam in arte, quae imitatur naturam in opere non-naturali, quam in illa quae perficit opus naturae et iuvat naturam in opere naturali."

[26] Albertus Magnus, *Quaestiones super libris "De animalibus"*, XVIII, q. 4 (ed. Filthaut, 1955, 299: 67–76). See also Albertus Magnus, *Quaestiones super libris "De animalibus"*, XVIII, q. 5 (ed. Filthaut, 1955, 299: 57–62): "Nam quaedam sunt agentia, quae in suis operationibus non recipiunt impedimenta, ut superiora, et quaedam sunt, quae impedimenta recipere possunt, sicut omnia quae agunt ex suppositione materiae, quia ex inoboedientia materiae potest agens frustrari debito fine."

[27] For a detailed exposition of Albert's account of monsters, see Miteva, Chap. 5 in this volume, pp. 73–90; see also Zuccolin (2019, 144–157).

matter.[28] There are other causes as well. For instance, as Albert states referring to Ptolemy's *Tetrabiblos*, a hostile disposition of the stars and planets may disrupt the regular transmission of the form.[29] When the principles of the species present in sperm or in the matter of a being to be generated is damaged or disturbed by an unfavourable cause, monstrous generation occurs.[30] Because sperm carries the formative virtue, its corruption, whether partial or total, affects the transmission of the form and, thus, the generation of beings, possibly producing monsters. Such occurrences are more frequent in animals than in plants, as the seed of the former is more fragile and their generation far more complex than that of other natural organisms.[31]

The corruption of the complexion of sperm may also cause infertility as another reproductive anomaly. The sperm that has a bad constitution (*male complexionatum*), i.e., unbalanced inner qualities, can indeed impede the generative power. In a passage in the *De animalibus* which is highly reminiscent of Avicenna's *Canon*,[32] Albert points out that, when moisture (*humidum*) is insufficiently digested (*incompletum*) or dryness (*siccitas*) is too dominant, they affect and are present also in the sperm. Given this, simple physiological causes, such as the consumption of sour food, may indeed alter the complexion of the body and prompt the production of ill-balanced sperm.[33] The reason why very young or old people, as well as gluttons and heavy drinkers, are incapable of procreation is that the sperm they produce is

[28] Albertus Magnus, *Physica*, II, tr. 3, cap. 3 (ed. Hossfeld, 1987, 136: 67–70; 137: 20–24).

[29] Albertus Magnus, *Physica*, II, tr. 3, cap. 3 (ed. Hossfeld, 1987, 138: 38–47). See Ptolemaeus Claudius, *Tetrabiblos* III, cap. 9–10 (ed. Vuillemin-Diem, Steel, De Leemans, 2015, 245–248).

[30] Albertus Magnus, *De animalibus*, XVIII, tr. 1, cap. 6 (ed. Stadler, 1920, 1214: 46): "His ita determinatis, consequens est dicere de causa partuum monstruosorum [...]. De monstris ergo et causis eorum hoc primum advertendum est quod quando principia moventia et motus ipsorum qui in spermate sunt ex natura speciei et ex materia individui, destruuntur per aliquam causam fortiter in oppositum agentem, non remanebit nisi virtus universalis generis, quae tolli non potest materia salvata: aliter enim nihil generaretur ex tali materia. Manente autem hac universali virtute in materia quae habet aequalitatem complexionis, erit generatio monstruosa: quoniam virtus universalis generis movet ad vitam et causa specialis in oppositum speciei agens, inducet aliam formam: et hoc praecipue causatur ex duobus, ex virtute caelesti videlicet, et ex materia concepta." This somewhat intricate passage dwells on the situation in which the capacity of sperm to transmit formal features is partially corrupted, leading to the generation of monsters: sperm may lose some of its formal powers meant to secure features proper to the individual or features proper to the species; at the same time, it may retain the formal aspect of the genus, which is purely universal and therefore highly undetermined. On the topic of *virtus universalis*, see Draelants (2003).

[31] Albertus Magnus, *Physica*, II, tr. 3, cap. 3 (ed. Hossfeld, 1987, 139: 5–12): "Plus monstra in animalibus apparent quam in plantis [...]. Haec autem est, quod mollius est semen animalis et plura exiguntur ad terminationem eius quam plantae et animal difficilis est generationis et planta similis et facilis generationis, et ideo facilius fit, ut corrumpatur aliquod principium generationis animalis quam plantae."

[32] Avicenna, *Canon medicinae*, III, sent. 21, doctr. 1, cap. 8 (Venetiis 1507, 362ra–b).

[33] Albertus Magnus, *De animalibus*, X, tr. 2, cap. 1 (ed. Stadler, 1916, 746: 39): "Causa autem sterilitatis, quae est in spermate, est quod male sit complexionatum et ideo impediat virtutem generativam calido vel frigido excellenti, sive per naturam hoc habeat sive per accidens infirmitatis: si enim illae qualitates diu infuerunt, resperguntur in sperma. Similiter autem dico de humido incompleto aut siccitate nimia: virtus enim generativa vult esse in eo quod completum et aequale est, si debeat formare bonum et perfectum natum in vita humana proficientem. Causa autem illius corruptionis aliquando, est ex cibis inconvenientibus et praecipue ex acetosis: haec enim infrigidant et exsiccant."

undigested.[34] Alterations in the complexion of sperm bring about infertility and confound the processes of embryo formation, nutrition, and generation.[35] Generally, Albert reasserts the principle of *ex sano sanum et ex infirmo infirmum*,[36] meaning that healthy beings—i.e., beings with a good complexion[37]—generate healthy offspring, because they produce well-digested and balanced semen, which is perfectly capable of transferring the formative virtues of the species.

To sum up, all the irregular reproductive phenomena, such as the generation of monsters and infertility, have a common explanation. Since sperm is the main vehicle of the form, any alteration of it—whether caused by a peculiar constellation of the stars, by changes in the bodily complexion, the availability and quality of matter, or by affections coming from the mind[38]—may hinder and complicate the process of generation. As Albert states, the fathering of foolish children by philosophers is ascribable to the insufficient digestion of their semen, which is one of the many possible obstacles to the natural process of generation.

6.5 Heat and Moisture in Albert's Physiology: Old Men and Generation

The reason why philosophers have hapless progeny—as explained in the *Quaestiones*—lies in the fact that the fathers' semen proves unsuitable for procreation. Their sperm is not arranged and shaped well (*terminatum*), thus being incapable of transmitting the disposition of their souls to the unborn children. Their virtues, such as aptitude for learning and wisdom, are not passed on because of the physiological imperfection of their insufficiently digested seed, which is unable to gather and transfer the paternal features.

[34] Albertus Magnus, *De animalibus*, X, tr. 2, cap. 1 (ed. Stadler, 1916, 747: 41): "Adhuc autem sperma non proficiens ad generationem est sperma infantis aut nauseativam satietatem patientis frequenter aut ebrii frequentius aut senis: omnia enim haec indigesta sunt et non constantia spisse, et ideo generationi non proficiunt."

[35] Albertus Magnus, *De animalibus*, X, tr. 2, cap. 1 (ed. Stadler, 1916, 746: 39): "Siccitas autem complexionis spermatis quidem aliquando non impedit generationem, sed facit eam difficilem et corrumpit alimentum embryonis, quod sugitur de matrice."

[36] Albertus Magnus, *De animalibus*, X, tr. 2, cap. 1 (ed. Stadler, 1916, 747: 41): "Similiter etiam est de spermate eius cuius non sunt sana membra corporis, eo quod sperma fluit ex omnibus membris, et est ex sano sanum et ex infirmo infirmum, et quamvis aliquando generet, tamen non proficit natus, et esset aliquando melius talem non natum esse."

[37] On the notion of complexion in the Middle Ages, see Thorndike (1958); Jacquart (1997); Klemm (2012); Chandelier and Robert (2013); Köhler (2008, 320–374).

[38] Besides the effects of thinking and meditation on the formation of sperm, several accidents of the soul, such as sadness, worry, and fear, are also listed among the causes of infertility in the *De animalibus*. See Albertus Magnus, *De animalibus*, X, tr. 2, cap. 1 (ed. Stadler, 1916, 746: 39).

In Albert's philosophy, the interaction between heat and moisture plays a crucial role as the basis for all processes in nature.[39] Although Albert distinguishes between animate and inanimate beings, their generation and corruption, their operations, functions, and inner structures are ruled by the relation between these two principles. Central to this is digestion defined as a process in which active heat and passive moisture shape, transform, or modify matter according to the form and the species that is to be generated.[40] Digestion subsumes both chemical processes (such as the formation of minerals) and the processes of biological life (such as nutrition and growth).[41] Thus, the dynamics of digestion encompass the generation of beings, the assimilation of victuals, the production of humours, the cooking of food, and even artificial operations, such as the transmutation of metals.

In the generation of animals, the material and moist principle of the female is digested by the formal and hot principle of the male. Although Albert's position is poised between Aristotelian and Galenic-Avicennean interpretations of the role that the sperm of both sexes has in generation, the German master does not question the peripatetic dichotomy of masculinity/form and femininity/matter.[42] Following the Aristotelian paradigm, he attributes the role of the material principle to females and the formal function to males. This framework is extended onto other entities. For example, sulphur and quicksilver—which, as has been transmitted by Arabic sources, are considered to be the prime matter of metals—are described as the father and mother of all metals; and plants are said to have soil for the mother and the sun for the father.[43] The male principle is always associated with the formal principle and at least carries heat, which Albert's physics perceives as the material quality closest to the formal principle.[44] Sperm, which is the outcome of the digestion of

[39] On the notion of *digestio* in Albert's works, see Cadden (1980); Reynolds (1999).

[40] Albertus Magnus, *Meteora*, IV, tr. 1, cap. 3 (ed. Hossfeld, 2003, 214: 29–33): "Subiecta enim activarum virtutum in rebus mixtis naturalibus sunt passivae virtutes, quae sunt humidum et siccum, ex quibus subiectis calidum et frigidum generant res mixtas, quando proportionaliter vincunt materiam humidi et sicci, hoc est, quando agunt permutando ista secundum exigentiam formae et speciei eius, quod generatur; ita tamen quod aliter agat generando agens calidum et aliter agens frigidum, quia unum agit ad speciem et alterum coadunando et continendo partes materiae, ut supra diximus."

[41] Cf. Loconsole (2020).

[42] See Demaitre and Travill (1980); Jacquart and Thomasset (1981, 76–82); Cova (2017).

[43] Albertus Magnus, *Mineralia*, IV, tr. 1, cap. 1 (ed. Borgnet, 1890, 63b): "Sulphur enim est quasi pater, et argentum vivum mater: quod convenientius dicitur si sulphur dicatur esse in commixtione metallorum quasi substantia seminis paterni, et argentum vivum sicut menstruum quod coagulatur in substantiam embryonum." See also Albertus Magnus, *De vegetabilibus*, I, tr. 1, cap. 7 (ed. Meyer, Jessen, 1867, 25): "Terra mater est plantarum, materiam per modum patientis ministrans, et sol pater, per modum agentis generans."

[44] Albertus Magnus, *Meteora*, IV, tr. 1, cap. 2 (ed. Hossfeld, 2003, 212–213).

blood,[45] is a pale and subtle humour produced by animal bodies whose ultimate goal is the conservation of the species.

In his commentary on the *Metereologica*, Albert claims that when heat is not strong enough to overcome moisture, digestion is incomplete,[46] meaning that the ongoing process in which matter is moulded is not concluded. Indeed, if heat is responsible for the closure of the formative process of the species,[47] coldness weakens the power of heat, prevents its action,[48] and thwarts the digestion of matter.[49] Albert explains in the *De animalibus* that since the balance between the active and the passive qualities is fundamental in the generation of animals, the sperm which acquires its virtue through coldness (*secundum frigum*), or in a similar way, forestalls digestion and paralyses the process of generation.[50] Given this, the infertility of the sperm of older men must ensue from its insufficient digestion caused by the excessive coldness of their nature. Such undigested sperm is unable to bear all the formal virtues necessary for generation. Therefore, the transmission of hereditary traits and the development of the foetus's form and body may be hindered or even

[45] Albertus Magnus, *De animalibus*, III, tr. 2, cap. 8 (ed. Stadler, 1916, 343: 151): "Humiditates vero posterius generatae in corporibus animalium sunt quae pertinent non ad salutem individui, sed ad permanentiam speciei, et sunt sperma generationis, et lac quod est nutrimentum iam generati in his quae generant similia. Sed differentia est, quia sperma generatur ex sanguine optime digesto quarta digestione et subtili."

[46] Albertus Magnus, *Meteora*, IV, tr. 1, cap. 3 (ed. Hossfeld, 2003, 214: 36–43): "Cum vero agentia calidum et frigidum non vincunt proportione dicta materiam, quae est humidum et siccum, tunc planum est, quia fit secundum aliquam partem humidi, quod non terminatur per activas molinsis, hoc est indigestio. Quia si bene digereretur humidum, quod est materia generati, ipsum susciperet speciem mixti, quando calidum agit."

[47] Albertus Magnus, *Meteora*, IV, tr. 1, cap. 2 (ed. Hossfeld, 2003, 212: 67–69): "Multipliciter calidum est activum et multiplicius quam frigidum, quia calidum ducit ad formam speciei."

[48] Albertus Magnus, *Meteora*, IV, tr. 1, cap. 1 (ed. Hossfeld, 2003, 210: 49–51): "[Frigiditas] permutat autem alterando et calidi effectum impediendo, sicut patebit in omnibus indigestionis speciebus."

[49] Albertus Magnus, *Meteora*, IV, tr. 1, cap. 16 (ed. Hossfeld, 2003, 229: 77–81; 230, 1–2): "Indigestio enim est incompletio, quae fit propter indigentiam sive defectum naturalis caloris, qui humidum non potest alterare ex passionibus sive qualitatibus contraiacentibus. Causa autem indigentiae sive defectus caloris frigiditas aliqua est, sive proveniat ex nimia abundantia humidi sive ex alia causa secundum omnem diversitatem rerum, in quibus est digestio."

[50] Albertus Magnus, *De animalibus*, XVI, tr. 1, cap. 13 (ed. Stadler, 1920, 1097: 68–69): "Omnino necesse est ponere in spermate aliquas virtutes quae faciunt ipsum conveniens generationi, ut sit perfectum ab anima ignobili vel nobili aut nobilissima. Hae autem virtutes praecipue sunt secundum illud in spermate quod calor naturalis vocatur et generativus eius quod concipitur. Non enim est haec virtus secundum frigus aut aliquid simile frigori: eo quod frigus sistit motum et facit indigestionem sicut in libro quarto *Meteororum* ostensum est."

discontinued.[51] Thus, the disproportion between heat and moisture is always deleterious. The abundance of moisture or the action of coldness precludes proper digestion insofar as a moist substrate remains that is not digested by heat. For its part, excessive heat or the lack of a moist substrate cause the desiccation of matter.[52]

6.6 Two Faces of Melancholy: Aristotle and Galen

A cold complexion, such as that of an aged man, produces undigested semen that only poorly conveys the formative powers necessary for reproduction. The question remains what factors, if any, produce such a cool complexion in philosophers and preclude their procreative success. The reason why the semen of philosophers is undigested can be deduced from Albert's reception of Aristotle's *Problemata*. Containing a mixture of Aristotle's and later authors' (e.g., Theophrastus') theories, the *Problemata* were translated by David of Dinant as early as in the second half of the twelfth century. His translation might have been only partial, and in all likelihood, only excerpts circulated until a new translation was made by Bartholomew of Messina at the court of King Manfred of Sicily (1258–1266).[53] Albert remarked that he did not have the entire text at his disposal when drafting *De somno et vigilia*,[54] which can be dated to the late 1250s.[55] *Problem* XXX.1, in which the Platonic idea of frenzy is combined with the purely medical notion of melancholy,[56] conveys the idea that all eminent philosophers, politicians, poets, and artists, including Socrates and Plato themselves, were evidently melancholic.[57] Albert quotes this text several

[51] Albertus Magnus, *Super Lucam*, 1: 18 (ed. Borgnet, 1894, 35b): "Senex autem frigidus est et siccus, et ideo ad generandum inhabilis: quia frigiditas sistit motum generativae, et siccitas abscindit semen generationis." See also Albertus Magnus, *De animalibus*, IX, tr. 1, cap. 1 (ed. Stadler, 1916, 677–678): "Quidam tamen sunt, praecipue mares, quibus propter calorem et siccitatem non accidit spermatizare in illo tempore, et in processu aetatis virilis quando tepescit calor et humectatur humore senectutis complexio, incipit in eis desiderium coitus et potentia fervens coeundi. Est autem in quibusdam feminis similiter. Et quidam sunt quibus ex frigiditate et siccitate complexionis in annis pubertatis non accidit semen, sed illis numquam redibit potentia coeundi in aliqua aetate, sed potius ex frigiditate impotentes erunt in aeternum: et in feminis quibusdam est similiter, sed est valde rarum." See also Albertus Magnus, *Meteora*, IV, tr. 1, cap. 13 (ed. Hossfeld, 2003, 227: 17–21): "In senectute abundat frigiditas cum sicco et in ultima aetate abundat frigiditas cum humido frigido quod non nutrit, sed tantum extrinsecus humefacit partes et est phlegmaticum humidum et indigestum."

[52] See Albertus Magnus, *Meteora*, IV, tr. 1, cap. 18 (ed. Hossfeld, 2003, 232: 18–35).

[53] See De Leemans, Goyens (2006, X).

[54] Albertus Magnus, *De somno et vigilia*, I, tr. 2, cap. 5 (ed. Borgnet, 1890, 145b): "[…] in *libro de Problematibus* ab Aristotele, qui liber non ad me pervenit, licet viderim quaedam excerpta de ipso."

[55] On the chronological order of Albert's works on the natural science, see Weisheipl (1980); Anzulewicz (2011).

[56] Klibansky et al. (1979, 17).

[57] Klibansky et al. (1979, 15–40); Schutrümpf (2015).

times, highlighting the melancholic nature of great men and repeatedly emphasising the relation between learning and melancholy.[58]

Besides this idea of melancholy as an exceptional condition, the Galenic humoral understanding of melancholy as a kind of disposition and part of the doctrine of the four temperaments,[59] is also pivotal in Albert's texts. The four humours are viewed as the material causes of certain types of physical and mental constitution. According to the humoral theory, various bodies are dominated by various humours: yellow bile (*cholera*), which is hot and dry like fire; blood, which is hot and moist like fire; phlegm, which is cold and moist like water; an melancholy, which is cold and dry like earth.[60] These qualities shape complexions, which influence the attitudes and inclinations of human beings: those who are sanguine are well-mannered and calm; those in whom choler prevails are usually smart but ireful; and melancholy people are inclined to be sad and studious.[61]

In particular, melancholy—or black bile—is perceived as the driest and coldest of humours,[62] and, accordingly, the melancholic complexion is the driest and coldest of complexions. Albert claims that old men's coldness is caused by their melancholic complexion, which makes them sad and fearful.[63] Because of the natural coldness of melancholy, the places where processes of digestion and nutrition take

---

[58] See Albertus Magnus, *De motibus animalium*, I, tr. 2, cap. 5 (ed. Borgnet, 1890, 277a); Albertus Magnus, *De somno et vigilia*, III, tr. 2, cap. 8 (ed. Borgnet, 1890, 206a); Albertus Magnus, *Summa theologiae*, II, tr. 12, q. 72, membrum 3 (ed. Borgnet, 1894–1895, 39a); Albertus Magnus, *Ethica*, IV, tr. 2, cap. 10 (ed. Borgnet, 1890, 313a); Albertus Magnus, *Ethica*, VII, tr. 2, cap. 5 (ed. Borgnet, 1890, 511b).

[59] In this respect, see Klibansky, Panofsky, Saxl (1979, 1–126). For the reception and the dissemination of the theory of the temperaments, see Jacquart (1997, 2013).

[60] Albertus Magnus, *De causis proprietatum elementorum*, I, tr. 1, cap. 2 (ed. Hossfeld, 1980, 52, 16–31): "Animalium autem corpora recedunt maxime et ideo non generantur ex elementis proxime, sed oportet elementa commisceri in humores et ex humoribus constitui animalium corpora. Humores tamen, licet sint actu humidi omnes, propter quod et humores vocantur, sunt tamen virtutem obtinentes cuiuslibet elementi; invenitur enim commixtio, in qua dominatur cholera citrina, quae est calida et sicca, sequens proprietates ignis. Et invenitur alia in qua dominatur sanguis, qui est calidus et humidus, sequens proprietates aeris. Invenitur etiam alia quae est ex phlegmate, sequens proprietates aquae in hoc quod est frigida et humida. Et invenitur alia quae est ex cholera nigra, quae melancholia vocatur, quae est frigida et sicca, virtutes habens terrae. Humorum autem istorum generationem et diversitatem determinare habet medicus."

[61] Albertus Magnus, *Summa theologiae*, II, tr. 12, q. 72, membrum (ed. Borgnet, 1894–1895, 39a): "[Principium] inclinativum est complexio: sicut dicimus sanguineos esse bonorum morum et quietos, cholericos autem acutos et furibundos, melancholicos autem tristes et studiosos, ut dicit Aristoteles in *Problematibus*."

[62] See Albertus Magnus, *De somno et vigilia*, III, tr. 2, cap. 4 (ed. Borgnet, 1890, 201b): "Melancholia autem est in extremo complexionis humanae frigida et sicca"; Albertus Magnus, *Ethica*, VII, tr. 2, cap. 5 (ed. Borgnet, 1890, 511b): "Melancholia enim naturalis, ut dicit Galenus, frigida est et sicca."

[63] Albertus Magnus, *De motibus animalium*, I, tr. 2, cap. 5 (ed. Borgnet, 1890, 277b): "Si autem sit spiritus paucus et grossus et obscurus et frigidus, hic erit de conditione melancholiae et senii: et hoc tantum excedere potest in his, quod numquam gaudebit, sed semper tristabitur: et si non excedit multum, tunc tamen non facile gaudebit, sed facile tristabitur, et erit pavidus ex qualibet causa, et fugiens, et etiam, ut ita dicam, formidans: et ideo senes tristes sunt et melancholici de melancholia naturali."

place become cold, which then leads to inadequate digestion.[64] Many physiological factors, such as the lack of *spiritus* and an aversion to bodily delight, foster the predisposition for study and learning in melancholics. In stark contrast to this, the abundance of clear, subtle, hot, and moist *spiritus*—the animating substance—in the heart makes sanguine people prone to cheerfulness, pleasure and delight, eliminating sadness and affliction. While the sanguine disposition hinders the study of sciences, the dry complexion of melancholics, which makes them prone to illnesses, inclines them to solitude, suppresses their desire for corporeal pleasures, and appears to promote the study together with the exercise of virtues and the intellect.[65]

When addressing the conundrum of philosophers generating foolish children in the *Quaestiones*, Albert neither explicitly evokes the doctrine of the humours nor discuss the theory of complexions in terms of the inclinational physical equilibrium of the individual. Nevertheless, he refers to the inadequate digestion of sperm, which is usually a symptom of the cold complexion proper to melancholy people (and, likewise, old men), and triggers the cooling of the reproductive organs. The idea that philosophers produce undigested sperm, because their digestive power is deficient as a result of their melancholic, cold, and dry complexion, is compatible with Albert's notions in *quaestio* 4 of the *Quaestiones de animalibus* and may be construed as his solution to the riddle.

6.7 What Is Unnatural Melancholy?

Nevertheless, a certain discrepancy inheres in Albert's works. It is true that Albert certainly noticed the overlap of the two pseudo-Aristotelian and Galenic traditions on melancholy. A passage in the *Ethica*, in which natural melancholy and unnatural

[64] Albertus Magnus, *De somno et vigilia*, I, tr. 2, cap. 8 (ed. Borgnet, 1890, 152a): "Senes exsiccati et frigidi graves et severi: nigra enim cholera, quae melancholia est, cum naturaliter frigida sit, locum nutrimenti et digestionis frigidum facit, similiter omnes alias partes ubicumque erit secundum potestatem excellentem huiusmodi superfluitas nigrae cholerae." The insufficient digestion of the semen due to coldness is also the cause of the birth of females, in keeping with the principle that males are always hotter. See Albertus Magnus, *De animalibus*, XVI, tr. 1, cap. 14 (ed. Stadler, 1920, 1100: 72): "Si autem indigestionem patiatur aliquam humidum quod formatur, erit frigidum quod est causa indigestionis: et haec complexionalis frigiditas et humiditas erit causa sexus feminini. […] Per omnem eundem modum accidit indigestionem aliquam pati sperma modo praedicto et tunc generat feminam et si non patiatur eam, procreat masculum." On the topic, see a synoptic article by Gabriella Zuccolin: Zuccolin (2017).

[65] Albertus Magnus, *De motibus animalium*, I, tr. 2, cap. 5 (ed. Borgnet, 1890, 277a): "Dicamus igitur, quod si in corde est multus spiritus, clarus, subtilis, calidus, et humidus temperate, hic est aptissimus gaudio et delectationi: et non tristabitur nec angustiabitur nisi ex magna causa. Et haec est optima dispositio quae quasi vitam florentem facit: et tamen hoc propter coniunctam sibi dissolutionem impedit studium scientiae et virtutis, sicut Aristoteles in libro de *Problematibus* dicit, asserens quod omnes excelsi in sapientia et heroicas virtutes habentes, fere fuerunt melancholici. […] Si quis autem est pauci spiritus et tenuis et non clari, sicut sunt spiritus convalescentium ex aegritudinibus et macilentorum siccorum, hic non gaudet de facili: eo quod spiritus vix sufficit principio vitae: et ideo in corde retinetur, et non effunditur in delicias corporis."

melancholy (called *adusta* or *accidentalis*) are set apart, exhibits the difference between the Galenic idea of melancholy as a dry and cold humour and the Aristotelian idea of the melancholic genius in *Problem* XXX.1.[66] Yet, the idea of a naturally melancholic humour and complexion and the idea of exceptional melancholy cannot always be easily reconciled. In fact, according to Albert, intellectual activities are sometimes related simply to melancholics in purely medical terms, as having a cold and dry complexion (as shown above). At other times, however, like in *Problem* XXX.1, a kind of unnatural melancholy is mentioned, which is hot, vaporous, and similar to red wine. Albert observes that Aristotle himself pointed out that melancholy may be both the hottest and the coldest of the humours under some conditions.[67] Moreover, his own conception of unnatural melancholy is directly borrowed from Avicenna[68] and frames it as *melancholia adusta*, i.e., a kind of burned melancholy produced by the heating of one of the four humours and believed to be different from any of these. In Albert's work, this kind of unnatural melancholy is sometimes cited as the physiological condition of virtue and philosophising, which is reminiscent of Aristotle's idea of genius.[69]

The physiological features of burned melancholy—i.e., being hot and vaporous—are more redolent of the sanguine type than of natural melancholy with its proper qualities of coldness and dryness. This is why the fine exercise of thought and the production of *melancholia adusta* are sometimes said to be proper to the sanguine rather than the melancholic complexion.[70] Albert sometimes also associates the medical concept of the melancholic temperament with serious flaws, such

---

[66] Albertus Magnus, *Ethica*, VII, tr. 2, cap. 5 (ed. Borgnet, 1890, 511b): "Melancholici de melancholia accidentali ex acumine humoris morsa habeant corpora, tamen dicit Aristoteles in *Problematibus*, quod omnes hi qui fuerunt heroicarum virtutum, Hector, et Priamus, et alii, in hac melancholia laborabant: eo quod haec melancholia rubei vini quod vaporosum est, habet similitudinem: et quia gravitatem habet, constantiam facit: quia vero vaporosa, virtutem erigit ad operationem. Melancholia enim naturalis, ut dicit Galenus, frigida est et sicca."

[67] See Schutrümpf (2015, 362).

[68] Albert's digression on *melancholia innaturalis* or *adusta* can be found in Albertus Magnus, *De animalibus*, III, tr. 2, cap. 3–4 (ed. Stadler, 1916, 329–334). Albert's notion derives from Avicenna; see Avicenna, *Canon medicinae*, I, sent. 1, doctr. 1. As in Avicenna, this particular type of humour is produced in the combustion of one of the main four humours. See also Klibansky et al. (1979, 87–88).

[69] See, for example, Albertus Magnus, *Ethica*, VII, tr. 2, cap. 5 (ed. Borgnet, 1890, 511b). See also Albertus Magnus, *De animalibus*, III, tr. 2, cap. 3 (ed. Stadler, 1916, 330: 121): "Et ideo dicit Aristoteles in libro de *Problematibus*, quod omnes maiores philosophi sicut Anaxagoras et Tales Mylesius et omnes illi qui virtutibus praecellebant heroycis, sicut Hector et Eneas et Priamus et alii, de tali erant melancholia [*sc.* melancholia adusta]. Dicit enim, quod talis melancholia habet naturam vini rubei, quod fumosum est et confirmatorum et stabilium spirituum generativum."

[70] See Albertus Magnus, *De animalibus*, XX, tr. 1, cap. 11 (ed. Stadler, 1920, 1304: 60): "Quaecumque autem grossi sunt sanguinis et calidi, immixtam in sanguine habent choleram adustam vel aduri incipientem quae est quoddam genus melancholiae: et haec satis sunt stabilia et constantis audaciae et multorum spirituum mediorum inter grossos et subtiles: propter quod etiam talis complexionis existentes homines, stabilis sunt animi et fortis et non praecipites. Talis enim melancholia est quasi de complexione vini rubei sicut dicit Aristoteles in libro de *Problematibus*."

as suicidal tendencies, depression, malice, and kleptomania, which cannot be properly attributed to philosophers.[71] As thinking or being studious often involve heat and vapour, unnatural melancholy enhances the ambiguity of the philosopher's identity: Is the philosopher a dry and cold melancholic or a person permeated by this unnatural humour?

Despite all these hesitations, Albert's texts embrace the idea that natural (cold and dry) melancholy and the burned (hot and vaporous) melancholy are complementary in the development and exercise of intellectual faculties. In fact, Albert believes that natural melancholy may produce the finest type of burned melancholy, because the gentle heat of natural melancholy does not burn violently (*alterat ipsa sed not incinerat*), thus sustaining the vaporousness of burned melancholy.[72] Moreover, Albert claims in his *De somno et vigilia*[73] that coldness and dryness, proper to melancholy, may interact optimally with the warmness of an act of thinking. Coldness mitigates the action of heat, as a result of which a subtler, vaporous *spiritus* is produced; meanwhile, dryness is able to sustain images.

Consequently, melancholics are capable of forming many images, remaining focused, and producing strong, persistent thoughts (*multum fixas habent cogitationes et fortes*).[74] In fact, due to their dry nature, melancholic people boast good memory, which is able to retain images and forms;[75] contrary to them, phlegmatic people, whose nature is moist, possess poor memory.[76] Melancholy *non multum adusta*, i.e., not completely burned, but burned with the gentle heat of natural melancholy, indi-

[71] Cf. Albertus Magnus, *De animalibus*, XX, tr. 1, cap. 11 (ed. Stadler, 1920, 1305: 61): "Et homines aliquando nihil delectationis apud se invenientes et malae suspicionis etiam existentes ad alios, occidunt se ipsos, et sunt nec diligentes nec diligibiles, solitudinem quae malitia vitae humanae est amantes, et in sordibus esse delectabile est eis, et alia multa mala contingunt eis et sunt frequenter fures etiam quando non indigent de re quam furantur, et multum sunt insomnes propter complexionis siccitatem et frigiditatem."

[72] See Albertus Magnus, *De animalibus*, III, tr. 2, cap. 3 (ed. Stadler, 1916, 330: 121): "In hominibus autem etiam diversimode participatur humor iste [*sc.* melancholia adusta]: quoniam si non sit multum adusta, sed adustio aliqua cum melancholia naturali quae alteret ipsam et non incineret, ita quod fumare incipiat, tunc illa melancholia erit habens multos et stabilitos et confirmatos spiritus: quia calidum eius bene movet et humidum eius cum hypostasi terrestri non incinerata optime movetur propter quod tales habent stabilitos conceptus et ordinatissimos affectus et efficiuntur studiosi et virtutum optimarum."

[73] Albertus Magnus, *De somno et vigilia*, III, tr. 2, cap. 4 (ed. Borgnet, 1890, 201b): "Calidum ergo paucum quod est in melancholicis, elevat vaporem et spiritum debilem. Quod si multus esset, consumeretur debilis spiritus: et quia calor non commiscet huiusmodi spiritum, ideo imagines bene tenent se in ipso: siccum autem melancholici complexionale est, et hoc in actu humidum et potentia siccum et virtute. Adhuc autem non tollitur vaporabilitas: quoniam potius vaporat, et vapor ex siccitate tenet fortiter figuras imaginum: frigus etiam non omnimode est, sed conditionale. Hoc etiam facit ad bonam imaginum retentionem. [...] Cum igitur ista coniuncta fuerint, sequitur quod melancholici multarum imaginum sunt, et multum in eis permanent."

[74] Albertus Magnus, *De spiritu et respiratione*, III, tr. 2, cap. 3 (ed. Borgnet, 1890, 249b).

[75] See Albertus Magnus, *De somno et vigilia*, III, tr. 2, cap. 4 (ed. Borgnet, 1890, 201b).

[76] Albertus Magnus, *De animalibus*, XX, tr. 1, cap. 11 (ed. Stadler, 1920, 1305: 61): "Et est eorum ingenium malum et memoria mala: ingenium quidem propter frigus quod non movet spiritus neque adducit, sed abscidit eos: memoria autem propter humorem aqueum qui non retinet."

cates that natural melancholy is somehow capable of offsetting the hot and vaporous nature of unnatural melancholy. If this the case, the philosopher's or, more generally, the learned man's nature is best equipped to produce and handle unnatural or burned melancholy, precisely because of their natural melancholic complexion.

Relevantly to my argument, Albert also emphasises the similarity between the two humours of burned melancholy and wine,[77] conceived as two kinds of *spiritus*,[78] described in the *Problemata*.[79] A highly vaporous and hot humour, wine cools the lower part of the body and may extinguish the heat of the human body.[80] Burned melancholy acts in the same way: its moving heat reaches the upper part of the body and extinguishes the natural heat of the living body that produces it. As with wine, the reproductive organs in the lower part of the body become cooler, which hinders the digestion/production of well-balanced semen. These interrelations among study, melancholy, and coldness can explain the frequently observed foolishness of philosophers' sons. In Albert's writings, the phenomenon is situated at the intersection of several philosophical traditions, including Aristotle's *Problemata*, Galen's theory of humours and temperaments, and Avicenna's *Canon* and *De animalibus*, and it is bound up with the process of digestion. The complexion of philosophers—whether cold and dry by nature or made such by the exercise of intellectual activities—weakens their natural virtue. The excessive coldness of their reproductive organs hinders the formative action of heat, as a result of which their semen is unbalanced, undigested, and consequently ineffective in transmission of the formative powers.[81] This commonly prevents the transfer of some dispositions, such as the capacity to acquire knowledge and wisdom, onto philosophers' sons.

6.8 James of Viterbo's *Quodlibet* IV, Question 23

If Albert's solution to the riddle of philosophers having foolish offspring must be gleaned from his various writings, James of Viterbo provides a much more compact and organised answer in his *Quodlibet* IV, question 23.[82] Explicitly focused on the

---

[77] See, for example, Albertus Magnus, *De somno et vigilia*, III, tr. 2, cap. 8 (ed. Borgnet, 1890, 206a): "Sunt enim melancholici, ut dicit Aristoteles in libro de *Problematibus*, studiosi, praecipue si patiantur melancholiam quae ex corpore per incinerationem facta est: illa enim est fumosa sicut vinum rubeum, et est adhaerentium multum phantasmatum, circa quae profundatur intellectus et coniicit ea."

[78] For more on the notion of *spiritus*, see Panarelli, Chap. 7 in this volume, pp. 115–132.

[79] See Ps.-Aristoteles, *Problemata*, XXX.1, 953b20–27.

[80] See Ps.-Aristoteles, *Problemata*, XXX.1, 954b35–40. On the relation among drunkenness, reproduction, and melancholy, see Fortenbaugh (2015, 100–123).

[81] In the Salernitan question quoted above, the bad complexion of sperm is attributed to its excessive dryness, which produces hard brains (*duri cerebri*) unable to use reason properly.

[82] Jacobus de Viterbo, *Disputatio quarta de quolibet*, q. 23 (ed. Ypma, 1975, 79–82). On the dating and context of James's *quodlibeta*, see Schabela and Courtenay (2007, 545–568).

case of wise men, the Augustinian master takes into account the idea of the "hereditariness" of features, the distinction between specific and accidental dispositions, the concept of inclination (*idoneitas*) to a certain ability, and several hindrances to the transmission of traits. James responds to the question *Utrum pater sapiens ut plurimum generet stultos filios* by outlining the basic principles behind the natural transmission of features from parents to offspring. As far as their specific form is concerned, parents always bring forth an individual resembling them: it will be of the same species—human beings certainly only generate human beings—and will share all the species-related proprieties.[83]

Nevertheless, the transmission of accidental features, both corporeal and those proper to the soul, may not always run as smoothly. The parent-child transfer of accidental corporeal features, such as height, colour, or size, can only take place if the generative power is strong enough and matter is inclined to be moulded (*non indisposita nec repugnans*). Basically, James refers to the same principles as Albert and similarly claims that the perfect process of generation takes place when both the formal and material forces involved are in harmony.[84] For their part, accidental features proper to the soul are not directly transmissible, because they accrue over time. For example, knowledge is acquired through education and virtue through practice. To illustrate it vividly, people versed in grammar cannot simply generate offspring equally competent in this respect at the time they are born. As a matter of fact, a wise man generates neither a dumb son nor a smart one, since no new-born possesses either of these features, although they may develop them later in life.[85]

However, James observes that certain kinds of accidents of the soul are somehow transmissible to the offspring; these are called aptitudes or inclinations. One person may have an aptitude for generosity, and another for avarice, temperance, or prudence. Likewise, an individual may be inclined towards mathematics, mechanical arts, or even foolishness.[86] All such inclinations are inheritable because, even though considered accidents of the soul, they originate from the complexion of the body. As such, they can be transferred from parents to offspring under ideal conditions, the way corporeal features are. Yet if a formal or material deficiency occurs, a wise man may father a foolish child.[87]

---

[83] Jacobus de Viterbo, *Disputatio quarta de quolibet*, q. 23 (ed. Ypma, 1975, 79: 7–10): "Genitum assimilatur generanti semper in forma speciei et in proprietatibus consequentibus speciem. Semper enim homo generat hominem et risibilis risibilem et disciplinae perceptibilis disciplinae perceptibilem."

[84] Jacobus de Viterbo, *Disputatio quarta de quolibet*, q. 23 (ed. Ypma, 1975, 79–80: 10–20).

[85] Jacobus de Viterbo, *Disputatio quarta de quolibet*, q. 23 (ed. Ypma, 1975, 80: 21–37).

[86] Jacobus de Viterbo, *Disputatio quarta de quolibet*, q. 23 (ed. Ypma, 1975, 80: 38–44): "Quaedam vero sunt accidentia animae indita vel innata, ut idoneitates; quaedam sunt aptitudines spirituales ad scientias et virtutes, vel ad horum opposita. Sicut in quibusdam est naturalis quaedam aptitudo ad liberalitatem, in quibusdam ad avaritiam, in aliquibus ad temperantiam, in aliis ad oppositum, in quibusdam est aptitudo ad prudentiam, in quibusdam ad stultitiam et ut sint facile deceptibiles, in quibusdam ad scientias mathematicas, in aliis ad artes mechanicas; et sic de aliis."

[87] Jacobus de Viterbo, *Disputatio quarta de quolibet*, q. 23 (ed. Ypma, 1975, 81: 45–57).

James specifically inquires why such deficiencies in the transmission of traits are so frequent in wise men.[88] He answers that it is because wise men are highly fatigued by their study of the sciences. As a result, their inner sensitive powers are debilitated and exhausted by the exercise of intellect, phantasy, estimative faculty, and memory. The weariness of studious people may be so extreme that their deep speculation turns into madness and insanity. More generally, study cripples the regular process of nutrition, compromising the digestive power (*vis digestiva*) and having an adverse effect on all the bodily activities dependent on it, such as the generative power (*vis generativa*): sperm, as seen in Albert's texts, is indeed a humour produced by the digestion of blood. Hence, prolonged intellectual effort may trigger malfunctions of the human powers and alterations in the bodily balance, which blight the disposition of the organs and the functions of powers responsible for nutrition and generation, impacting offspring as well (*redundent in prolem*).

6.9 Conclusion

The question why philosophers or, more generally, wise and studious men generate foolish children has highly relevant implications. Although rather peculiar and only marginal in the scientific debate on the generation of living beings, the subject revives several considerations. Hence, Albert's and James's physiological approaches to medicine-related questions are significant for an inquiry into the relationship of mind and body.

First, the explorations prompt concluding that some mental states may interfere with bodily processes just as corporeal causes do. In Albert, the insufficient digestion of the male semen, which weakens its generative power, can be caused not only by physical circumstances, such as the excessive consumption of certain foods and beverages, but also by specific mental conditions. Health and the functions of some corporeal principles may be marred by an excessive weariness of intellectual faculties, impairing the physical performance of the individual.

Second, while the influence of the mind on the body is no surprise, given that the soul is designed to control and animate the body, the opposite idea—that the corporeal sphere may prevail over the intellectual and formal spheres—is also seriously considered. Albert states that certain mental properties are subject to the influences of corporeal principles. Indeed, the transmission of dispositions of the soul, such as wisdom and knowledge, may be entirely thwarted if the complexion of the father and, consequently his semen, are altered by a material deficiency. For all his caution or simply accuracy in talking of resemblance and of the transmission of inclinations, rather than of dispositions of the soul, James believes that the inclination for studying is directly derived from the inner and corporeal structure of the individual;

[88] Jacobus de Viterbo, *Disputatio quarta de quolibet*, q. 23 (ed. Ypma, 1975, 81–82: 58–83).

personal intellectual attitudes stem from the person's complexion, which may be transmitted just like biological features. This position echoes Albert's account.

Albert's peculiar notion of *melancholia adusta* also seems instrumental in exploring the physiological and material dimension of intellectual activity. Thinking is posited to have a direct physiological counterpart, a humour produced by the digestion and material modification of certain corporeal humours. Analogically, acts of thinking assume elemental qualities, such as heat and *vaporabilitas*, which link them to particular activities, humours, and operations of the body.

Notably and curiously, the puzzle of wise men siring foolish children is a site of a clash between medical and philosophical discourses. The ideal conditions for using the intellect and thinking are at the same time revealed to be detrimental to both health and reproduction. This marks an unsolved paradox in which medical knowledge openly contradicts the philosophical tradition, as the physical and intellectual natures seem to represent two contradicting models. While the exercise of intellectuality is the highest goal of human nature according to philosophers, the perspective of the physicians reveals that, against expectations, *salus* and *sapientia* are not complementary.

References

Primary Literature

Albertus Magnus. 1867. In *De vegetabilibus libri VII*, ed. Ernst Meyer and Carl Jessen. Berlin: Reiner.

———. 1890a. *De motibus animalium*. In *Alberti Magni opera omnia 9*, ed. Auguste Borgnet. Paris: Vivès.

———. 1890b. *De somno et vigilia*. In *Alberti Magni opera omnia 9*, ed. Auguste Borgnet. Paris: Vivès.

———. 1890c. *De spiritu et respiratione*. In *Alberti Magni opera omnia 9*, ed. Auguste Borgnet. Paris: Vivès.

———. 1890d. *Mineralia*. In *Alberti Magni opera omnia 5*, ed. Auguste Borgnet. Paris: Vivès.

———. 1890e. *Ethica*. In *Alberti Magni opera omnia 7*, ed. Auguste Borgnet. Paris: Vivès.

———. 1894a. *Super Lucam*. In *Alberti Magni opera omnia 22*, ed. Auguste Borgnet. Paris: Vivès.

———. 1894b–1895. *Summa theologiae sive de mirabili scientia dei*. In *Alberti Magni opera omnia* 31–33, ed. Auguste Borgnet. Paris: Vivès.

———. 1916–1920. *De animalibus*, ed. Hermann Stadler (Beiträge 15–16). Münster: Aschendorff.

———. 1955. *Quaestiones super "De animalibus"*. In *Alberti Magni opera omnia 12*, ed. Ephrem Filthaut. Münster: Aschendorff.

———. 1980. *De causis proprietatum elementorum*. In *Alberti Magni opera omnia 5/2*, ed. Paul Hossfeld. Münster: Aschendorff.

———. 1987. *Physica*. In *Alberti Magni opera omnia 4/1–2*, ed. Paul Hossfeld. Münster: Aschendorff.

———. 2003. *Meteora*. In *Alberti Magni opera omnia 6/1*, ed. Paul Hossfeld. Münster: Aschendorff.

Aristotle. 1965a. In *De generatione animalium*, ed. Hendrik J. Drossart Lulofs. Oxford: Clarendon Press.

———. 1965b–1970. *Historia animalium I–VI*, ed. Arthur L. Peck. London: Heinemann; Cambridge, MA: Harvard University Press.

———. 1991. In *Historia animalium VII–X*, ed. David M. Balme. Cambridge, MA: Harvard University Press.

Ps.-Aristoteles. 1991–1994. *Problemata physica*, ed. Pierre Louis. Paris: Les Belles Lettres.

Avicenna. 1961. *De animalibus*. In *Avicenne perhypatetici philosophi, ac medicorum facile primi opera in lucem redacta ac nuper quantum ars niti potuit per canonicos emendata*, per Bonetum Locatellum mandato Octaviani Scoti. Venetiis 1508; rist. anast. Frankfurt am Main: Minerva.

———. 1964. *Liber canonis Avicenne revisus et ab omni errore mendaque purgatus summaque cum diligentia impressus, apud Simonem Papiensem. Venetiis 1507; rist. anast.* Hildesheim: Olms.

Jacobus de Viterbio. 1975. In *Disputatio quarta de quolibet*, ed. Eelcko Ypma. Würzburg: Augustinus-Verlag.

Petrus Hispanus. 2015. *Quaestiones super libro "De animalibus" Aristotelis*, ed. Francisca Navarro Sánchez. Farnham: Ashgate.

Ptolemaeus Claudius. 1980. *Tetrabiblos*, ed. Frank E. Robbins. London: Heinemann; Cambridge, MA: Harvard University Press.

———. 2015. In *Ptolemy's Tetrabiblos in the translation of William of Moerbeke. Claudii Ptolemaei Liber iudicialium*, ed. Gudrun Vuillemin-Diem, Carlos Steel, and Pieter De Leemans. Leuven: Leuven University Press.

Secondary Literature

Anzulewicz, Henryk. 1999. Die Aristotelische Biologie in den Frühwerken des Albertus Magnus. In *Aristotle's Animals in the Middle Ages and Renaissance*, ed. Carlos Steel, Guy Guldentops, and Pieter Beullens, 159–188. Leuven: Leuven University Press.

———. 2009. Albertus Magnus und die Tiere. In *Tiere und Fabelwesen im Mittelalter*, ed. Sabine Obermaier, 29–54. Berlin/New York: de Gruyter.

———. 2011. Zeittafel (Chronologie nach derzeitigem Forschungsstand). In *Albertus Magnus und sein System der Wissenschaften*, ed. Hannes Möhle, 28–31. Münster: Aschendorff Verlag.

De Asúa, Miguel. 1997. Peter of Spain, Albert the Great and the "Quaestiones de animalibus". *Physis* 34: 1–30.

———. 1999. Medicine and Philosophy in Peter of Spain's Comentary on "De animalibus". In *Aristotle's Animals in the Middle Ages and Renaissance*, ed. Carlos Steel, Guy Guldentops, and Pieter Beullens, 189–211. Leuven: Leuven University Press.

———. 2013. War and Peace: Medicine and Natural Philosophy in Albert the Great. In *A Companion to Albert the Great*, ed. Irven M. Resnick, 269–297. Leiden/Boston: Brill.

Bazán, Bernardo C., John W. Wippel, Gérard Fransen, and Danielle Jacquart. 1985. *Les questions disputées et les questions quodlibétiques dans les facultés de théologie, de droit et de médecine*. Turnhout: Brepols.

Berti, Enrico. 2010. Natura e generazione degli animali in Aristotele. *Kriterion: Revista de Filosofia* 51 (122): 489–504.

Bowler, Peter. 1989. *The Mendelian Revolution. The Emergence of Hereditarian Concepts in Modern Science and Society*. Baltimore: The Johns Hopkins University Press.

Burnett, Charles. 1990. The Planets and the Development of the Embryo. In *The Human Embryo: Aristotle and the Arabic and European Traditions*, ed. Gordon R. Dunstan, 95–112. Exeter: University of Exeter Press.

Cadden, Joan. 1980. Albert the Great's Universal Physiology: the Example of Nutrition. In *Albertus Magnus and the Sciences*, ed. James A. Weisheipl, 321–339. Toronto: Pontifical Institute of Mediaeval Studies.

———. 1993. *The Meanings of Sex Difference in the Middle Ages. Medicine, Science, and Culture*. Cambridge: Cambridge University Press.

Cova, Luciano. 2017. Seme e generazione umana nelle opere teologiche di Alberto Magno. In *Summa doctrina et certa experientia*, ed. Gabriella Zuccolin, 237–256. Firenze: Sismel-Edizioni del Galluzzo.

Chandelier, Joël, and Aurélien Robert. 2013. Nature humaine et complexion du corps chez les médecins italiens de la fin du Moyen Âge. *Revue de Synthèse* 134: 473–510.

Coles, Andrew. 1995. Biomedical Models of Reproduction in the Fifth Century BC and Aristotle's "Generation of Animals". *Phronesis* 40 (1): 48–88.

Demaitre, Luke, and Anthony Travill. 1980. Human Embryology and Development in the Works of Albertus Magnus. In *Albertus Magnus and the Sciences*, ed. James A. Weisheipl, 405–440. Toronto: Pontifical Institute of Mediaeval Studies.

Draelants, Isabelle. 2003. La *virtus universalis*: un concept d'origine hermétique? Les sources d'une notion de philosophie naturelle médiévale. In *Hermetism from Late Antiquity to Humanism*, ed. Paolo Lucentini, Ilaria Parri, and Vittoria Perrone Compagni, 157–188. Turnhout: Brepols.

Fancy, Nahyan. 2018. Generation in Medieval Islamic Medicine. In *Reproduction: Antiquity to the Present Day*, ed. Nick Hopwood, Rebecca Flemming, and Lauren Kassell, 129–140. Cambridge: Cambridge University Press.

Fortenbaugh, W. William. 2015. On "Problemata" 3: Wine-Drinking and Drunkenness. In *The Aristotelian "Problemata Physica" Philosophical and Scientific Investigations*, ed. Robert Mayhew, 100–123. Leiden/Boston: Brill.

Gadelrab, S. Sherry. 2010. Discourses on Sex Differences in Medieval Scholarly Islamic Thought. *Journal of the History of Medicine and Allied Sciences* 66: 40–81.

Goldstein-Préaud, Tamara. 1981. Albert le Grand et les questions du XIIIe siècle sur le "De animalibus" d'Aristote. *History and Philosophy of the Life Sciences* 3 (1): 61–71.

Jacquart, Danielle. 1997. De *crasis à complexio*: note sur le vocabulaire du tempérament en latin médiéval. In *La Science médicale occidentale entre deux renaissances (XIIe–XVe siècle)*, ed. Danielle Jacquart, 71–76. Aldershot: Variorum Reprints.

———. 2013. La reconstruction médicale de la nature de l'homme aux XIe et XIIe siècles. *Revue de Synthèse* 134: 445–472.

———. 2014. *Recherches médiévales sur la nature humaine. Essais sur la réflexion médicale (XIIe–XVe s.)*. Firenze: Sismel Edizioni del Galluzzo.

Jacquart, Danielle, and Claude Thomasset. 1981. Albert le Grand et les problèmes de la sexualité. *History and Philosophy of the Life Sciences* 3 (1): 73–93.

Jacquart, Danielle, and Françoise Micheau. 1990. *La Médecine arabe et l'Occident medieval*. Paris: Maisonneuve et Larose.

Köhler, W. Theodor. 2008. *Homo animal nobilissimum. Konturen des spezifisch Menschlichen in der naturphilosophischen Aristoteleskommentierung des dreizehnten Jahrhunderts, Teilband 1*. Leiden–Boston: Brill.

Klemm, Matthew. 2012. Les complexions vertueuses: la physiologie des vertus dans l'anthropologie médicale de Pietro d'Abano. *Médiévales* 63: 59–74.

Klibansky, Raymond, Erwin Panofsky, and Fritz Saxl. 1979. *Saturn and Melancholy. Studies in the History of Natural Philosophy, Religion and Art*. Nendeln-Liechtenstein: Kraus Reprint.

Krause, Katia. 2015. Albert the Great on Animal and Human Origin in his Early Works. *Lo Sguardo* 18: 205–232.

Lawn, Brian. 1963. *The Salernitan Questions. An Introduction to the History of Medieval and Renaissance Problem Literature*. Oxford: Clarendon Press.

———. 1979. *The Prose Salernitan Questions Edited from a Bodleian Manuscript (Auct F. 3.10)*. London: Oxford University Press.

De Leemans, Pieter, and Michèle Goyens. 2006. *Aristotle's "Problemata" in different times and tongues*. Leuven: Leuven University Press.

Lefebvre, David. 2016. Le *sperma*: forme, matière ou les deux? Aristote critique de la double semence. *Philosophie antique* 16: 31–62.

Loconsole, Mario. 2020. A Theory on the Formation of Minerals. Albert the Great and the Constitution of Scientific Mineralogy. *Quaestio* 20: 369–388.

López-Beltrán, Carlos. 2008. Les *haereditarii morbi* au début de l'Époque moderne. In *L'hérédité entre Moyen Âge et Époque moderne*. In *Perspectives historiques*, ed. Maaike van der Lugt and Charles de Miramon, 321–354. Firenze: Sismel Edizioni del Galluzzo.

Navarro Sánchez, Francisca 2018. La obra médica de Avicena en las "Questiones super libro De animalibus Aristotelis" de Petrus Hispanus. *Enrahonar*. Supplement Issue: 387–398.

Perfetti, Stefano. 2017. La disseminazione del sapere sugli animali e l'iperaristotelismo di Alberto Magno. In *La zoologia di Aristotele e la sua ricezione dall'età ellenistica e romana alle culture medievali*, ed. Maria M. Sassi, Elisa Coda, and Giuseppe Feola, 269–297. Pisa: Pisa University Press.

Preus, Anthony. 1970. Science and Philosophy in Aristotle's "Generation of Animals". *Journal of the History of Biology* 3 (1): 1–52.

Reynolds, L. Philip. 1999. *Food and the Body. Some Peculiar Questions in High Medieval Theology*. Leiden: Brill.

Rossi, B. Pietro. 2017. L'entrata dei libri "De animalibus" nel Medioevo latino. In *La zoologia di Aristotele e la sua ricezione dall'età ellenistica e romana alle culture medievali*, ed. Maria M. Sassi, Elisa Coda, and Giuseppe Feola, 237–268. Pisa: Pisa University Press.

Schabel, Chris, and J. William Courtenay. 2007. Augustinian Quodlibeta after Giles of Rome. In *Theological Quodlibeta in the Middle Ages. The Fourteenth Century*, ed. Chris Schabel, 545–568. Leiden–Boston: Brill.

Schütrumpf, Eckart. 2015. Black Bile as the Cause of Human Accomplishments and Behaviors in Pr. 30.1: Is the Concept Aristotelian? In *The Aristotelian "Problemata Physica". Philosophical and Scientific Investigations*, ed. Robert Mayhew, 357–380. Leiden/Boston: Brill.

Thorndike, Lynn. 1958. De Complexionibus. *Isis* 49: 398–408.

Van der Lugt, Maaike. 2004. *Le ver, le démon et la vierge. Les théories médiévales de la génération extraordinaire*. Paris: Les Belles Lettres.

———. 2008a. L'animation de l'embryon humain et le statut de l'enfant à naître dans la pensée médiévale. In *L'embryon: formation et animation. Antiquité grecque et latine, traditions hébraïque, chrétienne et islamique*, ed. Luc Brisson, Marie-Hélène Congourdeau, and Jean-Luc Solère, 234–254. Paris: Vrin.

———. 2008b. Les maladies héréditaires dans la pensée scolastique (XIIe–XVIe siècles). In *L'hérédité entre Moyen Âge et Époque moderne. Perspectives historiques*, ed. Maaike van der Lugt, and Charles de Miramon, 273–320. Firenze: Sismel Edizioni del Galluzzo.

Van der Lugt, Maaike, and Charles de Miramon. 2008. In *Penser l'hérédité au Moyen Âge: une introduction. In L'hérédité entre Moyen Âge et Époque moderne. Perspectives historiques*, ed. Maaike van der Lugt and Charles de Miramon, 3–40. Firenze: Sismel Edizioni del Galluzzo.

Weisheipl, A. James. 1980. Albert's Works on Natural Science (*libri naturales*) in Probable Chronological Order. In *Albert the Great and the sciences*, ed. James A. Weisheipl, 565–577. Toronto: Pontifical Institute of Mediaeval Studies.

Wilberding, James. 2017. *Forms, Souls, and Embryos. Neoplatonists on Human Reproduction*. London/New York: Routledge.

Ziegler, Joseph. 2008. Hérédité et physiognomonie. In *L'hérédité entre Moyen Âge et Époque moderne. Perspectives historiques*, ed. Maaike van der Lugt and Charles de Miramon, 245–271. Firenze: Sismel Edizioni del Galluzzo.

Zuccolin, Gabriella. 2017. Questioni di genere tra medicina e filosofia nel Medioevo. Tendenze storiografiche recenti. *Revue des sciences philosophiques et théologiques* 101 (4): 585–610.

———. 2019. *I gemelli nel Medioevo. Questioni filosofiche, mediche e teologiche*. Como/Pavia: Ibis.

Chapter 7
Bodily Prerequisites of the Mind: The Spirit as the Highest Product of Digestion

Marilena Panarelli

Abstract One of the various meanings that the term *spiritus* has assumed throughout the history of philosophy is that of a vaporous bodily substance that serves the soul. This definition binds the notion of the spirit to the notion of the body itself. The aim of this chapter is to identify the major stages of this tradition, which begins with Aristotle and Galen, who applied the notion in philosophy and medicine. During the Middle Ages, the notion was rethought, first, in Costa ben Luca's *De differentia animae et spiritus*, in which the difference between soul and spirit was clarified by defining the spirit as a material principle managed by the soul. Then, Avicenna directly connected the spirit to the humours in his *Canon*, explaining that spirits emerge in the process of concoction. Subsequently, Albert the Great, who was strongly influenced by this tradition, defined the spirit as a vapour of radical moisture and assigned a key role to it in his description of the mechanisms of life in general and, more specifically, of the human body. Among the various powers of the soul, the spirit should serve the highest of these, pervading the inner senses, revealing a relation between the mental functions and the process of the concoction of the humours, e.g., digestion, in this tradition. This chapter analyses the steps and sources—both medical and philosophical—that led to the conception of the body as a process of becoming, involving ever more refined degrees of spirt.

Keywords Spirit · Digestion · Mind · Inner senses · Humours

7.1 Introduction

The concept of the spirit is one of the most ambivalent notions in the history of philosophy. Attempts to retrace the history of this concept resemble a winding road, forking in many directions, which correspond to the various philosophical traditions

M. Panarelli (✉)
Università Del Salento, Lecce, Italy

M. Gensler et al. (eds.), *The Embodied Soul*, Historical-Analytical Studies on Nature, Mind and Action 11, https://doi.org/10.1007/978-3-030-99453-2_7

that refer to this term.[1] Indeed, the concept of the spirit has been contested from the viewpoints of both immateriality and materiality. Straddling this antithesis, the spirit is one of the most frequently used ideas to explain the relationship between mind and body in the history of philosophy. When chancing upon a crossroads of the immaterial and the material along the path of its historical-philosophical development, the spirit has tended to turn in one or the other direction.

Nevertheless, there are stretches where these trails seem to run parallel or to touch. In one of the most comprehensive studies on the issue, James J. Bono has analysed the polyvalent meanings of the spirit, pointing out that a profound reformulation of the concept took place during the twelfth and thirteenth centuries, when Arabic medical texts were found to meet the needs of Christian theology. Bono has eloquently demonstrated how this reformulation in the medical field was borrowed from theology in order "to bridge the gap between the language of theology and the language of natural philosophy."[2] In this way, the notion of the spirit became deeply imbedded in the soul/body relation, as a result of which the philosophical perspective on the issue was broadened.

This chapter aims to investigate the idea of the spirit as a material principle by focusing on its medical roots and, more specifically, on its connection to the humoral theory. Conceived as a material principle directly linked to the humours, the spirit is understood as the most subtle part of the body capable of being governed by the soul, which is considered to be the immaterial principle. In order to produce this kind of historical account, it is necessary to outline significant steps in the development of the notion. Given this, I first retrace Aristotle's and Galen's positions, as the relevant frameworks of ancient philosophy, and then consider Costa ben Luca's *De differentia animae et spiritus* and Avicenna's *Canon*, which both mark turning points in the transmission of this classical tradition. Subsequently, I examine the contribution of Albert the Great, who summarised and rearranged the entire tradition, supporting his philosophical analyses with tools borrowed from the medical field. His approach shows how medical sources corroborated philosophical investigations, so that the operations of the inner senses could be given a physiological explanation.

Following this path, we realize that reconsidering the notion of the spirit involves a reassessment of the body as the basic concept to which it is linked. The fact that the notion of the spirit became an integral part of the humoral theory—so much so that it was seen as a product of the concoction of the humours at a certain point—greatly modified the concept of the body itself. The body came to be conceived not as a simple passive principle, but as a process of becoming, in which matter was transformed from the rawest humour to the most refined spirit. Moreover, in showing such articulated physiological mechanisms, all functions of the body, including the highest ones, were understood as the ongoing result of the concoction process. Given this, the inner senses, which were a prerequisite for any mental

[1] For more details, see Chenu (1957), Bono (1984, 91) and Walker (1984).

[2] Bono (1984, 102).

activity, even for the mental functions involving the immaterial principle of the intellect as such, also depended on the process of the concoction of the spirits.

7.2 Aristotle's Notion of *Pneuma*

The doctrine of *pneuma* (which is the Greek term for the spirit) as developed by Aristotle is particularly difficult to capture with tolerable accuracy.[3] Indeed, even though Aristotle often applies the theory of *pneuma* in his biological works, he never systematises it.[4] To further complicate the issue, in the Middle Ages, an apocryphal work entitled *De spiritu* was ascribed to Aristotle and served as a trusted source to philosophers of the day.[5] While medieval authors referred to a text erroneously attributed to Aristotle, any reliable account of his genuine position should primarily focus on relevant passages of his works on natural philosophy. One of the contexts in which Aristotle addresses the issue of *pneuma* is his discussion of the presence or absence of the soul in the seed, which represent the state of matter prior to the soul. Interestingly, Aristotle claims that matter is potentially alive in seeds because of the presence of the *pneuma*, which is a precondition for the soul. He conceives of the *pneuma* as the bearer of the soul. Similarly, in *The Generation of Animals*, Aristotle raises the question of whether semen possesses the soul:

> In all cases the semen contains within itself that which causes it to be fertile—what is known as 'hot' substance, which is not fire nor any similar substance, but the *pneuma* which is enclosed within the semen or foam-like stuff, and the natural substance which is in the *pneuma*; and this substance is analogous to the element which belongs to the stars.[6]

This passage indicates that *pneuma* is the proper bearer of the soul. In this regard, Aristotle states that "the activity of the soul of every kind has to do with some physical substance which is different from the so-called 'elements' and more divine than they are."[7] Thus, Aristotle links identifies *pneuma* with the astral element, that is, with ether. Based on this passage, Abraham P. Bos states that "*Pneuma* in Aristotle's view is the fifth element in disguise, incognito."[8]

Having admitted the presence of *pneuma* in seeds, Aristotle can conclude that:

> We have now determined in what sense fetations and semen have soul and in what sense they have not. They have soul potentially, but not in actuality.[9]

---

[3] Düring (1966, 343–344), Solmsen (1957), Nussbaum (1978) and Freudenthal (1995).

[4] Bos (2013, 417).

[5] For the debate on the attribution of this work to Aristotle, see Bos and Ferwerda (2008).

[6] Aristoteles, *De generatione animalium*, II.3, 736b30–737a1 (transl. Peck, 1942, 171)

[7] Bos and Ferwerda (2008).

[8] Bos (2013, 135–146).

[9] Aristoteles, *De generatione animalium*, II.3, 737a16–18 (transl. Peck, 1942, 173).

Thus, seeds do not contain soul but *pneuma*, which is the vehicle of the soul. In Aristotle's account, the connection between *pneuma* and the soul is so close that the former is a precondition of the latter.

As the well-known definition of the soul in *De anima* 412a5 holds that the soul is the first actuality of a "natural body that has life potentially", it becomes clear that having life potentially implies that *pneuma* is present. According to Aristotle, *pneuma* is a precondition of life. Moreover, as a precondition of life, it is not only bound up with the elements, but it also possesses something divine, namely ether.[10] While Aristotle investigates *pneuma* mostly with reference to semen and seeds, he does not provide a detailed explanation of its behaviour in the living body. Nevertheless, his interest in this issue is borne out by various writings, for instance, *De motu animalium* 703a10, where he insists that the function of *pneuma* in living beings should be further investigated. Hence, although Aristotle's argument is partial and not systematic, it contains indispensable insights into the mechanisms of life.

7.3 Galen's Pneumatology

Galen offers a systematised depiction of the mechanisms regulating the functions of *pneuma* and proposes a pneumatological doctrine that constitutes a fundamental prerequisite for understanding his physiology. Detailing Galen's complex elaboration of pneumatology lies beyond the scope of my argument, but it is relevant to show how his intricate, methodical, and meticulous framework impacted later tradition. One of his most comprehensive treatments of *pneuma* is to be found in *The Doctrines of Hippocrates and Plato*, where he asserts that:

> [T]he soul's first instrument for all the sensations of the animal and for its voluntary motions as well, is [...] pneuma; and therefore, when the pneuma has escaped [when the ventricles of the brain have been cut], and until it is collected again [i.e., the cut has been closed], it does not deprive the animal of its life but renders it incapable of sensation and motion. Yet if the pneuma were itself the substance of the soul, the animal would immediately die along with the escape of the pneuma.[11]

According to Galen, *pneuma* is able to spark life in the body instantaneously.[12] In fact, the life or death of animals depends on its presence or absence. Besides defining *pneuma*, Galen classifies its varieties, although it is rather unclear whether distinguished two or three kinds of it:

> Accordingly, the pneuma in relation to the arteries is called vital, and that in regard to the brain is psychic, not that it exists as the substance of the soul, but rather as the first instrument of the soul which resides in the brain, whatever may be its substance. And just as vital

[10] Lloyd (2007, 141).

[11] Claudius Galenus, *De placitis Hippocratis et Platonis*, V (ed. De Lacy, 1978–1984, 605–606).

[12] Rocca (2008, 248).

> pneuma is generated in the arteries as well as the heart, obtaining the material for its generation from inhalation and the vaporization of the humours, so the psychic pneuma is generated by a further elaboration of the vital. For it was necessary that this pneuma by all means be changed in precisely the correct fashion. If Nature, needing to fashion semen and milk with precision, even though they are far inferior in power to psychic pneuma, nevertheless arranged for each a lengthy period in the organs of concoction and for that reason provided for semen the spiral vessels for the testes and for milk the length of the vessels that go to the breasts, so naturally also that when elaborating psychic from vital pneuma in the brain it constructed close to the brain a complex labyrinth, as it were, the retiform plexus.[13]

In this passage, Galen distinguishes between vital and psychic *pneuma*. Vital *pneuma* originates in the heart as a direct product of the inhalation of the air by the lungs and the "vaporization of the humours." It is thus possible to glean a connection between pneumatology and the humoral theory in Galen's thought, but this aspect has not been explored particularly well so far. What should be pointed out first is Galen's tendency to conceive of the human body as a whole in which the highest and most sophisticated functions can be explained in the same way as the lower simpler ones.

More precisely speaking, the action of *pneuma* is envisaged as developing from the most basic physiological function to the highest mental one. The first form of *pneuma* emerges in the heart from the encounter of the inhaled air with vapours fuelled by the vital heat. Afterwards, progressing refining process caused by concoction leads to the production of the brain's psychic *pneuma*. Psychic *pneuma* is a fundamental notion Galen uses in his explanation of the anatomy of the brain, whose ventricles form a labyrinth in which *pneuma* can circulate. It is only because of the presence of *pneuma* that the soul can exercise its highest functions.[14] To make this possible a conversion of vital *pneuma* into psychic *pneuma* takes place in the body:

> From the outside air, pneuma is drawn in by the rough arteries and receives its first elaboration in the flesh of the lungs, its second in the heart and the arteries, especially those of the retiform plexus, and then a final elaboration in the ventricles of the brain, which completes its transformation into psychic pneuma.[15]

Pneuma is first engendered in the heart and then in the brain. Moreover, its elaboration is envisioned as concoction that analogously explains the genesis of both vital and psychic *pneuma*.[16] This process became a fundamental point of reference for future authors, as will be demonstrated below.

---

[13] Claudius Galenus, *De placitis Hippocratis et Platonis*, V (ed. De Lacy, 1978–1984, 608–609).

[14] Boudon-Millot (2016, 241).

[15] Claudius Galenus, *De usu partium*, III (ed. Helmreich, 1968, 541–542).

[16] Rocca (2008, 255).

7.4 Costa ben Luca's Definition of Spirit

The notion of the spirit was crystallised for medieval authors in Costa ben Luca's *De differentia animae et spiritus*, in which the Syrian philosopher explained how the physiology of the spirit worked in detail.[17] Contrary to Galen's works, which were not easily available in the Middle Ages, this text became pre-eminent for the circulation of the doctrine. It disseminated the Galenic model, while synthesising it lucidly and clarifying the difference between the spirit and the soul, two concepts that had previously tended to be confused. The crucial point of Costa ben Luca's work was to reinforce this distinction by an analysis of the different ontological statuses of the two. At the beginning of his work, Costa ben Luca offers a definition that highlights the ontological dimension of his venture: he understands the spirit as a sort of subtle body,[18] whereas he considers the soul to be an immaterial entity. Then, he declares that the spirit has a dual origin, as it may originate in the heart or in the brain. These two possible origins sources of the spirit indicate its twofold nature. According to Costa ben Luca, what he calls *vital spirit* emerges in the heart, and he explains this in the same way as Galen did: the air that encounters the natural heat in the heart generates it. Life depends on the presence of this spirit, which is proven by the fact that it can be seen flowing out of the body when death sets in. Moreover, the speed or slowness with which the spirit leaves the body depends on the way in which an animal dies, specifically on whether its death is violent or not.[19]

For its part, the *animal spirit* has its origin in the brain. There is a direct genealogical relation between the two kinds of spirits: the vital one feeds the animal one.[20] In order to explain the function of the animal spirit, Costa ben Luca provides an anatomical division of the brain into anterior and posterior parts, each of them containing *ventricles*, through which the spirit passes.[21] Once again, the framework of these reflections is Galenic, and it is here that what Galen defined as the "labyrinth" is explained more elaborately. Namely, the vital spirit is purged in each of the ven-

---

[17] The theories of Costa ben Luca will have a great diffusion also thanks to Alfred of Sareshel's work De motus cordis. See Alfredus de Sareshel, *De motus cordis* (ed. Baeumker, 1923).

[18] Costa ben Luca, *De differentia animae et spiritus*, cap. 1 (ed. Barach, 1878, 121): "Spiritus est quoddam corpus subtile, quod in humano corpore oritur ex corde et fertur in assurient, i.e., in venis pulsus ad vivificandum corpus; operatur ei quoque vitam et anhelitum atque pulsum. Et similiter oritur ex cerebro et nervis et operatur sensum atque motum."

[19] Costa ben Luca, *De differentia animae et spiritus*, cap. 1 (ed. Barach, 1878, 123): "Causa autem, qua egreditur hic spiritus a corde, est causa mortis, vel causa velocitatis vel tarditatis eius exitus est levitas mortis sive gravitas et apparitio eius in quibusdam hominibus vel occultatio eius in aliis."

[20] Costa ben Luca, *De differentia animae et spiritus*, cap. 2 (ed. Barach, 1878, 124): "Spiritus vero, qui procedit ex cerebro et transit ad cetera membra corporis, nominatur animalis, cuius nutrimentum vel sustentatio est spiritus, qui fit in ventriculis cordis."

[21] Costa ben Luca, *De differentia animae et spiritus*, cap. 2 (ed. Barach, 1878, 124–125): "Cerebrum vero dividitur in duas divisiones, quarum una est anterior, quae est maior, et altera, quae est posterior; et in illa anteriori sunt duo ventriculi habentes introitum ad commune spatium, quod est medio cerebri. In posteriori vero est unus ventriculus faciens iter ad supradictum spatium, quod est commune utrisque ventriculis, qui sunt in anteriori parte cerebri."

tricles, and in this way it becomes increasingly subtler and, therefore, more and more capable of receiving the powers of the soul. The terms used by Costa ben Luca to explain the process are explicit and unambiguous: he talks of a progressive digestion of the spirit as it passes from one ventricle to another.[22] The direction of the movement of the spirit depends on what the mental activity takes place, for example, memorising or cogitation. The circulation of the spirit also depends on the position of the head during each of these activities: tilted forward if one is intent on thinking, and backwards if one wants to remember something.[23] Moreover, the animal spirit displays various qualities in various people: the clearer and more subtle the spirit is, the higher the aptitude for rational activity.[24] Therefore, the more purified and digested the spirit is, the more excellent the rational activity.

To sum up, according to Costa ben Luca, some cognitive activities, such as imagination, memory, and cogitation, depend on certain motions of the spirit in specific areas of the brain. Moreover, the movement of the spirit is also associated with the quality of the spirit itself. It is in this approach that ideas about the bodily underpinnings of the mind are outlined: mental activity depends on physiological processes within the body. However, while Costa ben Luca's short treatise usefully clarifies the terms, it does not explicitly explain the relationship between the spirits and the humoral theory. This challenge was undertaken by Avicenna.

---

[22] Costa ben Luca, *De differentia animae et spiritus*, cap. 2 (ed. Barach, 1878, 125): "Pulsus ergo subtiles, qui mittuntur a reti, quod est sub cerebro, ad interiora cerebri cum pervenerint, ad illum ventriculum, qui est in cerebro anteriori, perducunt spiritum vitalem, qui, inde transiens ad alterum ventriculum, fit ibi subtilior et purgatur atque aptatur ad recipiendum virtutem animae, et haec fit ei quasi digesto et converso in spiritum subtiliorem atque clariorem."

[23] Costa ben Luca, *De differentia animae et spiritus*, cap. 2 (ed. Barach, 1878, 125–127): "Cum ergo apertum fuerit foramen, transit spiritus de anteriori cerebro ad posterius, et hoc non fit nisi cum necesse fuerit recordari alicuius rei, quae tradita est oblivioni in tempore, quo fit cogitatio de praeteritis. Si vero foramen apertum non fuerit, nec transierit spiritus ad posterius cerebrum, nec recordatur homo, nec aderit ei responsio eorum, de quibus interrogatur. Illa vero apertio foraminis, quae fit per elevationem illius corporis, quod assimilatur vermi, est diversa in hominibus in celeritate et tarditate. Fit enim haec in quibusdam tardius, et ideo fiunt tardae memoriae et tardi ad respondendum multum cogitantes. Et ideo accidit ei, qui vult recordari alicuius rei, ut caput suum valde mergat et inclinando retro vertat et immotis oculis sursum aspiciat, ut haec positio vel figura fiat ei auxiliatrix ad aperiendum foramen praedictum, et ut ipsum corpus vermile possit sursum elevari. [...] Et ideo accidit ei, qui cogitat, caput mergere ad terram et multum aspicere eam, et incurvare, quasi scriberet aliquam scripturam et describeret in ea aliquas figuras, ut fiat ei hoc quasi auxilium ad deponendum illud corpus, quod diximus esse simile vermiculo, super foramen illius decursus, per quem transit spiritus ad posteriora capitis."

[24] Costa ben Luca, *De differentia animae et spiritus*, cap. 2 (ed. Barach, 1878, 127): "Spiritus autem, qui est in eo spatio, scilicet in ventriculo medio, est in hominibus diversus. In quibusdam enim est subtilis et clarus, et hic est rationabilis, cogitans dispositione et bonae cogitationis; est autem in quibusdam corruptus, eritque talis amens, irrationabilis, levis atque stultus."

7.5 Avicenna's Connection Between the Humours and the Spirits

In the *Canon*, Avicenna addresses the question of the spirit in the chapter dedicated to animal powers. Having previously listed the different kinds of humours and described their genesis, he explains that organs are made from their thickest part, while the spirit derives from their thinnest part.[25] In this sense, the spirit should be understood as the ultimate product of the concoction of the humours. In other words, the spirits mark the culmination of this universal physiological process and are its highest product. As the spirits perform sensitive and mental functions, the upper functions are shown to stem from and fall within the same process that regulates the entire human body. Consequently, the mental functions belong to the same process that regulates the entire human body, as the spirits perform sensitive and mental functions. As such, Avicenna's explanation of the genealogy of the spirits acquires a systemic character. The transformation of the elements into the humours and of the humours into either organs or the spirits is conceived as a gradual metamorphosis from the least to the most perfect forms. Each of these terms represents a step and, moreover, each term is connected to the other by a causal link.

The *Canon* inherited its epistemological framework from Aristotle's *Posterior Analytics*. According to Avicenna, the object of medical science is the human body. The aim of this science is practical insofar as it investigates the causes of the body's illness and health. In accord with the Aristotelian schema, there are four causes which medical science seeks to establish: formal, material, efficient, and final. In this framework, the elements, the humours, the spirits, and the organs of the body act as material causes by which every bodily transformation can be explained. The elements and the humours are the most remote material causes, while the organs and the spirits are the immediate material causes.[26] Consequently, all phenomena taking place in the human body can be directly linked to the spirit as their most proximate material cause. All the four material causes are perceived as various stages in the same process of transformation from the most remote elements to the most immediate spirits, in which moisture is concocted by natural heat. As such, the concoction process is essentially circular, with the spirits constituting its beginning and end, as will be shown below. All the steps of this process can be retraced by following the

[25] Avicenna, *Canon medicinae*, I, sen. I, doct. VI, cap. 4 (ed. Lugduni, 1522, f. 22vb): "Nos autem explanabimus summam hanc et dicemus quod quemadmodum ex humorum spissitudine secundum aliquam complexionem substantia generatur spissa, quae est membrum aut pars membri, ita ex humorum vaporibus et eorum subtilitate secundum aliquam complexionem substantia generatur subtilis, quae est spiritus."

[26] Sirasi (1987, 26); Avicenna, *Canon medicinae*, I, sen. I, doct. I, cap. 2 (ed. Lugduni, 1522, f. 4rb): "Causarum vero quattuor sunt species: materiales, efficientes, formales et finales. Causae vero materiales positae, in quibus sanitas et aegritudo mensurantur, sunt: subiectum quidem propinquum membrum aut spiritus et subiectum quidem longinquum humores et eis longinquiora sunt elementa."

index of the *Canon*.[27] Avicenna's analysis starts with the theory of the elements adopted from Aristotle, which reveals his reliance on Aristotelian philosophy. This is followed by an inquiry into the humours and their genealogy which provides a crucial reformulation of the humoral theory. Indeed, according to Avicenna, the humours are divided into primary and secondary types. Among the latter, Avicenna lists four kinds,[28] including fundamental radical moisture.[29] Radical moisture is the constitutive moisture of each organ, a "marker of individuality."[30] It derives from sperm and, as such, is present in the body from birth on. Indeed, the connection between the humours and the spirits can be first noted in sperm. That same connection can then be recognised in the fully-formed body, where these physiological principles perform their complete functions. Although the relation between radical moisture and the spirit is crucial, it has not received sufficient scholarly attention yet.

The formative power of sperm is due to the presence of the vital spirit, which makes the sperm frothy. Actually, the different kinds of the spirit are mixed and not distinguished in sperm.[31] Due to the formative power, sperm can spawn every organ, the first of which is the heart, insofar as it is the seat of the natural heat and the vital spirit. Thus, as already asserted by Aristotle, the spirit ages before the animal begins to live. As the most rarefied material principle, the spirit is a material part to which is imbued with formative power. Within the body, the spirit thus firstly resides in the heart from which it circulates and performs its functions owing to the continuous process of the concoction of moisture.[32] The transformation of the spirit takes place in the heart, the brain, and the liver. While each of these organs is the material sub-

[27] See Siraisi (1987, 19–40).

[28] Avicenna, *Canon medicinae*, I, sen. I, doct. IV, cap. 1 (ed. Lugduni, 1522, f. 8r): "Una est humiditas in foraminibus extremitatum parvarum venarum contenta, membris simplicibus propinquarum imbibentium ea, vel a quibus ipsa membra imbibunt. Alia est humor, per omnia simplicia transiens membra: sicut ros qui in nutrimentum converti est aptus, cum corpus nutrimento caret, et ut membra humectet, cum aliqua causa fortis motus, aut alia ea exiccaverit. Tertia est humiditas, quae in tempore proximo fuit coagulata. Et est nutrimentum, quod in substantia membrorum ex parte complexionis et similitudinis conversum est, sed ex parte essentiae completae nondum conversum fuit. Quarta est humiditas, quae est intus in membris simplicibus a principio nativitatis, per quam eorum continuitas existit: cuius principium est ex spermate."

[29] On the theory of radical moisture, see Hall (1971); Niebyl (1971); McVaugh (1974); Stolberg (1993); Reynolds (1999, 105–119); Crisciani (2005) and Pomata (2018).

[30] Pomata (2018, 199).

[31] Avicenna, *De animalibus*, IX, cap. 9 (ed. Venetia, 1500, f. 22v): "Et spiritus naturalis et vitalis et animalis non sunt segregati in spermate."

[32] Regard the origin of the spirit in the heart see also Avicenna, *De viribus cordis* (ed. Venetia, 1491), that is a fundamental work on Avicenna's elaboration on matter. Furthermore, some passages in Avicenna's *De animalibus* seem significant in this respect; see, in particular, Avicenna, *De animalibus*, XIII, cap. 4 (ed. Venetia, 1500, f. 28v–29r): "Et quia spiritus complet suam alterationem et completionem medullae et hoc per coctionem qua ibi decoquitur. Decoctio vero istius erit per motum et divisionem materiae in locis decoctionis secundum modum decoctionis in hepate, sed serra est magis aperta quam posterior est, quia fere est proportio sene ad senex, quae est membri ad membrum."

stratum of the various animal functions, the main powers of animals, which actually make organs function, are invested in the spirit (although organ activity primarily depends on the soul). Indeed, the material aspect of sensibility and movement in animals is linked to the action of the spirit.

All the powers that can be assigned to the spirit stem from the fundamental nutritive power (*virtus nutritiva*), which is the basic function necessary for life[33] and the source of all the other functions of living beings. The nutritive power is the first one to be conferred on the animate body by the spirit and, for this reason, it is also called the vital power (*virtus vitalis*). Although the spirit is the bearer of the vital power, this does not mean that it acts in the same way in every part of the body. In fact, its action is always adjusted to the complexion of the organ in which it acts. The relation between the spirit and complexion calls for some explanation. Indeed, the notion of complexion is one of Avicenna's most outstanding contributions to the Latin culture.[34] According to Avicenna, every animal power ensues from the penetration of the spirit into an organ which has its own complexion. Once the spirit passes through the brain, it becomes the so-called animal spirit,[35] enabling the body to perform the highest sensory activities.

Although the nutritive power is the origin of all the other animal powers, several of these complex animal powers can be assigned to the spirits. Among these, Avicenna dedicates a whole chapter to the sensorial powers, divided into manifest, i.e., those that can be linked to the five external senses, and occult powers. The example of sight helps understand how the spirit operates. Once the spirit has reached the eye through the optic nerves and assumed the complexion proper to the eye, it produces the sense of sight.[36] Among the occult sensorial powers, Avicenna lists the internal senses:[37] common sense, imagination, and cogitation. Since each of these faculties resides in one of the brain ventricles, which are activated by the transition of the spirit, they are the highest product of the process of concoction.[38]

---

[33] Avicenna, *Canon medicinae*, I, sen. I, doct. VI, cap. 4 (ed. Lugduni, 1522, f. 22vb): "Et si virtus nutritiva in quantum est nutritiva esset ipsa virtus, quae preparat ad sentiendum et movendum, vegetabilia preparata forent ad sensum et motum recipiendum. Restat ergo ut sit preparans alia res, quae propriam sequatur complexionem, quae vocatur virtus vitalis. Et ipsa quod est prima virtus, quae in spiritu provenit, cum spiritus ex subtilitate humorum procedit."

[34] For more details, see Köhler (2008, 305–340) and Zuccolin (2019).

[35] Avicenna, *Canon medicinae*, I, sen. I, doct. VI, cap. 4 (ed. Lugduni, 1522, ff. 22vb–23ra): "Praeterea dixerunt quod virtus haec praeter hoc ipsa est preparans vitam: et etiam motus substantiae spiritualis subtilis principium ad membra et principium constrictionis et dilatationis aeris attracti et pulsus. Affert enim adiutorium in comparatione vitae, quod est passio; et in comparatione operationum animae et pulsus praebet auxilium, quod est actio."

[36] Avicenna, *Canon medicinae*, I, sen. I, doct. VI, cap. 4 (ed. Lugduni, 1522, f. 22vb): "Harum vero virtutum operationes non ab hoc proveniunt spiritu in principio rei: sicut etiam sensus visus apud medicos non provenit per spiritum animalem qui est in cerebro, nisi ad chystalloidem penetret aut ad linguam aut ad alia."

[37] On Avicenna's theory of the internal senses, see Black (1993), Hasse (2000), Wolfson (1935).

[38] For more detail regard Avicenna's theory of internal sense, see Hernández-Rubio, Chap. 8 in this volume, pp. 133–161.

Subsequently, Avicenna explores the ontological status of the soul and the spirit[39] to conclude that the spirit is, as the most rarefied state of matter, an internal principle that is a precondition of the soul. In Avicenna's treatise, the "medieval language of life"[40] acquires a special complexity since his aim is to systematically explain the relationship among all its fundamental factors, such as complexion, the humour and the spirit.

Analysing the way in which the spirit interacts with the soul is a task that Avicenna ascribes to natural philosophy and not to medicine.[41] As already stated, the object of medical science is the human body; thus, the soul is excluded from the field of medical inquiry. Nonetheless, when Avicenna's *Canon* found its way to thirteenth-century Europe, its vocabulary became fundamental for natural philosophy.[42] An excellent case in point is Albert the Great, for whom the reformulation of the notion of the spirit, as proposed in the *Canon*, was indispensable.[43] Albert the Great made his mark as a scholar who, among other merits, synthesised this long and complex tradition, adapting it to the needs of Christian theology.

7.6 Albert the Great: The Spirit as the Highest Product of Digestion

Albert was aware of the many different meanings of the term "spirit."[44] In his *De spiritu et respiratione*, he lists an array of viewpoints on the spirit, including the beliefs of Heraclitus, who thought that the spirit was an immaterial principle (*incorporatissimus*). Albert entirely rejects this position, affirming that the spirit is something bodily.[45] When comparing Albert's and Costa ben Luca's definitions of the spirit, one immediately realises that Albert employs a more specific terminology. He

---

[39] Bono (1984, 96).

[40] Bono (1984, 97).

[41] Avicenna, *Canon medicinae*, I, sen. I, doct. IV, cap. 4 (ed. Lugduni, 1522, f. 23ra): "Huius autem virtutis declarationum rationem certificare et an sit una an plures quam una in sapientia continuetur naturali quae philosophiae pars existit."

[42] To this regard, see Jacquart (1985, 2002).

[43] See Siraisi (1980) and Bertolacci (2013).

[44] On Albert the Great see Evelina Miteva, Chap. 5 in this volume, pp. 73–90.

[45] Albertus Magnus, *De spiritu et respiratione*, I, tr. 1, cap. 3 (ed. Borgnet, 1890, 217): "Non autem stare potest quod dicit Heraclitus dicens ipsum spiritum esse incorporatissimum. Cum enim medium sit quo anima movet et vivificat corpus, non simplex, sed commixtum et complexionatum et compositum, non congrueret instrumentum motus cum eo quod movetur, si esset incorporatissimum omnium. Quod autem dicit hunc spiritum esse movens quiddam et motum, veritatem quidem habet, sed non sicut ille dicit ex seipso. Est autem motus spiritus ab anima vivificante et regente corpus, et movens per easdem virtutes corpus per omnia sicut faber et malleus movent et influunt materiae eandem formam: faber quidem ut artifex, malleus vero ut instrumentum. Similiter autem in eo verum dicit, quod spiritum vaporem esse dixit."

defines the spirit as the vapour of radical moisture,[46] whereby he reveals his profound knowledge of Avicenna's *Canon* see Takahashi (2008). This definition exemplifies Albert's capacity to bring together elements from different traditions. Albert connects the notion of the spirit inherited from Costa ben Luca to the physiology of moisture, which is of particular relevance to his natural philosophy. Radical moisture is one of the most important notions propounded by the re-elaborated humoral theory informed by Arabic science and medicine.[47] While in Avicenna's *Canon* the spirit was related to the humours in general, and not directly to radical moisture, Albert attempts to establish this fundamental connection. In Avicenna's medical system, radical moisture is a key concept, as it denotes the humour present in each single organ. The close relationship of the humour, the organ, and the spirit in Avicenna's account has been mentioned above. Admittedly the correct interpretation of the one required a profound understanding of the other already in Avicenna's account, but Albert advances this reasoning and states explicitly that the spirit coincides with the rarefied stage of radical moisture. Resulting from the action of heat on moisture, the generation of the spirit is defined as digestion, that is, as a universal physiological process in which the humour is concocted through heat.[48] All life can be explained by reference to this basic physiological process.[49] As the vehicle of the soul, the spirit is already present in the seed and, it guides the growth of the body up to its highest, pre-intellectual activities.

Albert distinguishes several kinds of the spirit by separating the spirit present in the seed or in the sperm, as Aristotle and Avicenna did, from the spirit present in the constituted body. Albert calls the spirit present in the seed *radical* (as in 'radical moisture') and divides the spirits present in the body into three kinds: *natural*, *vital* and *animal*. In animals, the spirit originates in the heart and then circulates in the body, which produces the other kinds of it. Although this framework is clearly Galenic, Albert clarifies that the spirit has just one source. From the heart, the *vital spirit* gives warmth to the body, and when it arrives in the liver, it changes into the *natural spirit*, which is able to digest food. Then, when it reaches the brain, it becomes the *animal spirit*, capable of regulating sensibility and pre-intellectual activities. The three kinds of the spirit represent three different kinds of perfection and, in this sense, are understood as three different stages of digestion. In this way, Albert emphasizes that the production of the spirit is a unified process. In addition, he establishes a parallel between the soul and the spirit. The rational soul incorporates the sensitive soul, which in turn incorporates the vegetative one, as claimed by

[46] Albertus Magnus, *De spiritu et respiratione*, I, tr. 1, cap. 3 (ed. Borgnet, 1890, 217b): "Est autem vapor humidi radicalis et cibalis spiritus."

[47] See n. 28.

[48] Albertus Magnus, *De vegetabilibus*, IV, tr. I, cap. 1 (ed. Meyer, Jessen, 1867, 215: 7): "Digestio enim est, ubi est humor sicut subiectum, et calor sicut agens. Et quando terminum proportionalem naturae rei consequuntur, tunc facta est digestio, quoniam tunc completum est ad esse naturae humidum, et calor retinetur in illo, sicut in subiecto sibi secundum naturam proportionali." See Takahashi (2008, 465–468).

[49] On the process of digestion in Albert's natural philosophy, see Cadden (1980).

Aristotle in *De anima* 412a, and the animal spirit likewise contains the natural and vital spirits, forming a continuous relationship.[50]

The three types of the spirit are arranged in a succession from the crudest, i.e., natural, spirit to the most refined, i.e., animal, spirit. This progressive refinement is manifested in the qualities of these spirits. The natural spirit represents the most elemental stage, at which the spirit does not differ much from the humour, as it is still hot and its vapour is heavy. For this reason, it can perform only the most basic vital functions: it is the instrument that the soul uses to bring life into the body.[51] The vital spirit constitutes the intermediate stage: it is neither as hot as the natural spirit nor as refined as the animal spirit. Compared with the natural spirit, it is drier and hence qualitatively more distant from moisture. This greater qualitative refinement enables it to perform more elaborate functions linked to the assimilation of nutrients.[52] Interestingly, the *animal spirit* does not designate just one, but many different kinds of spirits, which is evocative of Costa ben Luca. The animal spirits exemplify the most refined stage of the digestive process. Subtle, clear, and cold, they represent the most elevated state of bodily substance. The clearer and thinner they are, the more responsive they are to the forms they receive from the senses. In other words, being subtle and clear is a precondition of their receptivity. As they originate in the brain, they must be cold; if they were not, the sensitive forms would be confused by the heat.[53]

---

[50] Albertus Magnus, *De spiritu et respiratione*, II, tr. 2, cap. 5 (ed. Borgnet, 1890, 237b): "De anima enim diximus, quod sua determinatio est determinatio figurae; quoniam sicut in figuris trigonum est in tetragono, ita est vegetativum in sensitivo, et sensitivum et vegetativum in intellectivo. Omni eodem modo videtur vivificus esse spiritus in naturali, et naturali et vivificus in animali in omni eo quod habet has spirituum differentias."

[51] Albertus Magnus, *De spiritu et respiratione*, I, tr. 2, cap. 6 (ed. Borgnet, 1890, 239b): "Spiritus enim naturalis turbidus est et grossus et calidus quasi calore bulliente: propter quod etiam in vulneratis circa loca nutrimenti et superviventibus auditur spiritus quasi ollae bullientis. Tales enim proprietates sunt necessariae loco digestionis, qui fumosus est: et ideo turbans spiritum, neque retineret fortem caliditatem, nisi esset spissus, neque iterum nondum assimilatum nutrimentum converteret, nisi esset calidus valde: quando enim ex contraiacentibus dirigitur alimentum passionibus, oportet maioris vigoris esse agentia quam quando de simili efficitur similis. Nutrimentum autem quando primo cadit in vasa nutrientia, omnino dissimile est: et oportet quod spiritus naturales qui ad decoctionem deserviunt, sint calidiores et efficacioris operationis in convertendo et complendo massam cibalem."

[52] Albertus Magnus, *De spiritu et respiratione*, I, tr. 2, cap. 6 (ed. Borgnet, 1890, 239b): "Hi autem qui dicuntur vitae spiritus, oportet esse temperatiores et frigidiores et magis tenues et sicciores, et quasi semper expulsos, ita quod idem numero non retrahitur, sed alius similis sibi, et sunt magis albi spiritus isti, habentes tamen modicum tenuissimi sanguinis admixtum: sunt tamen temperatiores et frigidiores propter hoc quod refrigerium quoddam cordi dederunt. Tenues autem et subtiles sunt, ut undique cum vita quam advehunt, possint penetrare: sicciores autem sunt, ut sint velocioris motus: expulsi autem sunt, eo quod fuerunt cordis ventilabrum: et ideo fumosas in aliqua parte sui substantias admixtas habentes, non essent utiles, si ibidem statim retraherentur."

[53] Albertus Magnus, *De spiritu et respiratione*, I, tr. 2, cap. 6 (ed. Borgnet, 1890, 239b): "Animales autem subtilissimi, clarissimi et frigidiores omnibus sunt, et multiplices magis aliis, et diversi inter se huiusmodi qui vocantur spiritus animales. Subtilissimi autem sunt, ideo quia ad alteriorem locum sunt sublimati. Clarissimi autem sunt, ut susceptibiliores sint formarum et intentionum quae

Aware of Costa ben Luca's definitions, Albert explains the physiology of the spirit in the brain in the same way as the Syrian philosopher.[54] The higher activities of the internal senses (those which precede intellectual activities, which are in themselves immaterial), i.e., those of the memorative, imaginative and cogitative faculties, are made possible by the sublimation of the spirit. Each part of the brain has a specific function accorded to it, which is activated once the spirit reaches this part. Although the act of thinking involves to the immaterial principle of the intellect, it may only take place after the inner senses have systematised the *phantasmata*. Thus, thinking hinges on the movement of the spirit in the brain, so much so that there is a specific pose that promotes thinking. Specifically, if the head is tilted forward, the spirit can more easily reach the anterior part of the brain, where it produces cogitation. Meanwhile, if the aim is to remember something, the head should be tilted in the opposite direction, so that the spirit can move to the back of the brain. All this implies that, in Albert's account, mental activity is somehow fostered on the material level. The absolute immateriality of the intellect is retained, but the activity of thinking presupposes the inner senses triggered by the spirit. It is in this context that Albert proposes a precise, unambiguous definition of the ontological status of the soul and the spirit. If imagination, memory, and cogitation are the work of the soul, they are at the same time the powers (*vires*) of the spirit. The soul can therefore carry out these operations precisely because it possesses the spirit as an instrument.[55]

At the time when scholars were radically rethinking natural philosophy, Albert reformulated the doctrine of the spirit in the most up-to-date way. On the one hand, he adopted Costa ben Luca's insights and aims, and on the other, he repositioned this discourse by incorporating it into Avicenna's framework of the digestion of bodily moistures. While he relied on several other authors, it was exactly the

recipiuntur per sensus ex quibus perficiuntur operationes animales, et ut non commisceant rationes et formas, ablata est ab eis caliditas: quia si essent calidi, resolverent substantiam cerebri et permiscerent formas et operationes confunderent animales."

[54] Albertus Magnus, *De spiritu et respiratione*, I, tr. 2, cap. 4 (ed. Borgnet, 1890, 237a): "Sic igitur spiritus discurrit quidam a corde ad cerebrum et ibi opera animalia et animalium virtutem operatur transeundo de cellula in cellulam colatus et digestus et quasi sublimatus, sicut dictum est. Huius autem signum est, quod cum homo intente de aliqua re cogitat et meditatur, tunc elevat et deprimit caput, quasi ipso motu natura ostendat se laborare et adiuvare ad hoc ut facilius elevata caruncula vermiculari spiritus fluat de cellula in cellulam cum formis intellectus, imaginationis, et recordationis."

[55] Albertus Magnus, *De spiritu et respiratione*, I, tr. 2, cap. 4 (ed. Borgnet, 1890, 236b): "Spiritus igitur animalis in primis infunditur ab arteriis, et per quendam porum quem claudit caruncula figurae, et quantitas vermiculi, postquam porus reseratur, colatur in medio vacuum: et quia in posteriori parte qua dirigitur ad vacuum spatium, similis claudit caruncula, et ideo etiam cum illo porus reseratur, colatus amplius fluit in posteriorem capitis cellulam, et aliam perficit in omnibus illis quae sunt magis animae quam corporis. In primis quidem duabus sensum communem, et sensus proprios, et imaginationem influit. In tertia autem, ut dicit, memoriam, et influit in nervos motivos, sive oriantur a cerebro, sive a nucha quae vicarius est cerebri: influit enim spiritus motivus per spondiles dorsi per omnia paria nervorum, quae a spondilibus et nucha oriuntur, et dat eis motivam virtutem. Sentire autem et imaginari et memorari et intelligere animae vocantur opera: et ideo vires illae et spiritus animales vocantur."

combination of these two main sources that helped him advance a relatively novel notion of the spirit. In this framework, the spirit is defined as a derivative of radical moisture which is produced during the digestive process but acts as an instrument for the soul, proceeding from the most basic vital functions to the most complex ones, such as the activity of thought. Positing the material foundation of the mind, the concept of the digestion of the spirit offers a new image of the body. The body is conceived not as a static entity, but as an entity in becoming. The further digestion proceeds, the more rarefied the matter of the spirit becomes, in this way warranting the finest activity of thinking.[56] Indeed, even if thinking is an act of the intellect and, as such, immaterial, the very possibility to think resides in the spirit, because it regulates the inner senses, which are a sine qua non of thinking.

7.7 Conclusion

The present chapter has focused on a particular moment in the history of philosophical and medical thought when the notion of the spirit became intrinsically linked to the humoral theory, with the latter requisite for the former. The same process was believed to regulate the different mechanisms, from the simplest to the most complex bodily functions, including those performed by the inner senses. Such a reformulation was achieved through the contributions of various physicians and philosophers, whose main positions have been retraced above.

The primary aim of this study was to understand how the notion of the spirit arose in ancient philosophy and medicine by examining the positions of Aristotle and Galen. Then, the work of Costa ben Luca was considered to grasp the different meanings of the terms "soul" and "spirit." Costa ben Luca's work proposed that the spirit had a unifying role as a material entity that supported psychic functions. From there, the path led to Avicenna's *Canon*, in which material causes were directly considered in relation to one another: the elements, the humours, the spirits and the organs of the body were all regarded as products of the same process of concoction. The body thus regained its unity, as the doctrine of the spirit openly postulated the doctrine of the humours. Finally, Albert the Great combined these different approaches by adopting the complexity of Avicenna's medical terminology, while still pursuing the intentions of Costa ben Luca's work. With Albert, therefore, the notion of the spirit was put at the centre of natural philosophy, resulting in a more elaborate concept of the body.[57] The spirit was perceived as a product of a universal physiological process of transition from materiality to immateriality. Indeed, the spirit, to which the highest psychic activities were assigned, was deemed the most refined material substance.

---

[56] For more on the relation between intellect and spirit, see Loconsole, Chap. 6 in this volume, pp. 91–113.

[57] Regarding Albert contribution to natural sciences see Weisheipel (1980) and Zimmermann (1980).

To conclude, the concept of the body was profoundly reformulated, revealing a certain dynamic, primarily because the concept of the body developed in this tradition was basically inspired by medical inquiry. The body, closely investigated for medical purposes, gradually lost its passivity, and the explanation of its inner mechanisms prompted rethinking its relation to the soul. The reformulation of the notion of the spirt, as outlined in my argument, reveals a clear continuity. Indeed, the philosophical tradition analysed here certainly considered the spirit to be a bodily entity capable of becoming less corporeal the more it was digested. This, crucially, made it possible to theorise the body as having or being in different degrees and, as it were, bonding with the functions of the soul.

Furthermore, these observations reveal the speculative power of the digestion process. In Albert's reformulation, the most disparate natural phenomena—from the blossoming of plants to the capacity of imagining or remembering—could be explained by referring to the notion of digestion. Nature was thus terminologically reduced to a minimum, and arranging the terms in one way or another made it possible to explain the polyvalence of the phenomena of living. Although it is one of the most significant aspects of medieval natural philosophy, the connection between digestion and the spirit has not been thoroughly researched yet. The present study aims to address this gap.

References

Primary Literature

Albertus Magnus. 1867. *De vegetabilibus libri VII*, ed. Ernst Meyer, and Carl Jessen. Berlin: Reimer.

———. 1890. In *De spiritu et respiratione*, ed. Auguste Borgnet. Paris: Vivès.

Avicenna. 1491. In *De viribus cordis*, ed. Venetiis.

———. 1500. *Liber de animalibus*. Trans. Michael Scotus. Venetia: Apud Iohannem et Gregoruium de Gregoriis de Forlivio.

———. 1522. *Liber canonis de medicina*. Trans. Gerardus Cremonensis. Lugduni: Apud Simonem Papiensem.

Chenu, Marie-Dominique. 1957. Le vocabulaire de l'âme au XII e siècle. *Revue des Sciences philosophiques et théologiques* 41 (2): 209–232.

Costa ben Luca. 1878. In *De differentia animae et spiritus*, ed. Carl Sigmunt Barach. Innsbruck: Wagner Verlag.

de Sareshel, Alfredus. 1923. In *De motu cordis*, ed. Clemens Baeumker. Münster: Verlag des Wagnersc'en Universität Buchladung.

Freudenthal, Gad. 1995. *Aristotele's theory of material substance*. Oxford.

Galenus, Claudius. 1978–1984. *On the doctrines of Hippocrates and Plato*, ed. Philip De Lacy. Berlin: Walter de Gruyter.

———. 1968. In *De usu Partium*, ed. Georgius Helmreich. Leipzig: Teubner.

Hasse, Dag Nikolaus. 2000. *Avicenna's "De anima" in the Latin West: The Formation of a Peripatetic Philosophy of the Soul 1160–1300*. London: Warburg Institute.

Nussbaum, Martha C. 1978. 'The Sumphuton Pneuma and the De Motu Animalium's account of soul and body', in Aristoteles. In *De Motu Animalium*, ed. M. Nussbaum, 143–164. Princeton: Princeton University Press.

Solmsen, Friedrich. 1957. The vital heat, the inborn Pneuma and the aether. *The Journal of Hellenic Studies* 77: 119.

Secondary Literature

Bertolacci, Amos. 2013. Albert's Use of Avicenna and Islamic Philosophy. In *A Companion to Albert the Great. Theology, Philosophy, and the Sciences*, ed. Irven M. Resnick, 601–611. Leiden/Boston: Brill.

Black, Deborah. 1993. Estimation in Avicenna: The Logical and Psychological Dimensions. *Dialogue* 32: 219–258.

Bono, James J. 1984. Medical Spirits and the Medieval Language of Life. *Traditio* 40: 91–130.

Bos, Abraham P. 2013. "Pneuma" as Quintessence of Aristotle's Philosophy. *Hermes* 141 (4): 417–434.

Bos, Abraham P., and Rein Ferwerda. 2008. *Aristotle. On the Life-Bearing Spirit ("De spiritu")*. Leiden/Boston: Brill.

Boudon-Millot, Véronique. 2016. *Galeno di Pergamo. Un medico Greco a Roma*. Rome: Carocci.

Cadden, James. 1980. Albertus Magnus' Universal Physiology. In *Albert the Great and the Sciences*, ed. James A. Weisheipel, 321–399. Toronto: Pontifical Institute of Mediaeval Studies.

Crisciani, Chiara. 2005. Aspetti del dibattito sull'umido radicale nella cultura del tardo medioevo (secoli XIII–XV). *Arxiu de textos catalans antics* 23–24: 333–380.

Düring, Ingmar. 1966. Did Aristotle Ever Accept Plato's Theory of Trascendent Ideas? *Archiv für Geschichte der Philosphie* 48: 312–316.

Hall, Thomas. 1971. Life, Death and the Radical Moisture. A Study of Thematic Pattern in Medieval Medical Theory. *Clio Medica* 6 (1): 3–23.

Jacquart, Danielle. 1985. La réception du Canon d'Avicenne: Comparaison entre Montpellier et Paris aux siècles. In *Actes du 110e Congrès national des sociétés savantes*, 69–77. Montpellier.

———. 2002. Lectures universitaires du Canon d'Avicenne. In *Avicenna and His Heritage*, ed. Jules Janssen and Daniel De Smet, 313–324. Leuven: Leuven University Press.

Köhler, Theodor W. 2008. *Homo animal nobilissimum. Konturen des spezifisch Menschlichen in der naturphilosophischen Aristoteleskommentierung des dreizehnten Jahrhunderts*. Leiden/Boston: Brill.

Lloyd, Geoffrey. 2007. Pneuma between Body and Soul. *The Journal of the Royal Anthropological Institute* 13: 135–146.

McVaugh, Micheal R. 1974. The *Humidum Radicale* in Thirteenth-Century Medicine. *Traditio* 30: 259–283.

Niebyl, Peter H. 1971. Old Age, Fever, and Lamp Metaphor. *Journal of the History of Medicine and Allied Science* 26: 351–368.

Pomata, Gianna. 2018. Innate Heat, Radical Moisture and Generation. In *Reproduction. Antiquity to the Present Day*, ed. Nick Hopwood, Rebecca Flemming, and Lauren Kassell, 195–208. Cambridge: Cambridge University Press.

Reynolds, Philip L. 1999. *Food & the Body. Some Peculiar Questions in High Medieval Theology*. Leiden/Boston: Brill.

Rocca, Julius. 2008. Anatomy. In *The Cambridge Companion to Galen*, ed. Robert J. Hankinson, 242–262. Cambridge: Cambridge University Press.

Siraisi, Nancy G. 1980. The Medical Learning of Albertus Magnus. In *Albert the Great and the Sciences*, ed. James A. Weisheipl, 379–404. Toronto: Pontifical Institute of Mediaeval Studies.

———. 1987. *Avicenna in Renaissance Canon and Medical Teaching in Italian Universities after 1500*. Princeton: Princeton University Press.

Stolberg, Michael. 1993. Die Lehre vom "calor innatus" im Lateinischen Canon Medicinae des Avicenna. *Sudhoffs Archiv* 77: 408–428.

Takahashi, Adam. 2008. Nature, Formative Power and Intellect in Natural Philosophy of Albert the Great. *Early Science and Medicine* 13: 451–481.

Walker, Daniel P. 1984. Medical Spirits and God and the Soul. In *Spiritus. IV Colloquio internazionale. Roma, 7–9 gennaio 1983* (Lessico intellettuale europeao, 32), eds. M. Fattori and M. Bianchi. Roma: Pubblicazioni d'Ateneo.

Weisheipel, James A. 1980. *Albert the Great and the Sciences*. Toronto: Pontifical Institute of Mediaeval Studies.

Wolfson, H. A. 1935. The Internal Senses in Latin, Arabic, and Hebrew Philosophic Texts. *The Harvard Theological Review* 28 (2): 69–133.

Zimmermann, Albert. 1980. Albert le Grand et l'étude scientifique de la nature. *Archives de Philosophie* 43: 695–711.

Zuccolin, Gabriella. 2019. Tommaso d'Aquino sulla complessione corporea. *Revue des Sciences philosohiques et théologiques* 103.

Chapter 8
The Role of the *Intentio Individualis* in Albert the Great's Sense Perception Theory

Paloma Hernández-Rubio

Abstract According to Aristotle's *De anima*, human senses can recognize individual things *qua* individuals. This implies that they can apprehend individual forms, i.e., substances. However, substance is not sensible in itself. Although incidental perception accounts for the connection between intellectual and sensible properties, the question remains how intellect-lacking animals can perceive individual things. Avicenna advances an inner-sense theory to explain how animals interact with individuals without perceiving substance. Averroes's belief that the human senses grasp the *intentio individualis*, i.e., the individual substantial form, presupposes that animal and human senses are different. Albert the Great's solution, in his *Commentary on the De anima*, retains Averroes's *intentio individualis* but agrees with Avicenna's claim that rational and non-rational animals share identical sensitive apparatuses. Albert arrives at his solution in three main steps. First, he redefines the concept of abstraction from Avicenna's *Liber de anima*, part II, chapter 2, to separate the individual thing from its signification, i.e., its intention. Secondly, he applies his *forma totius* theory asserting that each of the five Porphyrian universals can be predicated of an individual thing because they refer to the totality of that thing. Thirdly, he uses the concept of *praedicatio per se tertio modo* from *Posterior Analytics* to argue that the Porphyrian universal accident refers to the totality of an individual thing. The resulting theory makes it possible to explain how the senses can apprehend the individual thing *qua* individual without the necessity of perceiving its substance.

Keywords Albert the Great · Avicenna · Averroes · Sense Perception · Intentionality

P. Hernández-Rubio (✉)
Universidad Autónoma Metropolitana, Mexico City, Mexico

M. Gensler et al. (eds.), *The Embodied Soul*, Historical-Analytical Studies on Nature, Mind and Action 11, https://doi.org/10.1007/978-3-030-99453-2_8

8.1 Introduction[1]

In his *Commentary on the De anima*,[2] Albert the Great offers a solution to a problem that Averroes and Avicenna faced before him: if non-rational animals and human beings share the same sense organs, they must be capable of apprehending the same cognitive content. In the *De anima*,[3] Aristotle says that the senses have the capacity to discern between different individuals. For Avicenna,[4] this means that humans and animals share very complex sensitive apparatuses, which account for this skill, and that the difference between them is that the human senses are related to the intellect while animals have no intellectual faculty. For Averroes, however, this means that the human senses can apprehend essential properties[5] mixed with sensitive ones by apprehending the *intentio individualis*, i.e., individual intentions. Since non-rational animals cannot apprehend intellect-related properties, Averroes makes a distinction between the animal senses and the human senses, but he fails to provide a sound explanation of it.[6] As we will see, Albert uses Avicenna's texts to prove that non-rational animals can grasp the individual intentions. Furthermore, Albert deliberately redefines many central Avicennian terms in order to develop an entirely novel theory, unrelated to either Avicenna or Averroes.

[1]The writing of this contribution was made possible by the PhD Scholarship CONACyT no. 398425, which enabled me to dedicate myself entirely to research at the Universidad Autónoma Metropolitana (UAM) in Mexico-City and at the Albertus-Magnus-Institut in Bonn. I want to thank everyone at both institutions who offered me their support. I am profoundly indebted to Henryk Anzulewicz and Jörg A. Tellkamp, whose academic guidance and encouragement were invaluable to me. Last but not least, my deepest gratitude goes to Monika Michałowska, Monika Mansfeld, Marek Gensler, Patrycja Poniatowska, and the anonymous reviewer for helping to bring some clarity into this text. Any remaining obscurities are my sole responsibility.

[2]Albertus Magnus, *De anima* (ed. Stroick, 1968b).

[3]Aristoteles, *De anima*, II.6 and III.3 (ed. Ross, 1961, 418a7–16; 428b21–22).

[4]Avicenna, *Liber de anima* (ed. Van Riet, vol. 1, 1968; vol. 2, 1972). Cf. Avicenna, *Liber de anima*, pars IV, cap. 1 (ed. Van Riet, 1972, 6). For more information, see Black (1991, 2000).

[5]For more on the problem of sensorial acquisition of the information required to produce universal concepts, see Pickavé, Chap. 11 in this volume, pp. 213–227.

[6]Averroes, *Commentarium magnum in Aristotelis De anima libros*, II, lectio 63 and lectio 65 (ed. Crawford, 1953). For more information, see Black (1996), Taylor (2000), and Wirmer (2004).

8.2 Albert the Great and the Individual Intentions

According to the search engines *Albertus Magnus Opera Omnia. Editio Digitalis*[7] and *Alberti Magni E-Corpus*,[8] the term *intentiones individuales* appears only twice in Albert the Great's works: once in *De somno et vigilia*,[9] and once in *Quaestiones super libris de animalibus* VIII, question 39: "Whether the camel and the horse distinguish their mothers from others" (*Utrum camelus et equus discernant suas matres ab aliis*). Albert presents two objections against it. One of them says: "It seems not. To distinguish is an operation of reason. But these animals lack reason. Therefore, they also lack the power of making a distinction."[10] In order to answer this objection, Albert cites very peculiar tales related by Aristotle in the *Historia animalium*:[11]

> For he says that a certain camel had coitus with his mother, and after she knew that she was his mother, she killed the one who had covered her head. And in the same way he tells of a certain noble horse in the northern region that had coitus with his mother and after he discovered it, he fled and cast himself off into a ditch and died.[12]

These episodes suggest two surprising findings: that some animals are able to be morally embarrassed, and that they have the power to make such distinctions, although they lack reason. If Albert's interesting explanation of the former goes beyond my purpose in this paper,[13] his position on the latter is certainly pertinent:

---

[7] Alberti Magni Opera Omnia. 2011. Editio Digitalis, https://www.albertus-magnus-online.de/#.

[8] Alberti Magni e-corpus, 2008–2018, http://www.albertusmagnus.uwaterloo.ca/.

[9] Albertus Magnus, *De somno et vigilia*, II, tr. 1, cap. 1 (ed. Borgnet, 1890 157a).

[10] Albertus Magnus, *Quaestiones super De animalibus*, VIII, q. 39 (ed. Filthaut, 1955, 200: 61–63). English translation taken from: Albert the Great, *Questions Concerning Aristotle's on Animals* (trans. Resnik, Kitchell, 2008, 298). "Et videtur quod non. Discernere est opus rationis; sed ista animalia carent ratione; ergo et discretione."

[11] Aristotle, *Histoire des animaux*, book IX, chapter 47, 630b31–631a7 (ed. and trans. P. Louis, tome 3, Paris: 1969, 136–137).

[12] Albertus Magnus, *Quaestiones super De animalibus*, VIII, q. 39 (ed. Filthaut, 1955, 200: 67–73). English translation taken from: Albert the Great, *Questions Concerning Aristotle's on Animals* (trans. Resnik, Kitchell, 2008, 298). "Dicit enim, quod quidam camelus coivit cum matre sua et postquam cognovit ipsam esse matrem, interfecit ipsum qui caput suum cooperuerat. Et idem dicit de quodam equo nobili in parte aquilonari, qui coivit cum matre sua et, postquam hoc cognovit, fugit et praecipitatus est in foveam et mortuus est." For more on *Quaestiones super De animalibus*, see Loconsole, Chap. 6 in this volume, pp. 91–113.

[13] The *solutio* to *quaestio* 39 gives us an answer: Albertus Magnus, *Quaestiones super de Animalibus*, VIII, q. 39 (ed. Filthaut, 1955, 201: 1–12): "Dicendum, quod sicut in ordine naturali causatum recipit influentiam et virtutem a sua causa, sic in via generationis generatum recipit virtutem a generante et non e contrario, et ideo verecundiam generatum magis debet generanti quam generans generato. Sed opus coeundi verecundiam inducit, et equus et camelus naturali instinctu ordinem naturae intendunt observare, et ideo filius naturaliter odit coire cum matre sua. Hoc tamen non est verum de omnibus equis, sed de equis nobilibus. Unde sicut homo odit coire cum matre, quia prohibitum est, sic equus, quia hoc ordini naturae repugnat."

> To the first argument one should reply that it is not an operation of reason to make universal distinctions among individual things. Rather, differentiation among individual intentions can be made by estimation, just as a bird differentiates between grain and a stone, and a sheep between a wolf and a lamb or a man. Thus, such differentiation can occur without reason.[14]

Even though Albert does not use the expression "individual intentions" very often, it is easy to recognize one of the major topics in Albert's psychology here, i.e., how non-rational animals are able to relate to their environment without having intellect or reason. As is well known,[15] Albert developed a theory of sense perception based on a complex sensory apparatus, mainly inspired by Avicenna's inner-sense theory, which explains the behaviour of rational and non-rational animals. In a nutshell, as long as they have a sentient soul, all animals have five exterior senses and four or five internal senses (depending on whether they are imperfect or perfect animals). These inner powers—common sense, imagination, estimation, phantasy, and memory—process the information that the five external senses convey to the soul.[16] In the version of the inner-sense theory that appears in Albert's *De anima*, the common sense grasps sensory inputs, which Albert calls forms, the imagination gathers them, and then the estimative power extracts (*elicit*)[17] intentions from them. Finally, intentions are stored in the memory, and the phantasy has the power of combining *ad libitum* forms with intentions. Although in question 39 Albert does not specify what an individual intention is, he recognizably addresses the same kind of intentions as in the *De anima*:

> We also experience in ourselves knowledge concerning sensory forms in three different ways. One of them deals with forms grasped by the senses in the manner in which they were sensed; another is concerned with intentions that have never been in the senses, but are also

[14] Albertus Magnus, *Quaestiones super De animalibus*, VIII, q. 39 (ed. Filthaut, 1955, 201: 67–73). English translation taken from: Albert the Great, *Questions Concerning Aristotle's on Animals*, (trans. Resnick, Kitchell, 2008: 299). "Ad primam rationem dicendum, quod discernere inter singularia universaliter non est opus rationis, immo per aestimationem potest fieri discretio inter intentiones individuales, sicut avis discernit inter granum et lapidem et ovis inter lupum et agnum vel hominem. Unde talis discretio potest fieri absque ratione."

[15] For more information on the relationship of Albert and Avicenna, see Wolfson (1935), Stenek (1974), Black (2000), Hasse (2000), Tellkamp (2013).

[16] The order and the number of inner faculties vary across Albert's works. The two principal accounts of them are to be found in *De homine*, *De anima sensibili, Sensus communis* (ed. Anzulewicz and Söder, 2008, 267–317); and in *De anima*, II, tr. 4, cap. 7 (ed. Stroick, 1968b, 156–158), and *De anima*, III (ed. Stroick, 1968b, 166–176). In *De homine*, the order is *sensus communis*, *imaginatio*, *phantasia*, *aestimativa*, *memoria*, and *reminiscentia* (only in human beings), while in *De anima* the order is *sensus communis*, *imaginatio*, *aestimativa*, *phantasia*, and *memoria*. For the number and order in Albert's *De spiritu et respiratione*, see Panarelli, Chap. 7 in this volume, pp. 115–132.

[17] In Albertus Magnus, *De anima*, II, tr. 4, cap. 7 (ed. Stroick, 1968b, 157–158), Albert calls these intentions extracted intentions (*intentiones elicitae*). In Albertus Magnus, *Super Ethica, Commentum et Quaestiones*, III, lectio 14, ad 3 (ed. Kübel, 1968a, 209) he explains that the estimative power extracts the intention of "convenience" and "inconvenience" from the form. For a theory about the meaning of *elicere* in this context, see Tellkamp (2013, 2016).

> not separated from their sensitive conditions, just as being convenient or inconvenient, or being a friend or an enemy, and being a son or not a son, a mother or not a mother, just as the sheep knows its own son, and extends its udder to its own young in order to feed it, and not to another; and flees from the wolf as from an enemy and follows the dog as a protector. [trans. P. H. R.][18]

The third way involves the phantasy.[19] As can be seen, in *De anima*, Albert explains what intentions are by means of an example similar to that in question 39. Moreover, the sheep in *De anima* shares with the camel and the horse of question 39 the skill of recognizing a given thing as *this* particular individual: its own son or its own mother. Throughout his work, Albert repeatedly uses examples which show the attitudes of sheep towards wolves and other sheep. The example is frequently restated in the same form as used by Avicenna in his *Liber de anima*,[20] where a sheep relies on its estimative power to recognize a wolf's hostility in its furry figure, though hostility itself is not a sensitive property. This example gradually evolves into many others in Albert's *De anima*: the sheep is able to distinguish between a hairy brownish thing that is a dangerous wolf and a hairy brownish thing that is a friendly dog.[21] Also, the sheep gives suck to its young and refuses to feed another lamb, because it can recognize the former as its own offspring.[22] Further on, the sheep sees its own offspring with compassion because it has the knowledge of both the individual lamb and the fact that this individual was born of her.[23] Finally, a sheep example appears in question 39 in the context of the ability of beasts to recognize individuals by telling one from another.

Avicenna's example of the sheep and wolf was originally intended to show that some non-rational animals are able to apprehend not only sensible forms but also some non-sensitive properties, e.g., the unsensed intentions (*intentiones non sensatas*) that enable them to act in a proper way. For Avicenna, the problem lies in how a creature without an intellect can apprehend a non-material property, such as hostility, which tells it when it should feel fear and flee. Albert's examples, however,

[18] Albertus Magnus, *De anima*, II, tr. 4, cap. 7 (ed. Stroick, 1968b, 157: 29–41): "Nos autem in nobis circa sensibiles formas triplicem experimur in genere cognitionem. Quarum una est, quae est circa formas acceptas per sensum hoc modo quo sensatae sunt; alia autem est, quae est circa intentiones, quae numquam in sensu fuerunt, sed tamen a sensibilium conditionibus non sunt separatae, sicut esse conveniens vel inconveniens, et amicum vel inimicum, et esse filium et non filium, matrem et non matrem, sicut ovis noscit filium et illi et non alii porrigit ubera lactando et fugit lupum ut inimicum et canem sequitur ut custodem." All translations with "[trans. P. H. R.]" are mine.

[19] Albertus Magnus, *De anima*, II, tr. 4, cap. 5 (ed. Stroick, 1968b, 157: 41–46): "Est autem et quaedam tertia, quae agit tam in formis sensatis quam in intentionibus componendo et dividendo, quae est quasi communis virtus, ad quam referuntur tam formae sensatae quam intentiones elicitae sicut sensata particularia ad sensum communem."

[20] Cf. Avicenna, *Liber de anima*, pars I, cap. 5 (ed. Van Riet, 1972, 86: 93–05), and Avicenna, *Liber de anima*, pars IV, cap. 3 (ed. Van Riet, 1968, 39: 39–52).

[21] Albertus Magnus, *De anima*, II, tr. 3. cap. 5 (ed. Stroick, 1968, 104).

[22] Albertus Magnus, *De anima*, II, tr. 4, cap. 7 (ed. Stroick, 1968, 156).

[23] Albertus Magnus, *De anima*, III, tr. 1, cap. 2 (ed. Stroick, 1968, 167).

show that his concern is different: he seeks to explain that non-rational animals are able to apprehend and recognize an individual thing and its meaningful properties.

8.3 The Inner Senses, Avicenna, and the Animal Problem

In general, Albert's concept of intention (*intentio*) comes from both Avicenna's and Averroes's theory of inner sensation. Whereas Avicenna thinks that non-rational animals can grasp intentions of certain kind, Averroes's theory of abstraction with its own concept of intention, which differs from Avicenna's, denies that non-rational animals can grasp any intentions at all. Albert was acquainted with Averroes's *Commentarium magnum in Aristotelis De anima libros*,[24] translated from the Arabic to Latin by Michael Scotus. According to the Commentator in this work, in grasping the sensitive characteristics of an external thing, human senses grasp an image, which is stored in the imagination. Richard Taylor explains that, subsequently, the cogitative power, "working with images resulting from sense perception, and through a process of discernment and distinction, distils individual intentions."[25] This implies that the common sense is able to grasp the individual intention. In the *Commentarium magnum in Aristotelis De anima libros* II, lectio 65, Averroes states that grasping the individual's intention (*intentio individui*) is specifically an act of the common sense of intelligent animals, i.e., human beings.[26] This clarification comes from Averroes's criticism of Avicenna's theory of sense perception in his *Liber de anima seu sextus de naturalibus*,[27] which Albert read in the Latin translation by Dominicus Gundissalinus and Avendauth. Because Averroes's theory stems from his criticism of Avicenna, and Albert follows Averroes's *Commentarium magnum in Aristotelis De anima libros* in his own *De anima*, at the same time extensively using the Avicennian theory and its animal examples, it is important to understand the basics of Avicenna's theory of inner sensation and the centrality of animals to it.

Avicenna's theory of sense perception builds on the Galenic model of the brain in order to solve a handful of classical Aristotelian problems of action and sense perception. According to the Galenic model, the brain is divided into three

---

[24] Averroes, *Commentarium magnum in Aristotelis De anima libros* (ed. Crawford, 1953). Richard Taylor has produced an English translation which assembles all the existent Arabic fragments. See Averroes (Ibn Rushd) of Cordoba, *Long Commentary on* De Anima *of Aristotle* (trans. Taylor, 2009).

[25] Taylor (2000, 120); cf. Averroes, *Commentarium magnum in Aristotelis De anima libros* III, lectio 6 (ed. Crawford, 1958, 415–416, and II, lectio 63, 225–226 and 228).

[26] Averroes, *Commentarium magnum in Aristotelis De anima libros*, II, lectio 65 (ed. Crawford, 1953, 228: 37–47): "Tamen videtur quod ista actio est sensus communis non secundum quod est sensus communis, sed secundum quod est sensus alicuius animalis, v.g., animalis intelligentis. [...] Videtur enim quod comprehensio intentionum individualium substantiarum, de quibus intellectus considerat, est propria sensibus hominis."

[27] Avicenna, *Liber de anima* (ed. Van Riet, vol. 1, 1968; vol. 2, 1972).

ventricles: anterior, middle, and posterior, and is penetrated by a corporeal, but very subtle, fluid, called *spiritus* (Greek *pneuma*; Arabic *ruhan*), which has a role in cognitive powers.[28] With this model, Avicenna tries to solve a range of problems. First, he wants to explain the relationship between the intellectual and sensitive content. Understanding this relationship is pivotal to action theory, to sciences, such as mathematics (especially geometry), and to fine arts, such as painting, because they are intellectual activities that always work with the products of sensation and imagination.[29] Another problem is how non-rational animals can perceive their environment and act on it in complex ways, although they are incapable of deliberate action, because they lack the intellect and rational concepts. To Avicenna, the Aristotelian imagination does not suffice to account for complex behaviour. The third problem is how to explain intemperance (Aristotelian *akrasia*). In order to adjust the Galenic model to solve the two latter problems, Avicenna introduces a new concept,[30] called the estimative power (Latin *aestimativa*; Arabic *wahm*), and coins the notion of what are translated into Latin as *intentiones non sensatas*, i.e., unsensed intentions. This concept is quite obscure in the *Liber de anima*, because Avicenna never properly defines it. He explains what unsensed intentions are only by saying how they are grasped by the sensitive powers and by giving some examples of them. These examples at least indicate what function these mental objects fulfil, but they do not say anything about their ontological status. Therefore, in order to understand how Avicenna uses this concept, we need to see how he defines the sensitive structure of the soul.

---

[28] I apply the term "Galenic model" to the sense-perception model proposed by Nemesius of Emesa in his *De natura hominis*. The original Galenic model, developed in the second century AD by Galen, included the tripartite division of the brain and the *pneuma*, the five exterior senses, and three inner perceptual powers: *phantastikón*, *dianoetikón*, and *mnemonikón*. Galen did not assign a definite location to each of these powers. This spatial relationship between the powers and the ventricles was only established in the fourth century, possibly by Porphyry, who may have been the source for Nemesius of Emesa (cf. Palanciuc and Chase, 2005). Nemesius is the oldest extant testimony of the inner senses located in the brain, and many of his definitions appear in Avicenna's the *Liber de anima*. For more information on the Galenic theory of inner senses, see Rocca (2003). For more information on *pneuma* theory, see Panarelli, Chap. 7 in this volume. Albert the Great read Nemesius in a Latin translation by Burgundio of Pisa (Némésius d'Émèse, *De natura hominis. Traduction de Burgundio de Pise*, ed. Gérard Verbeke, José Rafael Moncho, 1975). For more on Nemesius's role in the transmission of the Galenic theory, see Kijewska, Chap. 3 in this volume, pp. 37–56.

[29] Deborah Black has shown the importance of the estimative power in Avicenna, cf. Black (1991).

[30] In 1935, Wolfson published an influential paper on the "Internal Senses in Latin, Arabic and Hebrew Philosophical Texts." He attributed the invention of the estimative power to Al-Farabi. Nevertheless, Black (2000) shows that Wolfson misassigned to Al-Farabi the texts written by either Avicenna or by one of his followers (Black 2000, 60–70), and proves "that Avicenna is the originator of the internal sense tradition," using some of Averroes's assertions in his *Epitome of the Parva naturalia* and his *Tahafut al-tahafut*.

Avicenna identifies between five (or eight)[31] external senses and five internal senses. He says that the sensitive soul manages two different cognitive objects: forms (*formae*) and unsensed intentions (*intentiones non-sensatas*). He defines each cognitive object on the basis of the way it is apprehended: the form is a cognitive object apprehended first by the external senses and then by the internal senses, whereas the intention is apprehended by the soul, but only through the internal senses. Intentions are characterized as "unsensed" because they are not perceived by any of the external senses. Avicenna's most common example for explaining what these intentions are is that of a sheep's perception of a wolf. Once a sheep's senses have apprehended the forms of a wolf (for example, its smell, colour, and figure), the sheep's soul apprehends "that which allows the sheep to know that it must be afraid of the wolf and it must escape it," i.e., the wolf's hostility.[32] Hostility is an unsensed intention, because it is a meaningful property of the wolf, but is not sensible in itself.[33] As can be seen, forms are primary and secondary qualities, such as shape, colour, and other sensitive properties, while intentions are meaningful properties, which provide the sheep with information relevant to its practical behaviour.

The relationship of forms and intentions is fundamental to understanding how the entire internal sense apparatus works. As already stated, the brain is divided into three ventricles. The anterior and the posterior ventricles are both dual, which means that each of them has two different powers, whereas the middle ventricle houses a single power.[34] The common sense (*sensus communis*) and the imagination (*imaginatio*) are located in the anterior ventricle. The middle ventricle holds the imaginative power (*imaginativa*) in non-rational animals, and the cogitative power

---

[31] When Avicenna talks about the external senses, he always says "five or eight," because he ponders the long-standing question of whether the touch-sense is actually more than a single sense. Avicenna, *Liber de anima*, pars I, cap. 5 (ed. Van Riet, 1972, 84: 77–85: 87): "Et ex illis est tactus, qui est vis ordinata in nervis cutis totius corporis et eius carnibus, ad apprehendendum id quod tangit illium et afficit contrarietate permutante complexionem et affectionem compositionis. Videtur autem aliquibus haec vis non esse species specialissima, sed genus quattuor virium aut etiam plurium diffusarum simul in toto corpore. [...] Sed, quia collectio earum est in uno instrumento, ideo putantur esse una in essentia." For more on the multiplicity of touch in general, see Beneduce, Chap. 13 in this volume. For more on the same problem in John Blund, see Fedriga, Chap. 2 in this volume, pp. 17–35.

[32] Avicenna, *Liber de anima*, pars IV, cap. 1 (ed. Van Riet, 1968, 7: 82–83): "Inimicitia et malitia."

[33] Avicenna, *Liber de anima*, pars I, cap. 5 (ed. Van Riet, 1972, 86: 93–06): "Differentia autem inter apprehendere formam et apprehendere intentionem est haec: quod forma est illa quam apprehendit sensus interior et sensus exterior simul, sed sensus exterior primo apprehendit eam et postea reddit eam sensui interiori, sicut cum ovis apprehendit formam lupi, scilicet figura eius et affectionem et colorem, sed sensus exterior ovis primum apprehendit eam et deinde sensus interior; intentio autem est id quod apprehendit anima de sensibili, quamvis non prius apprehendat illud sensus exterior, sicut ovis apprehendit intentionem quam habet de lupo, quae scilicet est quare debeat eum timere et fugere, quamvis non hoc apprehendat sensus ullo modo. Id autem quod de lupo apprehendit primo sensus exterior et postea interior, vocatur hic proprie nomine formae; quod autem apprehendunt vires occultae absque sensu, vocatur in hoc loco proprie nomine intentionis."

[34] In the *Canon*, Avicenna enumerates only three internal senses: common sense, imagination and cogitative faculty. For more on Avicenna's inner sense theory in the *Canon* see Panarelli, in Chap. 7 in this volume, pp. 115–132.

(*cogitativa*) in human beings. The posterior ventricle is the locus of both the faculty called the estimative power (*estimativa*) and the memory (*memoria*). The powers in the anterior ventricle apprehend forms, and the powers in the posterior ventricle apprehend intentions. The power in the middle ventricle does not apprehend either forms or intentions, but divides or combines them. This yields the following pattern:

1. Common sense: it is linked to the five external senses. Its task is to apprehend the forms conveyed by the senses and discriminate between them; it makes simple judgments, for example, that something red is sour.[35] Among the sensitive powers, it is the one that is aware, therefore, when we are awake it perceives external stimuli, and when we sleep it senses internal stimuli; this enables us to hallucinate and dream.[36]
2. Imagination (or *formans*): its task is to compile the forms apprehended by the common sense.[37] It can present them to the estimation, to the intellect, or to the other psychological powers. These forms include the five Aristotelian common sense-objects, especially movement.[38]
3. Imaginative power/cogitative power:[39] its task is to combine and divide; in other words, it produces assertion-like and negation-like structures, i.e.,

[35] Avicenna, *Liber de anima*, pars IV, cap. 1 (ed. Van Riet, 1968, 6: 68–70): "Sensus vero communis et sensus exterioris discernunt aliquo modo et diiudicant: dicunt enim hoc mobile esse nigrum, et hoc rubicundum esse acidum."

[36] Avicenna, *Liber de anima*, pars IV, cap. 1 (ed. Van Riet, 1968, 5: 56–59): "Et haec virtus est quae vocatur sensus communis, quae est centrum omnium sensus et a qua derivantur rami et cui reddunt sensus, et ipsa est vere quae sentit." For more on the reception of Avicenna's theory of the inner sensation, please see Fedriga, Chap. 2 in this volume, pp. 17–35, and Mansfeld, Chap. 12 in this volume, pp. 229–251.

[37] Avicenna, *Liber de anima*, pars IV, cap. 1 (ed. Van Riet, 1968, 5: 60–63): "Sed retinere ea quae haec [sensus communis] apprehendit est illius virtutis quae vocatur imaginatio et vocatur formalis et vocatur imaginativa, et fortassis distinguunt inter imaginationem et imaginativam ad placitum: et nos sumus de his qui hoc faciunt."

[38] According to Avicenna, the perception of movement is produced through the consecutive impressions of the same object in different places of the imagination: Avicenna, *Liber de anima*, pars I, cap. 5 (ed. Van Riet, 1968, 88: 30–89: 43): "Cum autem volueris scire differentiam inter opus sensus exterioris et opus sensus communis et opus formantis [imaginationis], attende dispositionem unius guttae cadentis de pluvia, et videbis rectam lineam, et attende dispositionem alicuius recti cuius summitas moveatur in circuitu, et videbitur circulus. Impossibile est autem ut apprehendas rem aut lineam aut circulum nisi illa saepe inspexeris; sed impossibile est ut sensus exterior videat eam bis, sed videt eam ubi est; cum autem describitur in sensu communi et removeatur antequam deleatur forma sensus communis, apprehendit eam sensus exterior illic ubi est, et apprehendit eam sensus communis quasi esset illic ubi fuit, et quasi esset illic ubi est, et videt distensionem circularem aut rectam. Hoc autem impossibile est comparari sensui exteriori ullo modo, sed formans apprehendit illa duo et format ea, quamvis destructa sit res quae iam abiit."

[39] Avicenna, *Liber de anima*, pars IV, cap. 1 (ed. Van Riet, 1968, 6: 76–78): "[...] Et haec est virtus quae, cum intellectus ei imperat, vocatur cogitans, sed cum virtus animalis illi imperat, vocatur imaginativa."

quasi-propositional[40] content in non-rational animals, and propositional content in rational animals.

4. Estimative power: it can apprehend unsensed intentions related to the sensible objects perceived.[41] Its task is to make judgments that permit incidental perception.[42] In non-rational animals, it takes the ruling position held by the intellect in humans. In human beings, it may misappropriate the place of the intellect, and then it is responsible for akratic behaviour.[43]
5. Memory: its task is to store intentions and the propositional content generated by the work of the cogitative power (imaginative power in animals), the estimation and the imagination.[44] Memory is the name of both the power and action. The action involves the joint operations of all the internal senses.[45] Human beings also have reminiscence, which is a discursive type of memory linked to rationality.[46]

[40] Avicenna says that our soul has the power to divide and compound images in a way different from how we have seen them, *even if* (*quamvis*) we do not believe in them. That means that they have some kind of propositional structure in the way that Aristotle defines it. Cf. Avicenna, *Liber de anima*, pars IV, cap. 1 (ed. Van Riet, 1968, 6: 73–78). Nevertheless, Avicenna says that rational beings use these structures to make inferences that allow action, whereas the estimation, in non-rational animals, leads to action through immediate impulses. Cf. Avicenna, *Liber de anima*, pars IV, cap. 3 (ed. Van Riet, 1968, 35: 00–3): "Animalia autem et qui assimilantur eis homines non sequuntur in suis actionibus nisi hoc iudicium aestimationis, quod *non habet discretionem rationalem sed est ad modum adinventionis quae est in eius animo tantum*." (The italics are mine) That is why it is hard to talk about propositions *simpliciter*.

[41] Avicenna, *Liber de anima*, pars I, cap. 5 (ed. Van Riet, 1972, 89: 48–56): "Deinde est vis aestimationis [...], apprehendens intentiones non sensatas, quae sunt in singulis sensibilium."

[42] When we see a chocolate cake, we see a brown roundness, and simultaneously we perceive the cake's sweetness. This is incidental perception, and it results from a judgment made by the estimation. In this sense, we say that "we see sweetness." Avicenna states that the task of the estimation is to perceive intentions, such as hostility, which is non-sensitive in itself, as well as forms, such the sweetness in our example, that are not sensed in the chocolate-cake sense-perception act. As can be seen, the products of the estimation have a strong propositional content. Cf. Avicenna, *Liber de anima*, pars IV, cap. 1 (ed. Van Riet, 1968, 6: 79–7: 88).

[43] Avicenna, *Liber de anima*, pars IV, cap. 3 (ed. Van Riet, 1968, 35: 00–36: 06): "Animalia autem et qui assimilantur eis homines non sequuntur in suis actionibus nisi hoc iudicium aestimationis, quod non habet discretionem rationalem sed est ad modum adinventionis quae est in animo tantum, quamvis virtutibus hominis propter consortium rationis accidat aliquid propter quod virtutes eius interiores differunt a virtutibus animalium."

[44] Avicenna, *Liber de anima*, pars I, cap. 5 (ed. Van Riet, 1972, 89: 53–56): "Deinde est vis memorialis et reminiscibilis; quae est vis ordinata in posteriori concavitate cerebri, retinens quod apprehendit vis aestimationis de intentionibus non sensatis singulorum sensibilium."

[45] Cf. Avicenna, *Liber de anima*, pars IV, cap. 3 (ed. Van Riet, 1968, 39: 39–52).

[46] Elsewhere, Avicenna calls the "reminiscibilis" power the "recordatione" power, and identifies the brain as its physical location. See for example, *Liber de anima*, pars IV, cap. 3 (ed. Van Riet, 1968, 40: 59–63): "Id autem quo magis eget est memoria et sensus, sed forma opus es propter memoriam et recordationem. Memoria autem est etiam in aliis animalibus. Sed recordatio quae est ingenium revocandi quod oblitum est, non invenitur, ut puto, nisi in solo homine."

The major merit of Avicenna's inner-senses theory is that it can explain how sense-perception provides the soul with meaningful information about individual things dwelling in the world. This theory explains the learning process that happens through the senses, that is to say, it outlines the sequential process in which the soul first grasps sensual properties through the common sense, then accumulates them in the imagination, uses the imaginative or cogitative power to combine them with each other and with meaningful properties (i.e., unsensed intentions) recognized by the estimative power, and finally, reforges them into the propositional content (quasi-propositional content in animals), which is stored in the memory. Later, sentient beings use the stored propositional content to identify particular objects in a meaningful way, through the perception of just a single property, such as smell or colour. For example, when a sentient being sees a red roundness, all of its inner senses collaborate to help it recognize the object as a sweet apple. Because sweetness produces appetite, the sentient being undertakes action in order to eat this extramental thing. Avicenna's theory also explains how sentient beings can remember. Namely, when a horse craves something sweet, all of its inner senses work together in order to tell it which things can satisfy its appetite. For example, they tell it that it should seek a red roundness. At the same time, the human inner senses will tell a person craving something sweet that she should look for an apple. Essentially, Avicenna believes that the only considerable difference between animal and human sense perception lies in the relationship that the inner senses have—or do not have—with the rational powers of the soul. In all other respects, the animal and human inner senses are the same. It is in this context that accounting for the behaviour of non-rational animals is crucial.

According to Avicenna, non-rational animals display complex behaviour that far exceeds looking for food. In order to survive, they must be able to recognize dangerous things and safe environments, which involves grasping unsensed intentions through their estimative power. Avicenna insists that some animals learn from experiencing pleasure and pain associated with the things they perceive,[47] and that this is one way of grasping unsensed intentions. Nevertheless, without any previous experience, they can also instinctively flee from their natural enemies and recognize their own young to care for it. Avicenna calls this way of apprehending unsensed intentions "natural warnings" (*cautelas naturales*), and it is in this context that he uses his famous example of a sheep that fears a wolf. Although the sheep has never seen a wolf before, when it sees a brown, furry shape, it simultaneously perceives its hostility through its estimative power. Analogously, on spotting its fellow sheep, the sheep sees their colour and figure, and instantaneously perceives their friendliness. Avicenna claims that the intentions related to harm and benefit are mixed with the sensed things (*commixtae cum sensibilibus*). All instinctive behaviour, such as that

[47] Avicenna explains the internal process that happens in a dog when it learns to fear a stone and the role that the estimative power plays in it. Cf. Avicenna, *Liber de anima*, pars IV, cap. 3 (ed. Van Riet, 1968, 39: 39–41).

exhibited by non-rational animals, results from the unsensed intentions commingled with the sensed things.[48]

Expectedly, the origin and nature of unsensed intentions as posited by the Avicennian theory of sense perception is a controversial issue. Far from only being a problem for current scholarship,[49] it was already found problematic in the times of Al-Ĝhazālī and Averroes. Briefly, the question is whether the hostility/friendliness that the sheep perceives is a non-material but extramental property of an external thing, or whether it comes from another source.[50] Solutions given to the problem were so problematic that they did not survive in later Arabic authors. Nevertheless, although the ontological status and the origin of unsensed intentions in non-rational animals are far from certain, their function is clear: they are meaningful properties attached to pure sensory data that enable beasts to behave in ways promoting their self-preservation.

As already mentioned, for Avicenna, the human sensitive soul does not differ from other kinds of sensitive souls. The difference is due to the former's relationship to the intellectual realm. All the meaningful content of sense-perception unrelated to human non-rational behaviour comes from the intellect. Given this, the internal-senses structure linked to the intellectual powers can explain the capacity of rational beings to build propositions of scientific value based on sense-experience.[51] In other words, the structure that humans share with non-rational animals serves to explain how human beings are capable of developing a scientific empirical knowledge of the physical world. In *Liber de anima*, Avicenna says that when the imagination-related powers—especially the memory—collaborate with the human rational powers, they enable humans to bring together experiences (*experimenta*) and thoughts (*considerationes*) about singular things.[52]

The problematic origin of unsensed intentions in Avicenna's framework was one of the reasons behind Averroes's rejection of the concept. Another good reason was

[48] Avicenna, *Liber de anima*, pars IV, cap. 3 (ed. Van Riet, 1968, 38: 28–38).

[49] There is still no consensus on the nature and meaning of *intentiones non-sensatas* in Avicenna's *Liber de anima*. For example, Dag N. Hasse has argued that they are non-corporeal extramental entities attached to external things and, therefore, not just a mental content. On the ongoing dispute, see Hasse (2000); Black (2000); Kaukua (2007); Klubertanz (1952); López-Farjeat et al. (2010); and Wirmer (2004).

[50] This "other source" could be the sheep's soul itself or some supernatural entity.

[51] In Avicenna, *Liber de anima*, pars V, cap. 3 (ed. Van Riet, 1968) Avicenna shows four ways in which the sense-powers help the rational power to understand individual things in a rational way. The third way (Avicenna, *Liber de anima*, pars V, cap. 3, ed. Van Riet, 1968, 103: 10–19) is how rational beings acquire universal propositions about universal relationships between particular things from sense-experience.

[52] Avicenna, *Liber de anima*, pars IV, cap. 3 (ed. Van Riet, 1968, 36: 8–12): "Et eius [hominis] virtus imaginativa interior eiusmodi est quod valet ad scientias; et praecipue virtus suae memoriae valet multum ad scientias, eo quod confert nobis experimenta quae retinet memoria et considerationes singulorum et cetera huiusmodi." He will explain it in Avicenna, *Liber de anima*, pars V, cap. 3.

provided by an alternative conception of unsensed intentions that Averroes developed within his own theory of intellectual abstraction.

8.4 Averroes and the *Intentio Individualis*

Averroes's theory of inner sensation evolved somewhat between his early and late works.[53] Below, the principal focus will be on his *Commentarium magnum in Aristotelis De anima libros*,[54] because, in general, Albert uses it as a guide for his own commentary on the *De anima*, and, in particular, Albert's discussion of the nature of intentions—and the access that non-rational animals have to them—polemicizes with Averroes's position in this text. An outline of Averroes's theory of the inner senses and the relationship between intentions and images in his theory of perception will help better grasp this discussion.

According to Deborah Black, one of the most important changes Averroes made to Avicenna's inner-sense theory consisted in discarding the estimative power, both in human beings and in non-rational animals. For Averroes, the role attributed by Avicenna to the estimation is actually performed by the cogitative power,[55] whereas the estimative power is redundant in accounting for animal behaviour, because, according to Aristotle, imagination is enough "to explain the basic animal motions of pursuit and avoidance."[56] Thus, Averroes claims that there are only four internal senses in human beings: (1) the common sense, (2) the imagination, (3) the

[53] Apart from the *Great Commentary on the De anima*, Averroes develops his theory on sense-perception in his *Epitome on the De anima*, *Epitome on the Parva naturalia*, in the *Middle Commentary on the De anima*, and, as Black observes, in his *Tahafut al-Tahafut* (*Destructio destructiorum*). This late work was not known in the West in the thirteenth century. For Averroes's theory of the memory, see Black (1996). For more information on the change of the concept of *intentio/ma'na* in Averroes, see Wirmer (2004, 53–67).

[54] Averroes, *Commentarium magnum in Aristotelis De anima libros*. Importantly, Albert read Averroes's *Compendium libri Aristotelis De sensu et sensato* and believed that this work had been written by Al-Farabi. The *Compendium* is key to understanding the connection that Albert makes between intentionality and spiritualization in sense perception.

[55] Averroes does not mention the estimative power in his *Epitome on the Parva naturalia*, *Epitome on the De anima*, and *Middle and Great Commentaries on the De anima*, but he rejects it openly in his *Tahafut al-Tahafut*. See Black (1996, 2000). Black says that the role of Avicenna's estimation is actually played by Averroes's version of memory, because it can make the judgements ascribed to the estimation by Avicenna.

[56] Black (2000, 62; 69, n. 1; and 72, n. 32).

cogitative power, and (4) the memory,[57] whereas non-rational animals have just two inner senses: the common sense and the imagination.[58]

Like Avicenna, Averroes makes a distinction between two different cognitive objects in the soul, i.e., meaningful intentions and sensible images, but his approach differs from Avicenna's in two important respects. First, for Averroes, apprehended intentions concern the substantial, individuated form of the extramental thing.[59] Second, in apprehending the extramental thing, the external senses, the common sense and the imagination grasp simultaneously the image and the intention. For this reason, in human beings the task of the cogitative power is to extract the individual intention, i.e., the intention of the individual, the essential form, from the entire image that represents the sensible thing (*idolum imaginatum*).[60] That means that cognitive objects naturally supply the soul with enough data for the cogitative power to distil intentional meaningful information—the individual essence of the thing—from the perceived object. Finally, Averroes insists that, since non-rational animals have neither cogitative power nor memory, their souls cannot grasp intentions at all. These insights help understand the issues relevant to this argument in the sections of Averroes's commentary analysed by Albert.

In lectio 63–65 of the second book of the *Commentarium magnum in Aristotelis De anima libros*, Averroes comments on Aristotle's chapter 6, book II of *De anima*,[61] which is where Aristotle characterizes the three different types of objects of sense perception. Objects of two of these types are perceived in themselves, and objects of the third are perceived contingently. The first two types are handled by what we call essential perception and they are: the proper sensibles—objects that are in a one-to-one relationship with each of the external senses[62]—, and the five common

[57] Averroes, *Commentarium magnum in Aristotelis De anima libros*, III, lectio 6 (ed. Crawford, 1953, 415: 68–416: 74): "Et hoc aperte dixit Aristoteles in illo libro, cum posuit virtutes distinctivas individuales in quattuor ordinibus: in primo posuit sensum communem, deinde imaginativam, deinde cogitativam, et postea rememorativam. Et posuit rememorativam magis spiritualem, deinde cogitativam, deinde imaginativam, et post sensibiliem."

[58] Aristotle says that there are animals with memory, but, according to Averroes, the memory grasps intentions, and animals cannot do that. For a discussion of the problem of animal memory in Averroes, see Black (2000, 72, n. 33).

[59] Taylor (2000: 119–124).

[60] Averroes, *Commentarium magnum in Aristotelis De anima libros*, III, lectio 6 (ed. Crawford, 1953, 415: 62–64): "Declaratum est enim illic quod virtus cogitativa non est nisi virtus quae distinguit intentionem rei sensibilis a suo idolo imaginato." Cf. Averroes, *Compendium Memoria et Reminiscentia* in the *Epitome Parva Naturalia* (ed. Shields, Versio Vulgata, 56: 35–57: 43): "Et hoc erit quando sentiens senserit primo rem extra animam, deinde imaginaverit imaginans, deinde distinxerit distinguens intentionem illius formae a suo descripto, cuius est intentio, deinde recipit conservans illud quod distinguens distinguit." Cf. with the English translation of the Arabic version from Deborah Black (1996, 171).

[61] Aristoteles, *De anima*, II.6, 418a7–28 (ed. Ross, 1956, p. 40–41). For more on the reception of this Aristotelian distinction, see Mansfeld, Chap. 12 in this volume, p. 231.

[62] That is, colour is the proper object of sight because it can only be perceived by sight, and odour is proper to smell and it is essential because it belongs to the definition of smell. For its part, incidental perception is called thus because the relationship between the sense-power and the per-

sensibles[63]—objects that are perceived by more than one of the external senses. Objects of the third type are dealt with by so-called incidental perception. Averroes divides his comment into three parts: in lectio 63, he addresses proper sense-objects; in lectio 64, he discusses common sense-objects; and in lectio 65, he considers the objects of incidental perception. Lectio 63 and lectio 65 are relevant at this point, because they include Averroes's clarifications concerning the nature of individual intentions.

According to Aristotle, there is no room for error in the essential perception of proper sense-objects. The Aristotelian text quoted by Averroes says:

> but each of these [senses] judges those [each proper object of sense], and it does not make a mistake about which colour is which colour, nor about which voice is which voice, but it makes a mistake about what the coloured-thing [*colorato*] is and about where it is, and it makes a mistake about what the thing that is heard is and about where it is. [trans. P. H. R.][64]

Averroes explains that Aristotle does not mean that the senses can apprehend the essence of things when he says that we can make a mistake in saying "what the coloured-thing is," but rather that the senses apprehend individual intentions of things conjoined with their own proper sense-objects.[65] Examples of these individual intentions, says Averroes, are "the intention of this individual man" and "the intention of this individual horse." He adds that our senses can apprehend intentions of individuals under any of the ten Aristotelian categories.

The particularity of the individual intention does not depend on its relationship to the sensitive properties apprehended with it. As Richard Taylor observes,

> the cogitative power grasps the individual form which is responsible for the individual's existence and nature making use of the individual sensible characteristics revealed by the internal and external senses by way of the proper and common sensibles.[66]

But this does not mean that the individuality of the individual intention comes from these "sensible characteristics."[67] A few lines further, Averroes says that the task of

ceived property is not essential. For this interpretation of the terms "essential" and "incidental," see Cashdollar (1973).

[63] Magnitude, figure, motion, rest, and number.

[64] Averroes, *Commentarium magnum in Aristotelis De anima libros*, II, lectio 63 (ed. Crawford, 1953, 224: 10–14): "Sensus autem plures modos uno habet, sed unusquisque eorum iudicat ista, et non errat in colore quis color sit, neque in voce quae vox sit, sed in colorato quid est et ubi est, et in audito quid est et ubi est." Cf. Aristoteles, *De anima*, II.6, 418a14–16 (ed. Ross, 1961).

[65] Averroes, *Commentarium magnum in Aristotelis De anima libros*, II, lectio 63 (ed. Crawford, 1953, 225: 39–50): "Et cum dixit: sed in colorato quid est et ubi est, et in audito quid est et ubi est, non intendebat quod sensus comprehendit essentias rerum, […] sed intendebat quod sensus cum hoc quod comprehendunt sua sensibilia propria, comprehendunt intentiones individuales diversas in generibus et in speciebus; comprehendunt igitur intentionem huius hominis individualis, et intentionem huius equi individualis, et universaliter intentionem uniuscuiusque decem praedicamentorum individualium."

[66] Taylor (2000, 121).

[67] Black (1996, 169) believes that the individual intention allows all the sensed properties to conjoin in a coherent way and form an individual thing, instead of just as a bunch of sensations:

the cogitative power is to strip individual intentions of the sensitive properties imparted to them by the senses and then to place them in the memory. But even without its sensible properties, the individual intention remains individual. Moreover, the difference between the cogitative power and the imagination is that the latter perceives the individual intention attached to its sensitive properties, whereas the cogitative power perceives the individual intention as such. The cogitative power and the memory are individual entities, and, consequently, they can only grasp and retain individual entities. In this sense, distilling the universal essence is the task of the intellectual powers. Thus, it is clear, too, that the individuality of the individual intention is independent of the sensible properties attached to it.[68] In lectio 65, Averroes revisits this issue, when he explicitly states that the task of the senses is to apprehend differences among individuals *qua* individual things, especially substantial ones.[69]

But at this point, Averroes faces a major obstacle: if the external senses and the common sense can apprehend complex cognitive content, such as the intention, non-rational animals should also be able to do it, since these powers are shared by rational and non-rational animals. This conundrum prompts him to aver, in paragraph 63, that only the human senses possess this capacity and to marshal Aristotle's authority to prove this point.[70] In lectio 65, he reasserts that "it seems that the apprehension of the intentions of individual substances that the intellect regards is proper solely to the human senses" [trans. P. H. R.].[71] It is crucial to Averroes's abstraction theory that the senses can apprehend the individual intentions of substances, i.e., the individual form of extramental things, but, at the same time, he cannot offer a good argument to prove that grasping individual intentions is the exclusive prerogative of the human senses. Averroes concludes the passage about the relationship between the senses and the apprehension of individual intentions with a sentence that will be fundamental to Albert the Great: "And you should know that the apprehension of

"So the intention must represent some element within the ostensible individual that is not encompassed by its image, and that would seem to leave only its underlying individuality. [...] Thus, the perception of an intention for Averroes would seem to entail the recognition of an individual precisely in so far as it is individual."

[68] Crucially, the Averroistic interpretation of the individual intention as representing something like an ontological individual form fits well with the Aristotelian understanding of substance. What I mean is that all incidental properties are attached to the individual substantial form *as to a subject*.

[69] Averroes, *Commentarium magnum in Aristotelis De anima libros*, II, lectio 65 (ed. Crawford, 1953, 228: 41–45): "[...] quod accidit sensibus comprehendere differentias individuorum (secundum quod sunt individua) non secundum quod sunt sensus simplices, sed secundum quod sunt sensus humani; et praecipue differentias substantiales."

[70] Averroes, *Commentarium magnum in Aristotelis De anima libros*, II, lectio 63 (ed. Crawford, 1953, 225: 50–53): "Et hoc videtur esse proprium sensibus hominis; unde dicit Aristoteles in libro de Sensu et Sensato quod sensus aliorum animalium non sunt sicut sensus hominis, aut simile huic sermone."

[71] Averroes, *Commentarium magnum in Aristotelis De anima libros*, II, lectio 65 (ed. Crawford, 1953, 228: 45–47): "[...] Videtur enim quod comprehensio intentionum individualium substantiarum, de quibus intellectus considerat, est propria sensibus hominis."

the intentions of an individual belongs to the senses, whereas the apprehension of universal intentions belongs to the intellect" [trans. P. H. R.].[72]

8.5 Albert the Great: Paraphrases and *Intentio*

Albert's commentary on Aristotle's *De anima* does not have a traditional structure. Instead of commenting on each passage, Albert paraphrases Aristotle's text, expanding its excerpts with his own words in order to fill in the gaps or to shed light on obscure passages. One disadvantage of this technique is that it makes it hard to distinguish specific topics or problems from the commented text. In order to solve this problem, Albert inserts chapters, called "digressions," in between the segments of the paraphrased text. Most of Albert's original thought in his commentaries on Aristotle is found in chapters of this kind, which are thus key to understanding the Albertian approach to some topics and even some of his paraphrases. At the same time, Albert uses Averroes's *Commentarium magnum in Aristotelis De anima libros* as a guide for developing his own commentary, which means that he resumes topics and problems explained by Averroes and works out a solution of his own.

One of the problems posed by the Commentator in the *Commentarium magnum in Aristotelis De anima libros*, lectio 63 and lectio 65, is how the senses of human beings and non-rational animals can perceive the individual intentions of extramental things. The Averroistic assertion that the non-human senses cannot apprehend individual intentions is challenging to Albert, because his inner-sense theory is strongly influenced by Avicenna, with whom he shares the conviction that non-rational animals and human beings have the same sensitive apparatus.[73] Therefore, if the human senses are able to recognize individuals as individuals in the world, as Averroes contends, non-rational animals should be able to do so as well. However, this means that if non-rational animals cannot grasp the intentions of essences, the human senses cannot do so either. For this reason, Albert has to build a theory

[72] Averroes, *Commentarium magnum in Aristotelis De anima libros*, II, lectio 63 (ed. Crawford, 1953, 228: 47–50): "Et debes scire quod comprehensio intentionis individui est sensuum, et comprehensio intentionis universalis est intellectus."

[73] For Albert, the question of apprehending individuals is the same for rational and non-rational animals, because both have the same sense-perception apparatus. According to Jörg A. Tellkamp, for Albert "the physiological similarities between many animal species and human beings allow to establish that those similarities also hold on the cognitive level of sensory knowledge" (Tellkamp, 2016, 109). All animals, insofar as they are animals, have at least *sensus communis*, *imaginatio* and *aestimatio*, and perfect animals have *phantasia*, too. Human beings are animals as well. This means that both human beings and wolves apprehend individuals through the same physiological process. In rational and non-rational animals, the faculty involved in apprehending individuals is *sensus communis*. Both require another faculty in order to perceive the properties of the apprehended individual. Cf. Tellkamp (2016). For more on the problematic relationship between human beings and animals in Albert the Great, see Miteva, Chap. 5 in this volume, pp. 73–90. For more on the same matter in other medieval philosophers, see Trifogli, Chap. 10 in this volume, pp. 191–211.

capable of explaining what kind of sense-object is apprehended by sense-perception when it perceives individual intentions.

In order to solve this puzzle, Albert includes a chapter entitled "A digression explaining the four degrees and ways of abstraction" in *De anima* II.[74] The digression immediately precedes the section corresponding to Aristotle's *De anima* II, 6 and Averroes's *Commentarium magnum in Aristotelis De anima libros*, lectio 63. In this passage, Albert states that he will depict the operations of the apprehensive powers in order to make it easier to grasp the nature of proper sensibles, common sensibles, and incidental perception to be examined next.[75] He then proceeds to elaborate on the four degrees of abstraction or separation between the mental object and the material thing.

Almost half of the digression is a paraphrase of Avicenna's *Liber de anima* II, 2,[76] in which he classifies the respective sense-objects of particular sense-powers according to their degree of abstraction. Nevertheless, at the beginning of his gloss, Albert offers a series of clarifications which guide the understanding of Avicenna's two central terms: intention and form. Albert says:

> [...] that all apprehending consists in grasping the form of what is apprehended, but not according to the being that [the form] has in the apprehended thing, but rather in the sense that it is [the apprehended thing's] intention and species, under which any sensible or intellectual knowledge [*notitia*] of what is apprehended is contained. [trans. P. H. R.][77]

Albert makes these annotations because, as he observes, Avicenna uses "form" (*forma*) in more than one sense. Nevertheless, the main function of these clarifications is to modify the meaning of the Avicennian text in order to adapt it to his own purposes: Albert substitutes Avicenna's *intentio* with *species*, which functions as an intermediary between the act of knowing and the known thing, both at the sensorial level (*species sensibilis*) and at the intellectual level (the *species intellectualis*).

According to Albert, there are four degrees of abstraction. The first pertains to the common sense, which apprehends the form of its sense-object without its matter, yet requires the matter's presence. The second degree corresponds to the imagination, which apprehends the form of its sense-object without its matter and without any need for the matter, but without discarding the matter's appendages.[78] The third

---

[74] Albertus Magnus, *De anima*, II, tr. 3, cap. 4 (ed. Stroick, 1968b, 101–102): "Et est digressio declarans gradus abstractionis et modus."

[75] Albertus Magnus, *De anima*, II, tr. 3, cap. 4 (ed. Stroick, 1968b, 101: 57–61): "Sed ad faciliorem intellectum eorum quae dicturi sumus, faciemus capitulum breve de modo apprehensionis potentiarum apprehensivarum omnium; hoc enim perutile erit ad omnium sequentium notitiam faciliorem."

[76] Avicenna, *Liber de anima*, pars II, cap. 2 (ed. Van Riet, 1972, 120: 27–28).

[77] Albertus Magnus, *De anima*, II, tr. 3, cap. 4 (ed. Stroick, 1968b, 101: 62–66): "Dicimus igitur, quod omne apprehendere est accipere formam apprehensi, non secundum esse, quod habet in eo quod apprehenditur, sed secundum quod est intentio ipsius et species, sub qua aliqua sensibilis vel intellectualis notitia apprehensi habetur."

[78] Throughout Albert's work, two different distinctions between matter and matter appendages appear. On the one hand, it seems that Albert adopts the Avicennian position, where matter means

degree involves the estimative power, whose sense-object is the intention; and the fourth degree concerns the intellect, which is responsible for apprehending the universal form. Importantly, according to Albert,[79] the third degree of apprehension refers, in general, to any kind of incidental perception[80] and not only to the apprehension of intentions, such as hostility or friendship.

In the paragraph following the paraphrases, Albert again clarifies the meaning of intention and form, but this time he abandons the proper Avicennian meaning of form as primary or secondary qualities and the meaning of intention as a meaningful mental content. Instead, he uses "form" in its ontological sense, that is, as one constituent of composite substance, and he treats "intention" is a synonym of "species."

> We should note what the difference between the form of a thing [*forma rei*] and the intention of a thing [*intentio rei*] is; namely, that form is properly speaking that which by means of forming [*informando*] gives being in actuality to matter, and to the composite of matter and form. On the other hand, we call intention that which gives meaning to individual and universal things according to their degree of abstraction. And the intention does not give being to anything, neither to the senses, nor to the intellect, when it is in them, but generates [*facit*] a sign of the thing, and knowledge [*notitia*] of the thing. [trans. P. H. R.][81]

In this way, he frames the terms "thing's form" (*forma rei*) and "intention's form" (*intentio rei*) as opposites. Since the form is understood to give being to matter in actuality, intention is opposed to form, because *intentio* does not give being to that which receives it (the senses or the intellect), but instead it produces a sign and

---

the presence of an extramental thing at the moment of perceiving, and appendages are local and temporal properties (*situs, figura, tempus, hic et nunc*, etc.), which remain in the imagined object, even if the extramental thing is gone. On the other hand, in *De homine*, Albert says that, whereas individual extramental things are individuated by matter, the individual intentions of things are individuated by matter appendages, and the abstraction of the universal intention is abstraction not from matter but from the particular intention which is built with these appendages. For the former distinction, cf. Albertus Magnus, *De Anima III*, tr. 2, cap. 13 (ed. Stroick, 1968b, 195: 86–88); Albertus Magnus, *De IV coaequevis*, 4, q. 23, art. 1 (ed. Borgnet, 1895, 427a); Albertus Magnus, *De intellectu et intelligibili*, I, 2, 1 (ed. Borgnet, 1890, 490a); For the latter distinction, cf. *De homine, De anima sensibili, Sensus in communi* (ed. Anzulewicz, Söder, 2008, 264: 58–78).

[79] According to Avicenna, the third degree corresponds to the non-sensed intention, because we always apprehend it as mixed with matter, but in itself it is a representation of an immaterial entity.

[80] Albertus Magnus, *De anima*, II, tr. 3, cap. 4 (ed. Stroick, 1968b, 101: 90–102: 10): "Tertius autem gradus apprehensionis est, quo accipimus non tantum sensibilia, sed etiam quasdam intentiones quae non imprimuntur sensibus, sed tamen sine sensibilibus, numquam nobis innotescunt sicut est esse socialem et amicum, et delectabilem in convictu, et affabilem et his contraria, quae quidem cum sensibilibus accipimus, et tamen eorum nullum sensibus imprimitur. Et tale est, quod accipimus hunc esse filium Deonis, vel esse agnum vel hominem, aliud autem esse lupum vel leonem, secundum quod substantiales formae mediantibus sensibilibus et non separate ab eis apprehenduntur."

[81] Albertus Magnus, *De anima*, II, tr. 3, cap. 4 (ed. Stroick, 1968b, 102: 28–36): "Adhuc autem notandum est, quod differunt forma rei et intentio rei; forma enim proprie est, quae informando dat esse actu materiae et composito ex materiae et forma. Intentio autem vocatur id per quod significatur res individualiter vel universaliter secundum diversos gradus abstractionis; et haec non dat esse alicui nec sensui, quando est in ipso, nec etiam intellectui, quando est in illo, sed signum facit de re et notitiam."

knowledge (i.e., a representation) of the extramental compounded thing there. Albert concludes: "therefore, intention is not part of the thing as the form is, but rather it is the species, the knowledge of the totality of the thing" [trans. P. H. R.].[82] That means that the intention can represent the compound of the ontological matter and the ontological form.

In this way, Albert makes the most important assertion of this chapter-digression. Because the intention is the species of the thing, "and because the intention is abstracted from the whole thing and it is the meaning of the whole thing, it is predicated of the thing" [trans. P. H. R.].[83] This indicates how Albert has redefined abstraction to denote separating the signification[84] of a thing *from the signified thing*, i.e., separating the subject from its predicate, where the subject is an individual thing in itself and the predicate is any substantial or accidental property. This calls for showing what Albert understands by predication, that is how the subject and predicate are related to each other in his view. For example, the intention "human" is abstracted from one or several particular instances (Socrates, Zayd, or Marilyn Monroe), and therefore it contains all the *notitiae* of what corresponds to a human, including its matter and its form. In this sense, it is the meaning of the totality of the thing, and, as such, can be predicated of it, for example: "Marilyn Monroe is a human being." This in itself seems unproblematic; however, as soon as we start talking about sense perception, we run into trouble because sense perception deals with accidents. This is illustrated by one of Albert's own examples, in which he asserts that "the intention of the coloured thing, which is in the eye, conveys knowledge of the whole thing [*totam rem notificat*]" [trans. P. H. R.],[85] and continues: "Because not only the knowledge of the colour is perceived by the eye, but of the coloured thing, and its species in sight is the species of the coloured thing in so far as it is coloured, and judgement is made of the coloured thing in so far as it is coloured" [trans. P. H. R.].[86]

This example rehearses the problematic issue from Aristotle's *De anima* that Averroes did not know how to explain: how the animal senses, whether non-rational or rational, can apprehend the coloured thing. The intention that is in the eye, i.e., the colour, is that which is predicated of an individual thing. But what does it mean that not only the colour but also the coloured thing, as far as it is coloured, is in the

[82] Albertus Magnus, *De anima*, II, tr. 3, cap. 4 (ed. Stroick, 1968b, 102: 36–37): "Et ideo intentio non est pars rei sicut forma, sed potius est species, totius notitia rei."

[83] Albertus Magnus, *De anima*, II, tr. 3, cap. 4 (ed. Stroick, 1968b, 102: 38–40): "[...] Et ideo intentio, quia abstrahitur de toto et est significatio totius, de re praedicatur."

[84] Signification means here the mental content that some word or term generates, rather than what it refers to.

[85] Albertus Magnus, *De anima*, II, tr. 3, cap. 4 (Stroick, 1968b, 102: 39–42): "Intentio enim colorati, quae est in oculo, totam rem notificat, sicut et intentio, quae est in imaginatione particulari non praesente."

[86] Albertus Magnus, *De anima*, II, tr. 3, cap. 4 (Stroick, 1968b, 102: 47–51): "Non enim accipitur per visum notitia coloris tantum, sed colorati, et species eius in visu species est colorati, secundum quod coloratum est, et iudicium fit de colorato secundum quod coloratum est."

eye? To make sense of this, a look into Albert's *Organon*, where he defines *forma totius*, is in order.

8.6 Accidents, *Forma Totius*, and *Praedicatio per se Tertio Modo*

In *De sex principiis*, I, chapter 2,[87] Albert makes a distinction between what he calls ontological form and logical form. He refers to the ontological form, which gives being to individual substances in the world, as the "form of the part" (*forma partis*), because it is just one part of the totality of an individual substance, matter being its other part. He calls the logical form the "form of the totality" (*forma totius*). Albert defines the logical form as the intention that refers (*dicit*) to the formal being of the thing's totality, and as the knowledge (*notitia*) of the totality of the thing's being. Albert explicates that the form of the totality is what *Porphyry* called a universal,[88] and lists the four modes in which the Porphyrian universals can refer to the entire being of a thing: substantially, accidentally, potentially, and actually. For example, "animal," i.e., genus, refers to the whole being of a thing potentially, because in the genus the difference is contained potentially. "Rational," i.e., differentia, refers to the being of man in actuality, and "man," i.e., species, is a composite of both—animal and rational. "White," i.e., accident, refers to the whole being of a thing accidentally, and "risible," i.e., proprium, does the same but in respect to the accident that emanates from nature and species.[89] Each of these kinds of the form of the totality of a thing (*forma totius*) refers exactly to the same thing; they denote the same in different ways. For example, it is predicable of Socrates that he is animal, rational, human (rational animal), white, and able to laugh, because each of these predicates refers to the whole thing, i.e., to Socrates.

---

[87] Albertus Magnus, *De sex principiis* (ed. Meyer, 2006). For more information about *forma totius* in Albert the Great, see Piché (2013).

[88] These universals are genus, species, differentia, proprium, and accident. For Albert, everything predicable is a universal, and any universal is predicable. He defines the universal in *De quinque universalibus* (his commentary on *Porphyry*'s *Isagoge*). Albertus Magnus, *De quinque universalibus*, tr. 1, cap. 1 (ed. Santos Noya, 2003, 17: 39–43). "Universale autem est quod, cum sit in uno, aptum est esse in pluribus, ut in antehabitis ostensum est. Et per hoc quod in multis per aptitudinem est, praedicabile de illis est. Et sic universale est quod de sua aptitudine 'est in multis et de multis'." As can be seen, Albert considers mental content to have logical properties.

[89] Albertus Magnus, *De sex principiis*, tr. 1, cap. 2 (ed. Meyer, 2006, 4: 9–19): "Formam enim hic logica diffinimus diffinitione. Logice enim loquendo forma est intentio quae dicit esse formale totius et ipsa notio totius, ut dictum est in scientia *De Universalibus*. Dico autem quod dicit esse totius, sive hoc esse sit substantiale sive accidentale, sive sit esse potentiale, sive actuale, sicut 'animal' dicit esse totius potentiale et 'rationale' dicit esse totius hominis actuale, homo autem dicit esse totum compositum ex utroque, 'album' autem dicit esse totius accidentale, et 'risibile' dicit esse totius accidentale accidente naturae et speciei."

This raises the question about the way in which "white" can be the form of the totality of Socrates. Albert offers an explanation in *De praedicamentis*, II, 6,[90] where he discusses the ontological status of parts and wholes. Namely, an accident is not the form of the part of a thing but the form of the totality of a thing, because the subject in which the accident occurs is an entity complete in itself, and to be complete means to be a composite whole. However, an accident is not a constituent part of the totality of the subject, but rather its own being is constituted by the subject itself. Consequently, the subject in which the accident inheres appears in the definition of the accident "as the cause of the accident's being."[91] "Socrates is white" is a good example: "white" can be predicated of Socrates because it refers to its own cause, and its cause is the subject in which it occurs, which contingently happens to be Socrates.

This explanation corresponds to Aristotle's doctrine of *per se* predications in *Posterior Analytics*, 1, 4.[92] According to Aristotle, all definitions are expressed as *per se* predications, i.e., ones in which the relationship between the subject and the predicate is a necessary one, and there are four modes of them. The definition of substance—for example, "man is a rational animal"—is a *per se* predication of the first mode because the predicate is contained in the subject. The second mode of *per se* predication is famously illustrated by Aristotle's example of snubness, where the nose must be present in the definition of snubness, because snubness is exclusively a property of the nose. "Socrates is white" exemplifies the third mode of *per se* predication as colour is an accident and the subject of which it is an accident appears in its definition as a cause. In the definition of colour, the subject is the coloured thing as far as it is coloured. When sight receives the intention of the colour, since what it receives is the accident of a subject, it simultaneously receives the knowledge of the subject, but it can only know of the subject what it is in relation to the accident, i.e., being coloured. Finally, the fourth mode is causal and occurs when the cause of the predicate is co-signified in the subject, and the subject is referred to in an indeterminate way. Albert's example is that in saying that some beheaded is dead, the subject is individuated in an indeterminate way, and the cause of "is dead" is implicitly signified by "beheaded."[93]

---

[90] Albertus Magnus, *De praedicamentis*, tr. 2, cap. 6 (ed. Steel, Donati, 2013).

[91] Albertus Magnus, *De praedicamentis*, tr. 2, cap. 6 (ed. Steel, Donati, 2013, 33: 35–47): "Accidens enim non est forma partis sed totius, quia subiectum accidentis est ens in se completum; completum autem est totum compositum. Est autem adhuc etiam in hoc dissimilitudo, quod partes in toto, sive sint essentiales sive integrales, totum constituunt secundum esse, et ideo substantiae sunt totius. Accidens autem non est in toto sicut constituens ipsum secundum substantiam, sed sicut constitutum in esse quocumque per subiectum, et ideo in diffinitione accidentis recipitur subiectum sicut causa esse accidentis; in diffinitione totius compositi recipiuntur partes sicut principia cognoscendi totum, quia eadem sunt principia cognoscendi et essendi, sicut dicitur in principio *Physicorum*."

[92] Aristoteles, *Analytica Posteriora*, I.4 73a34–b24 (trans. Barnes, 1991, 7–8).

[93] Albert discusses the difference between the second, the third, and the fourth mode of predication in his commentary on the *Posterior Analytics*, *Libri 2 De demonstratione, id est, posteriorum analyticorum*, I, tr. 2, cap. 9 (ed. Borgnet, 1890, 40a–45b).

To return to *De anima*, Albert solves Averroes's problem by entirely altering the meanings of intention and abstraction, the terms that Avicenna uses in *Liber de anima*, II, 2. In apprehending their own proper sense-objects, the senses of animals, whether intelligent or not, can also apprehend the extramental individual thing. This happens because of the causal relationship between an accidental property and its subject: in apprehending colour, sight also apprehends the coloured thing. Once this riddle is solved, Albert goes on to explain the meaning of Averroes's quotation from Aristotle: "*universale est dum intelligitur; singulare vero, dum sentitur*":

> And this is what Aristotle says admirably well in the second book of his *De anima*,[94] that the senses deal with particulars, and he does not say that [they] deal only with some part of the form, but [with] the whole particular, just as the intellect deals with universals, which is not the knowledge and the species of one part, but of the whole, and therefore produces knowledge of the totality. [trans. P. H. R.][95]

Afterwards, Albert returns to his paraphrase of Avicenna and elucidates the proper meanings of abstraction and matter:

> When we say that we abstract something from matter, we should understand that matter is a certain particular which obtains its particularity from matter. For it is the first subject of form, and the particular thing subsists through it; and [it is also the first subject] of common nature, which is abstracted universally by the intellect, and with respect to the individuating forms whose intentions are abstracted by the other degrees of apprehension. [trans. P. H. R.][96]

Albert's reasoning deserves special attention at this point. He argues that matter is not to be understood as an undetermined potentiality but as a certain particular which plays the role of a subject standing in a certain relationship to the form, and this relationship is predication. From an ontological point of view, Albert adheres to Aristotle's hylomorphism, because what he calls a certain particular arises from the form's action on matter: it is through the form that matter obtains its particular way of being. But from a logical point of view, matter is also this certain particular that is the subject of the form from which it obtains its substantial essence, i.e., common nature, along with accidental properties of the individual thing, i.e., the individuating forms. Thus, the form can be predicated of matter. In order to clarify this new concept of matter understood as a certain particular, Albert refers readers to his

---

[94] Aristoteles, *De anima*, II.5, 417b22 (ed. Ross, 1961).

[95] Albertus Magnus, *De anima*, II, tr. 3, cap. 4 (ed. Stroick, 1968b, 102: 42–47): "Et hoc est, quod egregie dicit Aristoteles in secundo libro suo De anima, quod sensus sunt particularium, et non dicit, quod sint formae alicuius tantum, sed totius particularis, sicut et intellectus est universalium, quod non est notitia et species partis, sed totius; et ideo notitiam facit de toto."

[96] Albertus Magnus, *De anima*, II, tr. 3, cap. 4 (ed. Stroick, 1968b, 102: 53–60): "Similiter autem est, quando dicitur, quod abstractio fit a materia, intelligitur, quod materia est quoddam particulare quod particularitatem suam habet a materia. Hoc enim est primum subiectum formae, et per ipsam substat particulare et naturae communi, quae abstrahitur universaliter ab intellectu, et individuantibus formis, quarum intentiones ab aliis gradibus apprehensionis abstrahuntur."

commentary on Aristotle's *Physics*.[97] In this text, Albert talks of a *suppositum*, rather than of a certain particular. In his commentary on *Physics*, I, treatise 1, chapter 6,[98] Albert again explains two different meanings of the term "particular" (*particulare*):

> But it is necessary to know that the particular is considered in two different manners that are in it. There is the perfection of being in the particular according to nature, and it is this nature that is called the being of the species in the individual, that is, according to Boethius, that the species is the totality of the being of individuals. And also, in the particular there is a subject [*suppositum*] that underlies [*substat*] this nature, which is the being of this subject itself. [trans. P. H. R.][99]

As Albert spells out, "particular" denotes two different things. On the one hand, a particular is always a particular thing: a particular horse, a particular chair, or a particular human being, with each particular thing in the world always being a perfect instantiation of a species. On the other hand, a particular can also be understood as a *suppositum*. Albert's description of the *suppositum* overlaps with his description of "matter understood as a certain particular" in *De anima*. The *suppositum*, which is a grammatical technical term, can be comprehended as a "naked particular," i.e., a particular subject without any property of which all kinds of properties—both substantial and accidental ones—are predicable. In this sense, the *suppositum* is an individual in itself. A look at the context of the *Physics* passage quoted above will help better grasp what Albert means by *suppositum*.

In this section of the *Physics*, Albert sets out to explain the meaning of Aristotle's assertion that universal concepts are closest to sense perception in the physical sciences. What is perceived in the first place is the most general concept. This means that, for instance, we first see that a given individual is an individual substance; next, we are able to discern that it is a living being; then, that it is an animal; and, finally, that it is a sheep. Aristotle famously illustrates this by pointing out that children call all males fathers and all females mothers before they are able to distinguish between the respective concepts. According to Albert, a child in the Aristotelian example calls all men fathers, because it is first acquainted with the most general concept and learns to discriminate only later. Knowledge in the physical sciences works in the same way. Mixed concepts are more familiar to us, and general concepts have a wider extension than concrete concepts, so in a confused situation it is easier to obtain a more general concept first. Thus, according to Albert, the child example shows that the assertion that the universal is closer to us than the particular in the

---

[97] Albertus Magnus, *De anima*, II, tr. 3, cap. 4 (ed. Stroick, 1968b, 102: 60–63): "Sicut autem in principio *Physicorum* dictum est, communis natura aliquando accipitur ut confusa in particularibus et non separata ab eis."

[98] Albertus Magnus, *Physica*, I, tr. 1, cap. 6 (ed. Hossfeld, 1993, 11–12).

[99] Albertus Magnus, *Physica*, I, tr. 1, cap. 6 (ed. Hossfeld, 1993, 12: 67–74): "Sed oportet scire, quod particulare dupliciter consideratur secundum duo, quae sunt in ipso. Est enim in particulari perfectio esse secundum naturam, et haec est natura quae vocatur esse speciei in individuo, secundum quod dicit Boethius, quod species est totum esse individuorum. Et est in particulari suppositum, quod substat huic naturae quae est esse eiusdem suppositi."

physical sciences really means that concepts (*rationes*) are intermingled, and we cannot disentangle them. Nevertheless, at first sight, it seems strange that sense perception alone can grasp concepts at all. For this reason, Albert succinctly explains how sense perception works and how it is related to general and particular concepts. Specifically, our sensitive apprehension is threefold and takes place, first, according to the particular senses alone, secondly, according to the particular and common senses together, thirdly and finally, according to the particular and common senses together combined with "some cognition mixed with concepts in the senses or cognition that is in the place of reason, which some call estimation, and which is a part of the sensitive soul."[100] In his depiction of the third kind of apprehension, Albert elucidates:

> Thus, the apprehension of concepts—or estimations in the case of beasts—, mingled with sensations, is above the nature of things in which accidents are, which are sensations of each proper sense, and whose magnitude is the subject of this proper sensation, which is apprehended by the common sense. Through this [apprehension of concepts], the child grasps that a human male is a father and not a donkey, and the lamb, [through estimations], apprehends that a sheep is its mother and is not a wolf. [trans. P. H. R.][101]

Albert augments the Aristotelian example of the child with his favourite characters, namely the sheep and the wolf. Children and lambs apprehend proper and common sensations in the same way. They both apprehend something "above the nature of things," but children obtain concepts (*rationes*) from reason, while beasts obtain estimations from the estimation,[102] though the passage admittedly does not specify either where these estimations come from and or how they reach the estimative power.

Nevertheless, Albert spells out in *De homine* and in *De anima* that the estimative power is analogous to the practical intellect and the phantasy is analogous to the theoretical intellect.[103] In *De homine*, Albert observes that "the grasped intentions from the sensible things that are not perceived by the senses are apprehended in two ways." One of them is through universal reason, by which manner experience and the intellective powers extract them from sensible things. The other way, Albert

[100] Albertus Magnus, *Physica*, I, tr. 1, cap. 6 (ed. Hossfeld, 1993, 11: 54–59).

[101] Albertus Magnus, *Physica*, I, tr. 1, cap. 6 (ed. Hossfeld, 1993, 11: 65–71): "Acceptio autem rationis permixtae sensibus aut aestimationis in brutis est super rei naturam, in qua sunt accidentia, quae sunt sensata propriorum sensuum, et cuius est magnitudo subiecta sensibilibus propriis, quae accipitur per sensum communem. Et per illam accipitur puer, quod vir homo est pater et non asinus, et agnus accipit, quod ovis est mater et non lupus."

[102] It seems that Albert's usage of the terms "estimation" and "estimations" in the *Physica* is broader and different than in *De homine* and *De anima*, with "estimations" probably meaning "the grasped intentions from the sensible things that are not apprehended by the senses," which is discussed in the next paragraph.

[103] Cf. Albertus Magnus, *De homine, De anima sensibili, Partes apprehensivas deintus: aestimativa* (ed. Anzulewicz, Söder, 2008, 295: 70–73); Albertus Magnus, *De anima* III, tr. 1, cap. 2 (ed. Stroik, 1968b, 167: 58–63). Walter Burley has a non-identical but similar treatment of estimative power. For more information on Burley's treatment, see Mansfeld, Chap. 12 in this volume, pp. 229–251

explicates, is that the phantasy operates to extract (*elicere*)[104] true and false intentions (*intentiones veri et falsi*) from that which has been perceived by the senses, and the estimative power works to establish (*determinare*) what is noxious and what is beneficial in them. Ten years later, in *De anima*, Albert offers a very similar explanation.[105] Therefore Jörg A. Tellkamp certainly has a point concluding that estimations, which play a practical role in non-rational animals, are produced by the sentient soul itself.[106] In any case, it is clear that even non-rational animals receive and apprehend certain non-sensible properties, which are mixed with mere sense perceptions.

All of this means that both rational and non-rational animals apprehend sense-perceptual object *and* something "above the nature of things." This something is the meaning (*significatio*) of the perceived individual thing, that is to say, what is predicable of it. Human beings can apprehend that this thing is a substance, or that it is a man, or the son of Dion, while non-rational animals can apprehend that this thing is hostile, or that it is its own offspring. This entails that the species is the *forma post rem*, a universal, or, possibly, a concept. Human beings can grasp it because they are endowed with the intellect. Yet, in order to grasp the world meaningfully, human beings also need to apprehend a "naked" individual, i.e., a *suppositum* of which properties can be predicated. Our perception of the physical world lies in our capacity to attach predicated universals to naked particular subjects. Because beasts and human beings have the same apparatus, animals can do the same, though with their estimation rather than with their intellect.

8.7 Conclusion

Albert the Great explains how sentient beings are able to apprehend particular things in a meaningful way. "A meaningful way" means that in perceiving a thing, a sentient being can perceive what this particular thing is. We perceive horses, apples, cats, and chairs, and we know how to deal with each of them because we recognize what kind of things they are. At the same time, we believe, as Albert, Avicenna, and Averroes did, that non-rational animals cannot grasp things as

104 Albertus Magnus, *De homine, De anima sensibili, Partes apprehensivae deintus: imaginatio* (ed. Anzulewicz, Söder, 2008, 284: 29–34). "Secundum propriissimum autem modo accipiendi est imaginatio virtus tenens imagines sensibilium re non praesente, absque eo quod eliciat ex eis componendo vel dividendo intentiones aliquas non acceptas per sensum; hoc enim est phantasiae secundum Avicenna et Algazelem."

105 Albertus Magnus, *De anima* II, tr. 4, cap. 7 (ed. Stroik, 1968b, 157: 53–64): "Cum autem in nobis experiamur cognitionem esse intentionum elicitarum ex sensibilibus formis, oportet esse aliquid quod eliciat et agat illas intentiones; et illius erit quasi potentia activa agens intentiones illas ex sensibilibus, cuius perfectio in complexione sui organi erit calidum spirituale et formale valde. Reservatorium ergo formarum sensatarum vocaverunt Peripatetici imaginationem, et quibusdam placuit, quod vocaretur virtus formalis, eo quod servat apud nos formas; activum autem et elicitivum intentionum vocaverunt aestimativam."

106 Tellkamp (2016, 116).

complex as a universal intention, i.e., a concept. Nevertheless, they are able to interact successfully with a world full of individual things.

Averroes's solution to the problem of how we perceive the world in a meaningful way is that human beings are capable of directly perceiving individual intentions, which are the sensitive modes of being of individual forms. For example, when I see a horse, I can see several accidental properties attached to the individual "horseness," which makes a horse a horse. Averroes's position is straightforwardly Aristotelian, at least in the sense of the *Categories*, since each accidental property is necessarily attached to an individual substance. While such a framework helps Averroes support his own theory of intellectual abstraction, his solution is untenable in physiological terms because he fails to explain how the human sense organs apprehend individual substances, which are by definition non-sensible.

In contrast to Averroes, Albert's solution, which is indebted to Avicenna's inner-sense theory, retains a strong naturalistic view of sense-perception, because all its processes can be accounted for in terms of the physics of the sublunary world. Additionally, like Avicenna, Albert believes that non-rational animals can perceive individual things in a meaningful way, not as instantiations of general kinds, but rather under Albert's scheme of "connotational attributes," to use Dag N. Hasse's coinage for unsensed intentions.[107] Nevertheless, Albert's solution to the problem of individual perception starkly differs from Avicenna's. It is unclear what prevented Albert from following Avicenna's theory of sense perception.

How Albert's solution diverges from those offered by Averroes and Avicenna deserves a closer look. In contrast to Avicenna and Averroes, Albert does not make any distinction between meaningful mental contents and merely sensitive contents, and, unlike Averroes, he does not separate substantial properties from accidental ones at the sense-perception level. Instead, he differentiates between the properties of a thing and its individuality in itself. Consequently, the role of the sensitive apparatus is limited to the apprehension of the individuality in itself, and the individuality is always grasped as a potential subject of the properties of the apprehended thing. This subject is, of course, the *suppositum*. The sensorial apparatus, which is identical in rational and non-rational animals, apprehends the same kind of mental object: a *suppositum*. The difference between their respective sensorial apparatuses lies in whether they stand in a relationship to the estimative power or to the intellect. The *suppositum* is nothing new in Albert's theory of sense perception, as a similar notion appears in his theory of common sensation in *De homine*.[108] However, the use of the third way of *per se* predication in the explanation of how the inner sensation grasps the *suppositum* is a complete novelty.

---

[107] Hasse calls the unsensed intentions (*ma'na*) of Avicenna's *Liber de anima* connotational attributes because they are properties of the thing that can move the appetitive faculties thanks to their appetitive connotations, such as enmity, friendliness, etc. For more details, see Hasse (2000).

[108] See Albertus Magnus, *De homine* (ed. Anzulewicz, 2008, 267, 281). Henryk Anzulewicz says about abstraction in this section of *De homine*: "Der Intellekt abstrahiert vom Sinnenhaften das Allgemeine nicht wie von der Materie, sondern wie vom konkreten Ding, genauerhin von dem formalen Sinnesgehalt des Einzelnen, welcher vom Vermögen der sinnenhaften Seele als das der Materie Anhängende aufgenommen wurde." Anzulewicz (2002, 222).

References

Primary Literature

Alberti Magni Opera Omnia. 2011. Editio Digitalis. Aschendorff Verlag. https://www.albertus-magnus-online.de/#. Accessed 14 Dec 2019.

Alberti Magni e-corpus. 2008–2018. Bruno Tremblay. http://www.albertusmagnus.uwaterloo.ca/. Accessed 14 Dec 2019.

Albertus Magnus. 1890. *Libri 2 de demonstratione, id est, posteriorum analyticorum*. In *Albertus Magnus opera omnia 2*, ed. August Borgnet. Paris: Vivès.

———. 1891. *De somno et vigilia, libri tres*. In *Alberti Magni opera omnia 9*, ed. August Borgnet. Paris: Vivès.

———. 1955. *Quaestiones super de animalibus*. In *Alberti Magni opera omnia 12*, ed. Ephrem Filthaut. Münster: Aschendorff.

———. 1968a. *Super Ethica commentum et quaestiones*. In *Alberti Magni opera omnia 14*, ed. Wilhelm Kübel. Münster: Aschendorff.

———. 1968b. *De anima*. In *Alberti Magni opera omnia 7/1*, ed. Clemens Stroick. Münster: Aschendorff.

———. 1987. Super Ethica, Commentum et Quaestiones. In Alberti Magni opera omnia 14, ed. Bernhard Geyer. Münster: Aschendorff.

———. 1993. *Physica I–IV*. In *Alberti Magni opera omnia* 4/1 ed. Paul Hossfeld. Münster: Aschendorff.

———. 2003. *De quinque universalibus*. In *Alberti Magni opera omnia*, 1/1_1A, ed. Manuel Santos Noya. Münster: Aschendorff.

———. 2006. In *De sex principiis. In Alberti Magni opera omnia 1/1–2*, ed. Ruth Meyer. Münster: Aschendorff.

———. 2008. *De homine*. In *Alberti Magni Opera Omnia Tomus 27/2*, ed. Henryk Anzulewicz and Joachim Roland Söder. Münster: Aschendorff.

———. 2013. De praedicamentis. In Alberti Magni opera omnia 1, 1B, eds. Manuel Santos Noya et Silvia Donati. Münster: Aschendorff.

Albert the Great. 2008. *Questiones Concerning Aristotle's on Animals*. Trans. Irven M. Resnik, Kenneth F. Kitchell. Washington, DC: The Catholic University of America Press.

Aristote. 1969. Histoire des animaux. Tome III: Livres VIII–X. ed. Pierre Louis. Paris: Les Belles Lettres.

Aristoteles. 1961. In *De anima*, ed. David Ross. Oxford: Typographeo Clarendoniano.

———. 1991. *Complete works (Aristotle)*, ed. Jonathan Barnes. Princenton: Princenton University Press.

Averroes. 1953. *Commentarium Magnum in Aristotelis de Anima Libros*, ed. In *Stuart Crawford*. Cambridge, MA: Medieval Academy of America.

Averroes (Ibn Rushd) of Cordoba. 2009. *Long Commentary on* De anima *of Aristotle*. Trans. and notes Richard C. Taylor. New Haven/London: Yale University Press.

Avicenna. 1968. *Liber de anima seu sextus de naturalibus IV–VI*, ed. In *Simone Van Riet*. Brill: Louvain–Leiden.

———. 1972. *Liber de anima seu sextus de naturalibus I, II, III*, ed. Simone Van Riet. Louvain/ Leiden: Brill.

Némésius d'Émèse. 1975. *De natura hominis. Traduction de Burgundio de Pise*, ed. Gérard Verbeke, José Rafael Moncho, (*Corpus Latinum commentariorum in Aristotelem Graecorum* Suppl. 1). Leiden: Brill.

Secondary Literature

Anzulewicz, Henryk. 2002. Konzeptionen und Perspektiven der Sinneswahrnehmung im System Alberts des Grossen. *Micrologus* 10: 199–238.

Black, Deborah. 1991. Estimation (Wahm) in Avicenna: The Logical and Psychological Dimensions. *Dialogue* 32: 219–258.

———. 1996. Memory, Individuals, and the Past in Averroes' Psychology. *Medieval Philosophy and Theology* 5: 161–187.

———. 2000. Imagination and Estimation: Arabic Paradigms and Western Transformation. *Topoi* 19: 59–75.

Cashdollar, Stanford. 1973. Aristotle's Account of Incidental Perception. *Phronesis* 18 (2): 156–175.

Hasse, Dag N. 2000. *Avicenna's De Anima in the Latin West: The Formation of a Peripatetic Philosophy*. In *London: The Warburg Institute*. Turin: Nino Argano Editore.

Kaukua, Jari. 2007. *Avicenna on Subjectivity, a Philosophical Study*. Helsinki: University of Jyväsklä.

Klubertanz, George P. 1952. *The Discursive Power, Sources and Doctrine of the "Vis Cogitativa" According to St. Thomas Aquinas*. Carthagena: The Modern Schoolman.

López-Farjeat, Luis Xavier, Jorge F. Morales, and Ladrón de Guevara. 2010. El contenido cognitivo de la percepción: Avicena y McDowell. *Thémata* 43: 251–270.

Palanciuc, Chase. 2005. Némésius d'Émèse. In *Dictionarie des philosophes antiques*, ed. Richard Goulet, vol. 4, 625–654. Paris: CNRS Éditions.

Piché, David. 2013. La notion de forma totius chez Albert le Grand, ses contemporains et ses sources. In *Ad notitiam ignoti. L' 'Organon' dans la 'translatio studiorum' à l'époque d'Albert le Grand*, ed. Brumberg-Chaumont, 417–445. Turnhout: Brepols.

Sorabji, Richard. 1991. From Aristotle to Brentano: The Development of the Concept of Intentionality. In *Oxford Studies in Ancient Philosophy. Supplementary Volume: Aristotle and the Later Tradition*, ed. Henry Blumenthal and Howard Robinson, 227–259. Oxford: Oxford University Press.

Stenek, Nicholas H. 1974. Albert the Great on the Classification and Localization of the Internal Senses. *Isis* 65: 193–211.

Taylor, Richard C. 2000. Cogitatio, Cogitativus, and Cogitare: Remarks on the Cogitative Power in Averroes. In *L'élaboration du vocabulaire philosophique au Moyen Age: Actes du Colloque international de Louvain-la-Neuve et Leuven, 12–14 septembre 1998, organisé par la Société Internationale pour l'Étude de la Philosophie Médiévale*, ed. Jacqueline Hamesse and Carlos Steel, 111–146. Louvain: Brepols.

Tellkamp, Jörg A. 2013. Albert the Great on Perception and Non-Conceptual Content. In *Philosophical Psychology in Arabic Thought and the Latin Aristotelianism of the 13th Century*, ed. Luis Xavier López-Farjeat and Jörg A. Tellkamp, 205–221. Paris: Vrin.

———. 2016. Aping Logic? Albert the Great on Animal Mind and Action. In *Subjectivity and Selfhood in Medieval and Early Modern Philosophy*, ed. Jari Kaukua and Tomas Ekenberg, 109–123. Cham: Springer.

Wirmer, David. 2004. Der Begriff der Intention und seine erkenntnistheoretische Funktion in den De-anima-Kommentaren des Averroes. In *Erkenntnis und Wissenschaft, Probleme der Epistemologie in der Philosophie des Mittelalters*, ed. Mathias Lutz-Bachmann, Alexander Fidora, and Pia Antolic-Piper, 35–67. Berlin: De Gruyter.

Wolfson, Harry A. 1935. The Internal Senses in Latin, Arabic and Hebrew Philosophic Texts. *The Harvard Theological Review* 28 (2): 69–133.

Rocca, Julius. 2003. Galen on the Brain, Anatomical Knowledge and Physiological Speculation in the Second Century AD. In Leiden, Boston: Brill.

Chapter 9
Death, the Intellect and the Resurrection of the Dog: Geoffrey of Aspall's *Questions on the De Longitudine et Brevitate Vitae*

Michael W. Dunne

Abstract Geoffrey of Aspall was active in Oxford in the 1250s and may even have been a student of Adam of Buckfield. His scholarly activity ended in 1263 when he became a financial clerk at the court of Edward I until his death in 1287. This article examines Aspall's questions to *De longitudine et brevitate vitae* and specifically his comments in question 2 on whether the intellectual and sensitive souls can come to an end. (A list of the questions of Aspall's commentary and the text of question 2 are appended.) Of particular interest is Aspall's discussion of whether the vegetative and sensitive life of a dog, and even its body, could be educed from common matter through the action of an external agent. If that were the case, then a dead dog could live again, not as a resurrection of the dog, but rather a 'return' of the dog to the same species, yet not a 'return' of the individual. This discussion, found also in Aspall's questions to *De generatione et corruptione* (regarding Aristotle's views at 338b 1417) is of interest, not least because the general thesis is condemned in 1277.

Keywords Geoffrey of Aspall · Grosseteste · Life and death · Resurrection and return · First and last instants

9.1 Introduction

Geoffrey of Aspall was, up to comparatively recently, a rather unknown and somewhat neglected English author of the mid-thirteenth century. This, however, has been rectified by a flurry of publications and editions over the last ten years,[1]

[1] Galfridus de Aspall, *Quaestiones in De anima* (ed. Čizmić, 2010); Donati (2012, 2013); Galfridus de Aspall, *Quaestiones super librum in "De somno"* (ed. Ebbesen, 2014); Geoffrey of Aspall, *Questions on Whether the World in Generable and Perishable* (ed. Galle, 2012); Trifogli (2018).

M. W. Dunne (✉)
Maynooth University, Maynooth, Ireland
e-mail: michael.w.dunne@mu.ie

M. Gensler et al. (eds.), *The Embodied Soul*, Historical-Analytical Studies on Nature, Mind and Action 11, https://doi.org/10.1007/978-3-030-99453-2_9

especially with the recent edition by Silvia Donati and Cecilia Trifogli of his *Questions on the Physics*[2] which now surely becomes the point of reference for future editions and studies.

We can summarise what is known of Aspall's life and work as follows.[3] Geoffrey takes his name from the village of Aspall in Sussex which is nowadays best known in England as the origin of a popular brand of cider with the same name. In a letter of 13 December 1286, Archbishop John Pecham refers to Aspall as 'magister' but also 'conscholaris' from which, if we take it that Pecham was active in Oxford in the 1250s, we can infer that Aspall was also active in the same decade. Aspall's scholarly activity and production ended in 1263 when he began a busy career as financial clerk at the court of Edward I which lasted until his death on 11 June 1287. Geoffrey may have been a student of Adam of Buckfield (master by 1243) but the earliest reference to Aspall as 'master' is from 1262.[4] Although not such a great administrator of the finances of Queen Eleanor, Geoffrey was more successful in managing his own wealth, holding a plurality of benefices, so much so that the King had to write to the bishop of Chichester to defend him and to command that Aspall be not forced to take orders. At his death, he held six prebends, twelve rectories and had managed to become Archdeacon of Dublin.

The works of Aspall present more than the usual difficulties for the editor (to which I will return below). Questions of authorship, revisions, manuscript traditions abound. Do we have one work with one author or more than one work, the production of more than one author, revisions by the author or by someone else?

Aspall is regarded as having been part of the third and last phase of the reception of Aristotle at Oxford.[5] According to Callus, the first stage is represented by the work of John Blund and the second by the literal commentaries of Adam of Buckfield. The third phase is that of commentaries *per modum quaestionis* where there is no exposition of the text of Aristotle but rather questions are addressed which arise out of the text or are prompted by it. Sometimes these questions have little or nothing to do with the text at all. It may be the case that Aspall also composed literal commentaries and we can perhaps see remnants of this in our text (see below).

As regards the structure of the *Quaestiones*, I quote from the editors of the *Questions on the Physics*:

---

[2] Geoffrey of Aspall, *Questions on Aristotle's Physics* (tr. Ashworth, Trifogli, 2017).

[3] See Introduction to the *Questions on the Physics* (ed. Donati, Trifogli, 2017, xi–xiv).

[4] On the dating of Aspall's professional activity, see Geoffrey of Aspall *Questions on Aristotle's Physics* (ed. Donati, Trifogli, 2017, xi).

[5] For a general survey about the three phases of the reception of Aristotle's thought, see Gensler, Mansfeld, Michałowska, Chap. 1 in this volume, pp. 1–15; on Aspall, 00.

> The general structure is the standard one: the problem under discussion is stated in the title of the question; its actual discussion in the body of the question begins with the argument in favour of the view that is ultimately to be rejected; these arguments are then followed by arguments in favour of the contrary view (*ad oppositum*). Then comes the *solutio*, which contains Aspall's own reply to the question, followed by the reply to the arguments in favour of the contrary view. Within this general structure, however, there are many variations both in the specific form that an individual question may take as well as its length. One important general characteristic of Aspall's questions concerns the *solutio*: while in later commentaries *per modum quaestionis* the *solutio* is the *locus classicus* where the commentator expands and argues for his own view, this is not so in Aspall's questions, where the *solutio* is almost never given at great length, but is either stated in one sentence or may simply be implied by the replies to the initial arguments. The crucial elements of Aspall's view on a given problem are then to be found in his replies to the initial arguments, in which he addresses the objections to his own views. As a result, Aspall's treatment of the problem raised in the question is fragmentary and the difficult task of reconstructing his view from the scattered evidence is left to the reader.[6]

My focus is upon Aspall's *Questions on De longitudine et brevitate vitae (De morte et vita)* but we find here the same approach identified by the editors of the *Questions on the Physics* as can be seen below in the examination of Aspall's question on whether the intellectual and sensitive souls can come to an end and in what manner.

The overall text of the *Questions on the De longitudine et brevitate vitae* is very extensive both in number of words and in the number of issues addressed, as can be seen from the comparative list of questions in Appendix 1. Some brief comments in this regard should suffice here. Firstly, as expected, one can see from the headings in the Table that there is a pronounced disparity between the versions transmitted by the manuscripts. Secondly, although Aspall himself refers to numbered sections in other texts of his, the numbering in the manuscripts when present seems somewhat haphazard and they do not match up with each other. Finally, the latter part of the text preserves elements of a perhaps earlier exposition of the text *per modum commenti* which may owe something to the tradition of Adam of Buckfield; whereas the earlier part of the text is *per modum quaestionis*, perhaps a partial reworking of an earlier text which may have also been closer to a simple literal commentary.

---

[6] Geoffrey of Aspall, *Questions on Aristotle's Physics* (ed. Donati, Trifogli, 2017, xiii–xiv).

9.2 Manuscripts

Aspall's *Questions on De longitudine et brevitate vitae* (*De morte et vita*) are transmitted by three manuscripts which also transmit other works of Aspall including the *Questions on the Physics*. These are:

C = Cambridge, Gonville and Caius College, MS 509/386, ff. 276ra–286rb;
O = Oxford, Merton College, MS 272/304, ff. 282rb–294vb (incomplete);
T = Todi, Biblioteca Communale, MS 23 (163), 84ra–93va.[7]

Again, to quote the editors of the *Questions on the Physics*:[8]

> Each of the three manuscripts has a very high number of isolated readings that are not errors but variants, that is, they all result in acceptable texts, which often transmit what is essentially the same conceptual content but formulated in slightly different ways.

As already pointed out above, something of this can be seen from the Table of Questions given below in Appendix 1. As is clear from this Table and from the comparison given below for a section of the text, C and T are closer to each other than O is to either. Were more than one series of lectures given? With regard to the *Questions on the Physics* Donati and Trifogli argue for one, holding that all three versions arise from a common source and ultimately from two *reportationes*. The detailed examination they give of the various manuscript traditions[9] lead one to imagine that similar questions will have to be solved for a future edition of Aspall's *Questions on De longitudine et brevitate vitae* but the general approach and some of the difficulties involved are already apparent.

If we take two examples from the short extract of the text in Appendix 2 and compare them with MS O we can see an immediate divergence and convergence with C:

[7] A full description of the manuscripts is to be found in Geoffrey of Aspall, *Questions on Aristotle's Physics* (ed. Donati, Trifogli, 2017, xx–xxii).

[8] Geoffrey of Aspall, *Questions on Aristotle's Physics* (ed. Donati, Trifogli, 2017, xxii).

[9] Geoffrey of Aspall, *Questions on Aristotle's Physics* (ed. Donati, Trifogli, 2017, xxv–xxxv).

Oxford, Merton College, 272/304	Todi, Biblioteca Communale, 23	Cambridge, Gonville and Caius College, 509/386
(f. 286vb) 2.2. Item, omne quod corrumpitur corrumpitur in aliquod diminutum ex quo idem potest per agens educi. Si ergo vegetativa et sensitiva possunt corrumpi, corrumpentur in alia principia diminuta et eis iterum per agens possunt produci in actum quod non videmus accidere. 2.3. Item, corporeum et incorporeum quae sunt differentiae specificae, si dicamus ipsas corrumpi, hoc est per reductionem eorum ad eandem potentiam generis communem quae iterum potest per intellectum cedere in utrumque eorum. 2.4. Ergo a simili, si corpus et anima quae sunt formae eis corrumpentes in materia corrumpantur, reducentur ad eandem potentiam materiae ex qua iterum possent educi. Et ex hoc sequitur ex corpore possit fieri anima et e contrario – quod est impossibile.	(f. 85ra) 2.2. Praeterea, omne quod corrumpitur corrumpitur in aliquod <diminutum>; si igitur sensitiva et vegetativa corrumpuntur, corrumpuntur in aliud corpus quod est diminutum respectu illorum. Si hoc est verum, tunc vegetativa et sensitiva per agens extra possent educi ex aliquo diminuto, ergo vegetativa et sensitiva canis et etiam corpus peragens extra posset educi ex illo diminuto; tunc canis mortuus iterum viveret, quod falsum est in generantem post propagationem. 2.3. Hic tamen habetur dubium de substantia corporea et incorporea quae sunt in eadem potentia generantis remota, igitur primum. 2.4. Praeterea, sicut est in corruptione et resolutione <secundum rationem sic est in compositione et resolutione> secundum rem, sed ita est in compositione et resolutione secundum rationem quod, si corrumperentur diversae species, cederent in eandem potentiam communem generantis, quae iterum per rationem ex illa potentia communi possent educi per additionem duarum specialium illi potentiae communi. Sic igitur erit in re. Cum igitur sensitiva et vegetativa contineantur sub substantia incorporea et corpus sub substantia corporea, si corrumpantur, cedunt in potentiam generalem remotam quare iterum ex illa potentia remota communi utriusque, scilicet animae et corpori, potest iterum educi utrumque. Et etiam sequitur quod ex corpore possit educi anima et e contrario—ex quo illa potentia communis est istis.	(f. 277ra) 2.2. Praeterea, omne quod corrumpitur corrumpitur in aliquid; si igitur sensitiva et vegetativa corrumpuntur, corrumpuntur in aliquod corpus quod est diminutum respectu istorum. Sed si hoc est verum, tunc vegetativa et sensitiva per agens extra possent educi ex aliquo diminuto, <ergo vegetativa et sensitiva canis et etiam corpus peragens extra posset educi ex illo diminuto>; (f. 85rb) tunc canis mortuus (?) iterum viveret, quod falsum est in generantem post propagationem. 2.3. Hic tamen habet dubium de substantia corporea et incorporea quae sunt in eadem potentia generantis remota, ergo etc. 2.4. Praeterea, sicut est in corruptione et resolutione secundum rationem sic est in compositione et resolutione secundum rem, sed ita in compositione et resolutione secundum rationem quod, si corrumperentur diversae species generantis, cederent in eandem potentiam communem generantis, quae quidem iterum per rationem ex illa potentia communi possent educi per additionem duarum specialium illi potentiae communi. Sic igitur erit in re. Cum igitur vegetativa et sensitiva contineantur sub substantia incorporea et corpus sub substantia corporea. Si corrumpantur cedunt in potentiam generalem remotam quare iterum ex ista potentia communi remota utriusque scilicet animae et corpori potest iterum educi utrumque. Et etiam sequitur quod ex corpore posset educi anima et e contrario—ex quo illa potentia communis est istis.

(continued)

Ad primam quaestionem dico quod vivens vita intellectiva mori potest et anima intellectiva separari potest secundum quod dicit Aristoteles quod intellectiva separatur a corpore sicut incorruptibile a corruptibili.	Dicendum ad primum quod aliquid potest mori et etiam vivens vita intellectiva potest mori, unde intellectiva potest mori.	Dicendum quod aliquod potest mori et etiam vivens vita intellectiva potest mori, unde intellectiva potest separari.

It would seem that we have two versions which are not reducible one to the other but which deal with the same general ideas. The resolution of this matter must be left, however, to a future study. At present, and from the short text presented below, it would seem that the version in C and T is a later and more refined version of O. On the other hand, version CT never received a final edited form and like O presents elements of an earlier commentary by means of exposition of the text with features of a commentary by way of questions.

9.3 Authenticity

Much work has already been done to establish which works attributed to Aspall appear genuine.[10] The editors of the *Questions on the Physics* list the following commentaries of Aspall on the works of Aristotle: *Physics*, *De caelo et mundo*, *De generatione et corruptione*, *De anima*, *De sensu et sensato*, *De memoria et reminiscentia*, *De somno et vigilia*, *De longitudine et brevitate vitae*, and the *Metaphysics*. Aspall may have also commented on the *Meteors* and the *De vegetabilibus*.[11]

In addition to the fact that the commentary on the *De longitudine* is to be found in collections of other of Aspall's works, we have the attribution in the hand of the scribe of T (f. 286rb): "De Aspale quaestiones explicant." Another argument in favour of the text's authenticity is the fact that Aspall in his *Questions on De caelo* refers to his *Questions on De morte et vita*. In the former, in his question on the generability and perishability of the world, Aspall refers to a discussion in his question 9 of *De morte et vita* (Galle, 328; 352):[12]

[10] See Geoffrey of Aspall, *Questions on Aristotle's Physics* (ed. Donati, Trifogli, 2017, xi).

[11] Part of the commentary on the *Meteors* may survive, see Geoffrey of Aspall, *Questions on Aristotle's Physics* (ed. Donati, Trifogli, 2017, note 12). I have yet to examine the matter in detail but I wonder if the later sections of the commentary on the *De longitudine* dealing with the life of plants might have given rise to the suggestion of a commentary on the *De vegetabilibus*?

[12] See Geoffrey of Aspall, *Questions on Whether the World in Generable and Perishable* (ed. Galle, 2012, 328, 352).

> Dubium tamen est utrum aliqua creatura de se possit in pure nihil cedere. Sed de hoc sufficienter dictum est super librum *De morte et vita*, quaestione 9ª.

This question is to be found in the present text as the question "An aliqua creatura de sui natura possit cedere in non esse."

To conclude, at the present state of our knowledge the text dealt with here can be ascribed with reasonable certainty to Geoffrey of Aspall.[13]

9.4 On the Death of the Intellectual and Sensitive Souls

Geoffrey begins his commentary with asking such questions as what life is, what death is, etc. The initial arrangement of topics which he makes seems to have caused some difficulty for the reporters who were trying to follow the order Aspall gave, as can be seen from the Table of Questions. However, it is clear that the question of life is addressed first before moving on to the question of death. In any case, the question which interests us here is the fourth and is to be found between the third "An aliqua vita sit longa vel brevis" and the fifth "Utrum aliquod mixtum possit dissolvi."

The text of Question 4 begins with a straightforward definition of death: "Death is the privation of life in him who was born to have it." (It may be Aspall's original definition as I have not found it elsewhere so far. It could be noted here that "mors est privatio vitae" is found in both Augustine and Aquinas but not the second phrase.). Then there are two articles: (1) Whether something living through the intellectual life can die as such?; (2) Whether something living by the vegetative and sensitive life can die? Six arguments are proposed in the first article and four in the second. These are responded to and then a final argument introduces the topic of the elements mixed in a composite body, which will be the subject of the following question.

It is worth pointing out here that the structuring of the arguments is somewhat uneven and perhaps not even very well-developed at times, leading to difficulties in interpretation not least because the replies to the arguments sometimes only have a tenuous connection with them. On the other hand, the author comes across as not slavishly tied to the text and engaged in a genuine philosophical reflection on the points raised.

---

[13] On the other hand, one noticeable stylistic absence here is the phenomenon documented by S. Ebbesen in his edition of the many exhortations to the reader in the text, e.g., 'stude!'. See Galfridus de Aspall, *Quaestiones super librum in "De somno"* (ed. Ebbesen, 2014, 259) where he notes that this is to be found in other places in Aspall's works. It may be the case that a close study of all of the manuscripts of the *De longitudine* will yield similar examples, but I have come across none so far.

I remain somewhat unconvinced about the early date assigned to these texts by some which would place them in the 1250s.[14] If this were the case, then these texts would be the earlier and yet better developed of the commentaries *per modum quaestionis* which appear, generally speaking, a decade or two later. I, on the other hand, would tend to agree with Ebbesen who would place Aspall's activity somewhat later, Ebbesen suggests around 1260–1265, and also notes the surprising lack of any influence of Albert the Great which would also seem to be the case here.[15]

Turning to the first article, regarding intellectual life, the first argument reveals the interests of someone who commented on the *Physics* since it focuses on the role of the instant.[16] Geoffrey begins: If a living being with an intellective life can die, then the intellectual life can be separated from the body and its existence will be permanent outside of the body. Therefore, a) we have to assume that there is a first instant of its existence outside of the body and, b) there must be a last instant of its existence in the body. A problem arises as to whether it is the same instant or two separate ones and Aspall develops his argument by explaining why both parts of this alternative are false.

Aspall constructs the argument against the death of the life of the intellect on the basis of his understanding of the first and the last instants. Can the intellect be separated from the body and have a permanent existence outside of the body? At the moment of separation there would be the first instant of its separate existence. Thus, it must be assumed that there is the final instant of its existence in the body. Now are these one instant or two? If this is one and the same instant, then it could be concluded that the intellect would be both in the body and out of the body at the same time. If there are two distinct instants, then another problem arises since there must be an interval between the two instants; thus, an instant in which the soul would be neither in the body nor out of it. Since one can see that both of these positions are false, it could be concluded that it is impossible for the intellect to be separated from the body and so it remains perpetually in a body. We would then presumably have the absurdity that humans could not die (they do) or that the intellect moves from body to body as in the case of reincarnation (of which he drops a hint at the end of the next argument). Aspall, however, does not elaborate on these difficulties.

In the third argument a comparison is made between the movers of the heavenly bodies and the intellectual soul as the mover of the body. No separate substance is

[14] See the references given by Donati and Trifogli in Geoffrey of Aspall, *Questions on Aristotle's Physics* (ed. Donati, Trifogli, 2017, xi).

[15] Geoffrey of Aspall, *Quaestiones super librum in "De somno"* (ed. Ebbesen, 2014, 261). A similar point is made about Aspall's independent approach by Thomsen Thörnqvist in Chap. 4 in this volume, pp. 57–71.

[16] A similar theme but with a different emphasis is to be found in Aspall's Commentary on the *De somno et vigilia* (ed. Ebbesen, 2014, 274), q. I.4: *Utrum anima vegetativa possit esse sine sensitiva et intellectiva, neutra tamen sine ipsa*, arg. 1.1: "Et primo de intellectiva, quia secundum Aristotelem anima intellectiva separatur a corpore sicut incorruptibile a corruptibili, et ita patet quod ipsa manere potest sine vegetativa." There is no use of the term 'instans' in that text as far as I can tell.

idle, according to Aristotle, as each is given the task of moving some body or other. However, if the intellect were separated, then it too surely would be tasked with moving some other body and this would not happen unless it entered into another body. Thus, the intellect always enters into some body or other. So, it could be speculated that our separated human soul might move from our body and animate some other body—the corpse of a dog, for instance.

It is not my intention here to look in detail at all of the arguments (the reader can find these in the text appended in Appendix 2) but simply to point out those which might seem of greater interest. Another argument presented by Aspall (1.6) follows on from the one just looked at (1.3), namely that the end and beginning of the same thing are interrelated; but the soul at the beginning and before its infusion in the body did not have a separate existence from the body, so it will not exist after separation from the body but will always be with it. Does it then die with the death of the body? Again, the argument is not developed as seems to be Aspall's frequent practice both here and elsewhere.

Passing on to the second topic: Can sensitive life come to an end? Here we have a series of arguments concerning the death of the sensitive soul. The first argument is an unusual one perhaps owing something to Grosseteste's notion of the *lux spiritualis*.[17] Aspall says that according to (unnamed) philosophers, the sensitive life is called the *lux spiritualis* (I presume some kind of relationship to the *spiritus animales* yet there is no reference to that here). He further comments that this 'spiritual light' is related to the corporeal light in that both seem to work in much the same way and whatever we claim about one can be claimed about the other. Again, just as corporeal light is not destroyed in its passage here through its medium the same is true of the spiritual light. Aspall explains that corporeal light, if it were somehow to be destroyed, must resolve itself into the elements (which is not true, because it would have to be already composed of the elements), or resolve in nothing (which does not happen for several reasons described by Aspall), or it returns to its source.[18] Aspall regards all of these as impossible and his argument goes as follows: light could not be destroyed by the element of air since then it could be produced from the element of air, which is false because it is of the nature of a heavenly body; nor into nothing (because the nothing which is sometimes a being falls into nothingness—according to some again unnamed philosophers and perhaps anticipating some of the discussion in the later question

[17] Cf. Marone (1983, 196): "The *lux spiritualis* floods over intelligible objects and over the mind's eye (*oculus mentis*) and stands to the interior eye and to intelligible objects as the corporeal sun stands to the bodily eye and to visible corporeal objects"; Robertus Grosseteste, *Commentarius in Posteriorum Analyticorum libros* (ed. Rossi, I.17, 39–42): "Res autem dicuntur certe a comparatione quam habent ad cognitionem sive ad visum mentalem. Dico ergo quod est lux spiritualis quae superfunditur rebus intelligibilibus et oculo mentis, quae se habet ad oculum interiorem et ad res intelligibiles sicut se habet sol corporalis ad oculum corporalem et ad res corporales visibiles."

[18] MSS C and T lack this latter argument, but it is present in the version transmitted by MS O. See below.

"An aliqua creatura de sui natura possit cedere in non esse") and for this reason it exists perpetually.

The next argument is quite inventive and unique as far as I am aware, namely the case of the possible resurrection of the dog which appears in our title. The argument runs as follows: everything which is corrupted and passes away does so into a diminished mode of being, so if the sensitive and vegetative souls were corrupted they would pass away into a body or material state which is something less with respect to them. However, if this were true, then the vegetative and sensitive souls could be educed from a lesser state of being, namely from something non-living, through the action of an external agent. Therefore, the vegetative and sensitive life of a dog, and even its body, could be educed from a lesser state of common matter through the action of an external agent. If so, then a dead dog could live again. A long *dubium* follows with regard to how both the corporeal and incorporeal substance relate to a remote power of generation. He concludes that should they fall into a remote general power, then both soul and body can again be educed. It would also follow that the soul can be educed from the body and vice versa inasmuch as that power is common in both.

Aspall's last objection in the second part of the question concerns a famous dictum of Aristotle from the *De anima*: the sensitive power cannot be destroyed or removed because, if and old man could get the eye of a young man, he would see like a young man. In other words, the sensitive power remains intact behind the weakened or damaged bodily organ, in this case the power of vision.

The solution, as is normal for Aspall, is brief and states that something living can indeed die and even a living being living an intellectual life, therefore the intellective soul can die.

In his answer to the first argument Aspall points out that the two instants are not of the same kind: the first instant of the intellective soul outside of the body is permanent and no new instants are added, whereas the last instant of its existence in the body is not final as regards the body, since further instants occur in terms of the change and destruction of the body after death. We note briefly that in his response two further arguments and responses are added. The arguments are based on instants of change and here Aspall addresses the issue by giving a standard example of a change from being black to being white. We already had only a brief mention of the whiteness of Socrates in the relevant argument (1.1), so either the original argument was longer and referred to something which is being debated here, or in the course of his answer Aspall decided to make this digression. It should not come as a surprise since we are dealing with a *reportatio* after all. Aspall presents two objections and a solution to both; then he gives another objection and solution before reverting to the main argument again in response to the second argument. Are the two instants (the end of bodily life and the beginning of separated existence) simultaneous or not? As regards the order of existence they are but not, of course, as coexistent in the same individual, since in the latter one follows the other. With regard to the second point raised, namely that the intellectual soul should be permanently embodied,

Aspall makes the point that the soul has a natural inclination to be united with a body but only with a body which has an appropriate constitution and harmony between the parts which is essential to the union of soul and body. This union is habitual, the aptitude for living accidental, and so if the balance were to be destroyed then that which gave rise to the inclination for the intellective soul to be joined with the body is gone.

The response to the third argument is brief: Aristotle is, of course, talking about the separate Intelligences which are the movers of the spheres and not about the souls in this world. The response to the fourth argument takes the view that Aristotle, inasmuch as he is a philosopher, accepted the actual existence of an infinite number of separate souls but that he did not accept that the infinite division of a continuum can reach an actual infinity. He finishes with a theological response that the condition of the human creature is that unless it is impeded by sin, it goes from good to better. Since, therefore, it is better for it to be in itself than in another, initially it was in another so that it could gain or lose merit, and afterwards when it is separated it is in itself which is, of course, better.

In the reply to the second article, or 'problema' as he calls it, Aspall begins by stating that the vegetative and sensitive souls are destroyed with the body owing to a defect in the matter.[19] The replies to the two arguments expand firstly on Aspall's views on light and, secondly, on the harmonious proportion of the elements in an animal body together with its active powers or seminal reasons. Briefly, with regard to the first he states that it is not necessary that even if light falls into a transparent medium (*perspicuum*) that this is a lessened or diminished being of light. He embarks on a long discussion of how light is diffused from its source and irradiates the heavens without being lessened. Here again, it might be the case that the original argument was more extensive than has been preserved since the answer is clearly much more developed and wide-ranging. Aspall concludes that light is not corrupted as such but only in the way in which a habitus is corrupted through privation, giving the example of an image in the mirror which is 'corrupted' when the object is removed. The original point of the argument of speaking of the sensitive soul as a spiritual light seems to have been forgotten except maybe that we are supposed to infer that the sensitive 'light' is not corrupted as such but only when the bodily object is removed.

Finally, Aspall, somewhat echoing the words of Bonaventure,[20] states that the sensitive soul is a truly noble form arising from a noble harmony and proportion of the elements, but if this harmony is destroyed so is the sensitive soul. Thus, it would seem to follow that if an agent were to educe this harmony and proportion between the elements entering into the composite out of the potency of matter, in the end it would educe the act of life (spontaneous generation of life?). We then have Aspall's

---

[19] A similar point is made by Gilbert the Englishman, see Kijewska, Chap. 3 in this volume, pp. 37–56.

[20] See Bonaventura, *Breviloquium*, pars II, cap. 10, *Opera Omnia* V (Quaracchi: Collegium S. Bonaventurae, 1882–1902, 227–229); also Davies (2019, 83–84).

response to Aristotle's view that the soul is not a harmony: the harmony spoken about concerns the material elements and not the soul which is spiritual and incorporeal. Next Aspall treats of the notion of the seminal reasons, deriving ultimately from Augustine,[21] and states that the seminal reasons are found throughout the natures of the elements and are said to be active powers or diminished forms in matter. Thus, the vegetative and sensitive souls could be corrupted into their seminal reasons. Again, perhaps the body of the dog together with its vegetative and sensitive seminal reasons could again be sublimated and enabled by means of an external agent so that a dog might again be brought forth from them. This does not mean that the resurrection of the dog would follow because the 'return' of the dog would be to the same species, not a return of the individual.

9.5 Conclusion

Well, the poor faithful Fido might not return naturally as Fido but only as a 'dog' but perhaps all dog lovers might still hope for a miraculous intervention of the divine master!

We leave Aspall at a point which is highly interesting in terms of the later history of the development of Aristotle's idea of the cyclical renewal of the elements of the universe as put forward in *De generatione et corruptione*, II, ii. Aspall clearly distinguishes between 'reditio' and 'resurrectio', the return of an individual specimen of a species but not of the individual itself, but he has moved the discussion on from the renewal and return of an element such as water which transforms into air and then back into water to the death and return of a living thing, in a presumably likewise cyclical manner.

In a chapter entitled "Le retour numériquement identique" Dragos Calma provides us with a fragment of a commentary, probably transcribed by Pierre of Limoges, on the *De generatione et corruptione* of Siger of Brabant which opens with the question: "Quod idem numero postquam corruptum est, naturaliter possit generari et redire idem numero."[22] Calma wonders whether it was this text of Siger which lead to the censured thesis of 1277 "quod non contingit corpus corruptum redire idem numero, nec idem numero resurget" (n. 17). The challenges to Christian

[21] See, for example, Augustinus, *De Genesi ad litteram*, IX, cap. 17, 32 (*Opera Omnia*, Paris, 1836, V, 470): "Unde fit ut de grano tritici non nascatur faba, vel de faba triticum, vel de pecore homo, vel de homine pecus. Super hunc autem motum cursumque rerum naturalem, potestas Creatoris habet apud se posse de his omnibus facere aliud, quam eorum quasi seminales rationes habent, non tamen id quod non in eis posuit ut de his fieri vel ab ipso possit."

[22] Calma (2011, 233–263, 237). It is interesting to note that a compendium on the *De generatione et corruptione*, attributed to Siger in the edition by Bazán in the volume Sigerus de Brabantia, *Ecrits de logique, de morale et de physique* (ed. Bazán, 1974, 127–140), is an abbreviation of the text of Aspall's commentary.

notions of the resurrection were too severe to pass uncensured. The problem of the resurrection of the dog would be replaced by problems of far greater import.

To conclude, this brief examination of the text has allowed us to arrive at some provisional conclusions regarding the text and its content. Issues which one would look forward to developing further would be clearly the relationship between what is said here regarding the instant, the nature of light and a probable influence of Grosseteste and the very topical question of the origin of life from inanimate matter and its relationship to the discussions of others such as Siger and Boethius of Dacia. At present this will have to wait for a future study.

Appendices

Appendix 1

Geoffrey of Aspall, Table of Questions on the *De longitudine et brevitate vitae*

This table attempts to capture the somewhat evolving structure of a text which seemed to have gone through various revisions. At both the beginning and towards the end of the text we see features of a more traditional commentary. On the other hand, in all three manuscripts both marginal notes and numbering indicate that contemporary readers saw the text as mainly composed of questions. T is the most developed in this regard since it gives both titles for questions and numbers them continuously, albeit from a hesitant beginning; O begins numbering questions from 35, whereas C adopts the simplest approach simply by marking 'q^{o}' (*quaestio*) in the margin. Sometimes to our eyes they do not seem to be questions at all but simply arguments proposed in the course of a larger discussion. Again, the initial introductions sometimes list questions which do not appear later in the text or with a different title further on. I have tried to capture this below by marking the initial introductions in bold and then listing the questions as they appear in the text. The question titles and numbering supplied in the margin of T are supplied in square brackets. For the sake of clarity these are listed on their own in a separate table in Appendix 2.

Oxford, Merton College, 272/304	Todi, Biblioteca Communale, 23	Cambridge, Gonville and Caius, 509/386
(f. 285rb) In isto libro determinat Aristoteles de causis longitudinis et brevitatis vitae animalium et plantarum. Sit igitur prima quaestio de vita, secunda de morte. Item de vita: Primo quot modis dicatur vita ut sumatur illa de qua est ea; <secundo quid est vita de qua intenditur>; tertio quaeratur de vita an sit possibilis in aliquo inferiori; quarto an approprietur sibi aliquod membrum in quo operetur, ut cor vel cerebrum, etc. 1. Dicendum ad primum quod vita aut est in corpore aut extra corpus. 2. Secundo quaeritur: Quid est vita de qua intenditur? 3. An vita sit possibilis hic inferius. (f. 286rb) 4. An vita sit organum vel appropriet sibi membrum.	(f. 84ra) *De eo autem quod est longae vitae*, etc. Quaeritur quot modis dicitur vita, secundo quid sit vita proprie dicta, tertio quaeritur quid sit mors, quarto an aliquid inferius vivat, quinto an aliquid moriatur. [Prima quaestio an aliquid inferius vivat] 1. An aliquid inferius vivat. (f. 84va) [Secunda quaestio an vita appropriat sibi organum] 2. An vita appropriat sibi membrum vel organicum vel universaliter partem determinatam in corpore in quo primo reperiatur.	(f. 276ra) In isto libro determinat Aristoteles de causis longitudinis et brevitatis vitae animalium et plantarum. Sit igitur prima quaestio de vita, secunda de morte. Item de vita: quot modis dicatur vita et sumatur illa de qua ad praesens est intendere; secundo quid sit vita prout hic intenta de vita; tertio an vita sit possibilis in aliquo inferiori; quarto an appropriat sibi aliquod membrum in quo operetur, ut cor vel cerebrum, etc. 1. Dicendum quod vita aut est in corpore aut extra corpus. 2. Quid est vita hic intenta. 3. An vita sit possibilis hic inferius. (f. 276va) 4. An vita appropriat sibi membrum vel organum vel universaliter partem determinatam in corpore in qua reperitur.
(f. 286rb) An aliqua vita sit longa vel brevis.	(f. 84va) [Tertia quaestio an vita sit longa vel brevis.] 3. An aliqua vita sit longa vel brevis.	(f. 276va) An aliqua sit vita longa vel brevis.
(f. 286va) Habito de vita quaeritur de morte. Et primo quaerendum quid <sit mors>, secundo an aliquid possit mori. Et primo <an> aliquid vivens per animam intellectivam possit mori; secundo an vivens sensitiva vel vegetativa possit mori; tertio an mixtum possit mori.	(f. 84vb) [Quarta quaestio de morte] Habito de vita, consequenter de morte. Quaerendum primo quid sit mors, secundo an aliquid possit mori. Et primo an vivens vita intellectiva possit mori; secundo an vivens vita sensitiva vel vegetativa possit mori. [An vivens vita intellectiva possit mori] [Quinta quaestio an vivens vita sensitiva possit mori]	(f. 277ra) Habito de vita quaeretur consequenter de morte quid sit, secundo an aliquod possit mori. Et primo an vivens per intellectivam possit mori; et secundo an vivens sensitiva vel vegetativa possit mori; tertio an mixtum possit mori.

(continued)

(f. 287va) Tertio quaeritur utrum aliquod mixtum possit dissolvi.	(f. 85rb) [Sexta an omne mixtum ex elementis possit corrumpi.] An omne mixtum ex elementis potest corrumpi et resolvi in quattuor elementa.	(f. 277vb) An omne mixtum ex elementis possit corrumpi et resolvi in quattuor elementa.
(f. 288vb) Consequenter quaeritur de perpetuitate creaturarum. Et primo: Utrum aliqua creatura possit <cedere> ad non esse quantum est de se. (f. 289ra) Secundo quaeritur utrum de se possibile possit fieri necessarium per aliud.	(f. 86rb) An aliqua creatura de sui natura possit cedere in non esse. (f. 86rb) An illud quod de se est possibile tamen per aliud ut per virtutem naturalem fit necessarium.	(f. 278va) An aliqua creatura de sui natura possit cedere in non esse. (f. 278vb) An illud quod de se est possibile fiat per aliud necessarium.
(f. 289va) An possit esse complexio media secundum omnimodam aequalitatem. An inter contraria possit esse medium secundum aequalitatem.	(f. 86vb) An sit reperire inter contraria mediata aliquod medium aequaliter se habens respectu utriusque extremi, scilicet ex utroque aequaliter confectum. (f. 87rb) Consequenter quaeritur, si talis <esset> complexio, an sit corruptibilis.	(f. 279ra) An sit reperire inter contraria mediata aliquod medium essentialiter se habens respectu utriusque extremorum scilicet ex utroque aequaliter confectum. An aliqua sit complexio media aequaliter ex miscibilibus inter complexiones. (f. 279vb) Consequenter quaeritur, si talis complexio sit, an sit corruptibilis.
(f. 290rb) Utrum aliqua corruptio sit naturalis.	(f. 87vb) [Quaestio 11ª An aliqua corruptio sit a natura] An aliqua corruptio sit naturalis. [No Q. 12]	(f. 280ra) An aliqua corruptio sit naturalis.
(f. 290va) Consequenter, quia dicit Aristoteles in fine prooemii quod *speculandum est de causis vitae*, quaeritur utrum eas debeat determinare.	(f. 87va) [q. 13 An sit determinare in aliqua scientia de causa brevitatis et longitudinis vitae] An contingat determinare in scientia de causis brevitatis et longitudinis vitae.	(f. 280ra) An contingat in scientia determinare de causis longitudinis et brevitatis vitae.

(continued)

(f. 291ra) Consequenter quaeritur de hoc quod dicit Aristoteles quod scientia corrumpitur dupliciter scilicet a contrario et per corruptionem sui subiecti ut hominis vel animalis.	(f. 88ra) [q. 14 Utrum scientia in anima sit res et natura] Utrum scientia sit res vel natura vel species sola et intentio.	(f. 280va) Utrum scientia in anima sit res et natura, an species sola et intentio.
(f. 291rb) Utrum scientia corrumpitur in anima.	(f. 88ra) [q. 15 An scientia possit corrumpi in anima] Utrum scientia possit corrumpi in anima.	(f. 280vb) An scientia possit corrumpi in anima.
(f. 291rb) Utrum scientia corrumpitur a suo contrario.	(f. 88rb) [q. 16 An scientia possit corrumpi a contrario] An scientia possit corrumpi a contrario.	(f. 280ra) An scientia corrumpi possit a contrario.
(f. 291va) Utrum necessario corrumpatur scientia per corruptionem subiecti, ut hominis aut animalis.	(f. 88va) [q. 17 An scientia corrumpatur corruptione subiecti] An scientia corrumpatur corruptione subiecti, verbi gratia, an scientia corrumpatur corrupto homine.	(f. 281rb) An scientia corrumpitur corruptione subiecti, verbi gratia, an scientia corrumpatur corrupto homine.
(f. 291va) An subiectum per se corrumpatur <corrupto> suo subiecto.	(f. 88vb) [q. 18 An subiectum corrumpatur per se] An subiectum per se corrumpatur sive substantia.	(f. 281rb) An subiectum per se corrumpitur <corrupto> suo subiecto.
(f. 291vb) Consequenter quaeritur de hoc quod dicit quod contrarium coniunctum materiae passibilis est causa corruptionis … Primo quaeritur an contrarium sit causa corruptionis; Secundo an materia <sit> passibilis; Tertio an activum, et cetera.	(f. 88vb) [q. 19 An contrarium coniunctum materiae passibilis sit causa corruptionis] Consequenter quaeritur de hoc quod dicit quod contrarium coniunctum materiae passibilis est causa corruptionis … Primo quaeritur an contrarium sit causa corruptionis; Secundo an sit materia sic passibilis; Tertio an activum, et cetera.	(f. 281va) Consequenter dubitatur de hoc quod dicit quod contrarium coniunctum materiae passibilis est causa corruptionis … Primo quaeritur an contrarium sit causa corruptionis; Secundo an sit materia passibilis; Tertio an activum et passivum, et cetera.
(f. 292ra) Circa secundum quod materia non sit passibilis.	(f. 89ra) [q. 20 An materia sit passibilis] Circa secundum quod materia non sit passibilis.	Circa secundum quod materia non sit passibilis.

(continued)

(f. 292ra) Circa tertium quaeritur an activum et passivum.	(f. 89ra) [q. 21 An activum et passivum sint causa corruptionis vel an contrarium sit causa corruptionis inquantum activum et passivum.] Circa tertium quaeritur an activum et passivum est causa corruptionis.	Circa tertium an activum et passivum, et cetera.
(f. 292va) Dicit consequenter *impossibile est materiam habenti non habere contrarium.* Contra …	(f. 89rb) [q. 22 An omne habens naturam [*recte* materiam] habet contrarium] Dicit consequenter quod *impossibile est naturam habenti non habere contrarium.*	(f. 282ra) Dicit consequenter *impossibile materiam habenti non habere contrarium.*
	(f. 89va) [q. 23 – no text and not a question] *Si autem fuerit quodcumque utique* etc. Signat quod si unum duorum contrariorum fuerit activum et reliquum passivum, necessario est unum per se mutari et posse corrumpi. [q. 24 Utrum minor flamma corrumpatur maiore.] Dubitatur de hoc quod dicit minor flamma vincitur sive corrumpatur a maiori.	
(f. 293ra) *Oportet autem supponere*, et cetera. Dicit quod omne animal est calidum et humidum in natura. Primo quaeritur an hoc sit verum. Secundo an calidum et humidum sint causa longitudinis et brevitatis vitae. Tertio an frigidum et siccum sint causa brevitatis vitae. Quarto de complexione quae <est> nobilior et durabilior. Quinto quod calidum an igneum aereum <sit>. Sexto quod humidum an aereum an aqueum <sit>.	(f. 89vb) [q. 25 An animal sit calidum et humidum in natura] *Oportet autem supponere*, et cetera. Sic arguitur quod omne animal est calidum et humidum in natura. Primo quaeritur an hoc sit verum. Secundo an calidum et humidum sint causa longitudinis et brevitatis vitae. Tertio an frigidum et siccum sint causa brevitatis vitae. Quarto de complexione quae <est> nobilior et durabilior. Quinto quod calidum an igneum vel aereum <sit>. Sexto quod humidum an aqueum vel aereum <sit>.	(f. 282rb) *Oportet autem supponere*. Dicit Aristoteles quod omne animal est calidum et humidum in natura. Primo quaeritur an hoc sit verum. Secundo an calidum et humidum sint causa longioris vitae. Tertio an frigidum et siccum sint causa (f. 282va) brevitatis vitae. Quarto de complexione scilicet quae sit nobilior et quae sit durabilior. Quinto quod calidum scilicet an igneum vel aereum <sit>. Sexto quod humidum scilicet an aqueum vel aereum <sit>.

(continued)

(f. 293va) An calidum et humidum sint causa longitudinis vitae. (f. 294ra) Sed quaeritur ex quo in elemento ut igne constitit siccum sive humidum [...] et cetera.	(f. 90rb) [q. 26 An calidum et humidum sit causa longitudinis vitae] An calidum et humidum sint causa longitudinis vitae.	(f. 282vb) An calidum et humidum sint causa longitudinis et brevitatis vitae. Sed quaeritur ex quo in elemento ut in igne consistit siccum sive humido ut in igne propria terra non potest manere sine humido qualiter igitur potest ignis.
(f. 294ra) An calidum et humidum sit causa iuventutis.	(f. 90va) [q. 27 An calidum et humidum sint causa iuventutis] An calidum et humidum sint causa iuventutis.	(f. 283) An calidum et humidum sint causa iuventutis.
(f. 294rb) Quaeritur quae complexio nobilior sit.	(f. 90vb) [q. 28 Quaeritur quae complexio sit nobilior] Quaeritur quae complexio sit nobilior.	(f. 283rb) Consequenter quaeritur quae complexio sit nobilior. Et quod cholerica videtur sic …
(f. 294vb) Quaeritur quae complexio <est> durabilior.	(f. 91ra) [q. 29 Quae complexio sit durabilior.] Quaeritur quae complexio durabilior <est>.	(f. 283vb) Consequenter quaeritur quae complexio sit durabilior. Et quod melancholica sic …
(f. 295ra) [now begins to number in margin at 35] Consequenter de hoc quod dicit quod humidum et calidum est causa longae vitae. Et primo quaero utrum humidum de quo loquitur sit humidum aqueum an aereum. Secundo utrum calidum de quo loquitur sit igneum an aereum.	(f. 91rb) [q. 30 An humidum quod est causa longitudinis vitae sit aqueum an aereum] Consequenter quaeritur, cum dicit quod calidum et humidum sint principium longae vitae, quod calidum, quod humidum. Primo de humido an aqueum an aereum.	(f. 284ra) Consequenter quaeritur, ut dicit quod calidum est principium longae vitae, quod calidum et quod humidum. Sed primo de humido an humidum aqueum vel aereum.
(f. 295va) [36] Consequenter quaeritur de calido quod est causa.	(f. 91va) [q. 31 An calidum quod est causa vitae sit igneum vel aereum.] Consequenter <quaeritur> quod calidum, an <sit> igneum vel aereum.	(f. 284rb) Consequenter dicit quod calidum, et cetera. Quaeritur an <sit> igneum vel aereum.

(continued)

(f. 297vb) [37] *Sed quae habitant in calidis*, et cetera. Hic loquitur de causis extrinsecis confluentibus ad vitae longitudinem dicens quod quae habitant in locis calidis longius vivunt et sunt maioris corporis. Aquatica enim ut dicit minus vivunt quam animalia. Sic ergo prima quaestio: quaero quare non ponit duas causas extrinsecas.	(f. 91vb) [q. 32 An sint duae causae extrinsecae longitudinis vitae sicut intrinsecae.] *Sed quae sunt in locis calidi sunt longioris vitae*, et cetera. Quaeritur cum duae sunt causae intrinsecae longitudinis vitae scilicet calidum <et> humidum quare similiter non erunt <duae> extrinsecae respectu loci.	(f. 284ra) *Sed quae sunt in locis calidis sunt longioris vitae*. Quaeritur cum sint duae causae intrinsecae longitudinis vitae scilicet calidum <et> humidum quare similiter non erunt duae extrinsecae respectu loci.
(f. 296ra) [38] Secundo quaeritur de hoc quod dicit Aristoteles quod calidum extrinsecum confirmat.	(f. 92ra) [q. 33 An habitantes in locis calidis sint longioris vitae, et cetera.] An habitantes in locis calidis sint longioris vitae ut dicit.	(f. 284vb) An habitantes in locis calidis sint longioris vitae ut dicit.
(f. 296rb) [39] Consequenter quaeritur de tertio scilicet utrum animalia habitantia in locis frigidis et humidis, eo quod humidum est magis aquosum, sint parvi corporis et brevis vitae.	(f. 92ra) [q. 34 An habitantes in locis frigidis et humidis sint brevis vitae et parvi corporis.] Dicit quod habitantes in locis humidis et frigidis, eo quod humidum est magis aquosum et cetera, sunt parvi corporis et brevis vitae. Contra …	(f. 284vb) Item dicit quod habitantes in locis humidis et frigidis, eo quod humidum est magis aquosum et cetera, sunt parvi corporis et brevis vitae. Contra …
(f. 296rb) [40] *Aquatica autem et pedibus ambulantibus*, et cetera. Contra quaero an aquatica et habitantia in locis et frigidis et humidis sint brevis vitae sicut dicit.	(f. 92ra) [q. 35 – no text] *Aquatica autem pedibus ambulantibus minus longioris vitae sunt* propter humidum aquaticum quod velociter corrumpitur et condensatur, et cetera. Hoc dicit. Contra …	(f. 284vb) *Aquatica autem pedibus ambulantibus minus longioris vitae sunt* propter humidum aquaticum quod velociter corrumpitur et condensatur, et cetera. Verba sunt Aristotelis. Contra …
(f. 296vb) [41] Consequenter quaeritur de hoc quod dicit *alimentum enim habentia*. Dicit quod subtractio alimenti est causa extrinseca brevitatis vitae, quia calor naturalis, si non habeat alimentum agit in se ipsum, et hoc destruendo se ipsum. Circa hoc quaero …	(f. 92rb) [q. 36 – no text] *Augmentum enim non recipientia*, et cetera. Dicit quod subtractio alimenti est causa corruptionis in animalibus, quia calor naturaliter digestius, si non habeat in qua agat ut alimentum, *consumit se ipsum*. Contra …	(f. 285ra) *Alimentum enim non recipientia*, et cetera. Dicit quod subtractio alimenti est causa corruptionis in animalibus, quia calor naturaliter digestius, si non habeat in qua agat ut alimentum, *consumit se ipsum*. Contra …

(continued)

(f. 297ra) [42] *Sed quae sunt in plantis*. Hoc est tertium capitulum in quo determinat de causis longitudinis et brevitatis vitae in plantis. Et primo in generali [omnes hanc corruptionem] *plantae longioris vitae sunt quantum animalia* quam probat per duo media; primum est *quia minor aquositatis est*; secundum est *quia magis habent de pinguedine et dulcedine*. Primo quaero de corruptione in se et videtur quod sit falsa, quia …	(f. 92va) [q. 37 Utrum plantae sunt longioris vitae quam animalia] *In plantis*, et cetera. Dicit quod plantae sunt longioris vitae animalibus quibusdam, tum quia minoris aquositatis, tum quia magis habent de pinguedine et dulcedine quod animal longius vivat quam planta … Praeterea… [has a different structure, more like a commentary …]	(f. 285rb) *Sed in plantis*, et cetera. Dicit quod plantae sunt longioris vitae quibusdam animalium, tum quia sunt minoris aquositatis, tum quia magis habent de pinguedine et dulcedine <sicut habetur> primo *De generatione* quod animal longius vivat quam planta … Praeterea …
(f. 297vb) [43] Consequenter quaero de primo medio per quod probat per dictam corruptionem et est quod plantae minoris aquositatis sunt quam animalia.	(f. 92vb) [q. 38 An plantae sint minoris aquositatis quam animalia] Contra primam causam vel mediam sic: nobiliori formae nobiliora respondent principia corruptibilia et minus nobili ignobiliora; sed sensitiva <est nobilior> et vegetativa ignobilior ergo principium corruptibile ut humidum nobilius respondet sensitivae in animalibus ut aereum, et ignobilius ut aqueum vegetativae in plantis, quare et cetera. Praeterea …	(f. 285va) Contra primam causam vel mediam sic: nobiliori formae nobiliora respondent principia corruptibilia et minus nobili ignobiliora; sed sensitiva est nobilior et vegetativa ignobilior ergo principium corruptibile ut humidum nobilius respondet sensitivae in animalibus ut aereum, et ignobilius ut aqueum vegetativae in plantis, quare et cetera. Praeterea …
(f. 297vb) [44] Secundum medium est quod plantae magis habent de pinguedine et dulcedine quam animalia. Contra: pinguedo est a nobiliori complexione sanguinea … [text ends abruptly at the bottom of the folio]	(f. 92vb) [q. 39 An plantae habent plus pinguedinis quam animalia] Contra secundum medium sic vel secundam causam suae corruptionis sic Aristoteles in illo libro: pinguedo est a nobilissima complexione sanguinea, sed animalia habent nobiliorem complexionem quam plantae quaecumque, igitur habent plus de pinguedine. Praeterea …	(f. 285vb) Contra secundum medium sic vel secundam causam suae corruptionis sic Aristoteles in illo libro: pinguedo est a nobilissima complexione sanguinea, sed animalia plus habent nobiliorem complexionem quam plantae quaecumque, igitur habent plus de pinguedine. Praeterea …

(continued)

	(f. 93ra) [q. 40 – no text] De dulcedine sic: In lacte animalium feminarum reperitur dulcedo; in plantis non reperitur lac vel si sic haec non contingit reperire nisi per artificium ut patet in amigdalis …	(f. 285vb) De dulcedine sic: In lacte animalium feminarum reperitur dulcedo; in plantis non reperitur lac vel, si sic, haec non contingit reperire nisi per artificium ut patet in amigdalis …
	(f. 93ra) [q. 41 Utrum in animali possit esse innovatio membrorum sicut in planta innovatio ramorum.] *De eo autem multo tempore*, et cetera. Dicit quod arbores sunt longae vitae, quia habent innovationem ramorum et radicum post abscisionem, animalia non habent innovationem post abscisionem membrorum. De hoc dubitatur. Videtur enim …	(f. 285vb) *De eo autem multo tempore*, et cetera. Dicit quod arbores sunt longae vitae, quia habent innovationem ramorum et radicum post abscisionem, animalia non habent innovationem post abscisionem membrorum. De hoc dubitatur. Videtur enim …
	(f. 93rb) [q. 42 An pars plantae vivat post decisionem] De plantis sic: anima animalis perfecti <est> magis perfecta et minus ligata materiae quam anima plantae. Igitur maior est ratio ut ipsa existat ubique in corpore. Igitur, cum in animali perfecto anima existens tota ubique non potest post decisionem vivificare membra, igitur nec planta. Item …	(f. 286ra) De plantis ergo sic: anima animalis perfecti magis perfecta est et minus ligata materiae quam anima plantae. Igitur maior est ratio ut ipsa existat ubique in corpore. Igitur, cum in animali perfecto anima existens tota ubique non potest post decisionem vivificare membra, igitur nec planta. Item …
	(f. 93rb) [q. 43 – no text] Ex hoc sequitur secunda quaestio: Utrum vegetativa totius dividitur vel una cedit in duas quod dividitur sic sensitiva.	(f. 286ra) Ex hoc sequitur secunda quaestio scilicet utrum vegetativa totius dividitur vel una cedit in duas quod dividitur. (f. 286rb) Explicit. *In marg*. De Aspale quaestiones expliciunt.

Appendix 2

Table of question titles given in the margins of MS Todi, Biblioteca Communale, 23

(f. 84ra) Q. 1 An aliquid inferius vivat. (f. 84va) Q. 2 An vita appropriat sibi organum. (f. 84va) Q. 3 An vita sit longa vel brevis. (f. 84vb) Q. 4 De morte: (f. 84vb) Q. 4a An vivens vita intellectiva possit mori [not a question in T]. (f. 84vb) Q. 5 An vivens vita sensitiva possit mori.	(f. 85rb) Q. 6 An omne mixtum ex elementis possit corrumpi. [Q. 7 –unnumbered] An omne mixtum ex elementis potest corrumpi et resolvi in quattuor elementa.
(f. 86rb) [Q. 8 –unnumbered] An aliqua creatura de sui natura possit cedere in non esse. (f. 86rb) [Q. 9 – not numbered] An illud quod de se est possibile tamen per aliud ut per virtutem naturalem fit necessarium.	(f. 86vb) [Q. 10 – unnumbered] An sit reperire inter contraria mediata aliquod medium aequaliter se habens respectu utriusque extremi, scilicet ex utroque aequaliter confectum. (f. 87rb) [unnumbered] Consequenter quaeritur si talis <esset> complexio, an sit corruptibilis
(f. 87vb) Q. 11 An aliqua corruptio sit a natura. [No Q. 12]	(f. 87va) Q. 13 An sit determinare in aliqua scientia de causa brevitatis et longitudinis vitae.
(f. 88ra) Q. 14 Utrum scientia in anima sit res et natura.	(f. 88ra) Q. 15 An scientia possit corrumpi in anima.
(f. 88rb) Q. 16 An scientia possit corrumpi a contrario.	(f. 88va) Q. 17 An scientia corrumpatur corruptione subiecti.
(f. 88vb) Q. 18 An subiectum corrumpatur per se.	(f. 88vb) Q. 19 An contrarium coniunctum materiae passibilis sit causa corruptionis.
(f. 89ra) Q. 20 An materia sit passibilis.	(f. 89ra) Q. 21 An activum et passivum sint causa corruptionis vel an contrarium sit causa corruptionis inquantum activum et passivum.
(f. 89rb) Q. 22 An omne habens naturam [*recte* materiam] habet contrarium.	(f. 89va) [Q. 23] *Si autem fuerit quodcumque utique*, et cetera. (f. 89va) Q. 24 Utrum minor flamma corrumpatur maiore.
(f. 89vb) Q. 25 An animal sit calidum et humidum in natura.	(f. 90rb) Q. 26 An calidum et humidum sit causa longitudinis vitae.
(f. 90va) Q. 27 An calidum et humidum sint causa iuventutis.	(f. 90vb) Q. 28 Quae complexio sit nobilior.
(f. 91ra) Q. 29 Quae complexio sit durabilior.	(f. 91rb) Q. 30 An humidum quod est causa longitudinis vitae sit aqueum an aereum.
(f. 91va) Q. 31 An calidum quod est causa vitae sit igneum vel aereum.	(f. 91vb) Q. 32 An sint duae causae extrinsecae longitudinis vitae sicut intrinsecae.
(f. 92ra) Q. 33 An habitantes in locis calidis sint longioris vitae, et cetera.	(f. 92ra) Q. 34 An habitantes in locis frigidis et humidis sint brevis vitae et parvi corporis.

(continued)

(f. 92ra) [Q. 35] *Aquatica autem pedibus ambulantibus minus longioris vitae sunt.*	(f. 92rb) [Q. 36] *Augmentum* [recte *alimentum*] *enim non recipientia.*
(f. 92va) Q. 37 Utrum plantae sunt longioris vitae quam animalia.	(f. 92vb) Q. 38 An plantae sint minoris aquositatis quam animalia.
(f. 92vb) Q. 39 An plantae habent plus pinguedinis quam animalia.	(f. 93ra) [Q. 40] De dulcedine sic: in lacte animalium feminarum reperitur dulcedo; in plantis non reperitur lac vel si sic haec non contingit reperire nisi per artificium ut patet in amigdalis …
(f. 93ra) Q. 41 Utrum in animali possit esse innovatio membrorum sicut in planta innovatio ramorum.	(f. 93rb) Q. 42 An pars plantae vivat post decisionem.
(f. 93rb) [Q. 43 no text] Utrum vegetativa totius dividitur vel una cedit in duas.	

Appendix 3

Geoffrey of Aspall

Questions on Aristotle's *De longitudine et brevitate vitae*

Question 4: On Death [23]

[T f. 84vb]

Habito de vita, consequenter de morte. Quaerendum primo quid sit mors, secundo an aliquid possit mori. Et primo an vivens vita intellectiva possit mori; secundo an vivens vita sensitiva vel vegetativa possit mori.

Ad primum: Mors est vitae privatio in eo quod natum est habere illam.

1. An vivens vita intellectiva possit mori, ut homo.

Quod non videtur sic:

1.1. Si sic, tunc intellectiva possit separari a corpore, et suum esse est permanens extra corpus; igitur est sumere primum instans sui esse.

Item est sumere ultimum instans sui esse in corpore, cum sit magis permanens quam albedo Socratis.

Quaero: illa instantia aut sunt duo alia aut unum? Si unum, tunc simul et semel <sunt> in corpore et extra; si duo, tunc inter ea cadit tempus medium in quo nec est in corpore nec extra; quorum utrumque est falsum.

Et hoc sequitur ex separatione animae intellectivae a corpore; igitur impossibile est eam separari, quare manet perpetue in corpore.

[23] For our purposes here, I have followed MS T which is written in a fine clear English (or perhaps French) hand from the latter part of the thirteenth century. The version of the text which it transmits (together with C) seems to be a later, clearer and more considered version than that transmitted by MS O.

1.2. Praeterea, nulla substantia separata est otiosa secundum Aristotelem in decimo *Metaphysicae*,[24] quaelibet igitur deputatur operationem ad movendum corpus aliquod; ergo, si intellectiva [ergo] tua separetur, oportet quod deputetur ad movendum aliquod corpus aliud; sed hoc non contingit nisi ingrediatur aliud corpus, ergo ingreditur aliud corpus. Consequens falsum, ergo antecedens.

1.3. Praeterea, si separetur a corpore cum generatio processit in infinitum ex parte ante et <finitum> ex parte post secundum Commentatorem[25] et Aristotelis doctrinam[26] et qua ratione una <anima> separatur et alia, tunc infinitae sunt separatae et sunt actu substantiae; igitur sunt infinita in actu appositione. Quod est contra Aristotelem in *Physicis*.[27]

1.4. Praeterea, unio animae cum corpore est essentialis; convenientia miscibilium in mixto est accidentalis, et nullum accidentale perimit substantiale, ergo discoaequatio et improportio miscibilium in mixto non perimit illam emovere, et ita non facit animam separari.

1.6. Praeterea, finis et principium eiusdem rei proportionantur; sed anima in principio ante infusionem in corpore non existit separata, quare non existet post separationem eius, sed semper erit; quare et cetera.

2. De secundo: An vivens <vita vegetativa et> sensitiva possit mori.

Quod non videtur sic:

2.1. Sensitiva secundum philosophos dicitur lux spiritualis; sed lux spiritualis proportionatur luci corporali hic in medio. Si igitur lux corporalis <hic> in medio non corrumpitur, nec lux spiritualis ut[28] sensitiva. Sed quod lux in medio non corrumpatur videtur sic: si corrumpitur, aut in elementum aut in nihil aut regreditur in fontem suum. Sed quodlibet istorum est impossibile. Non enim corrumpitur [T f. 85ra] in elementum, quia tunc posset produci ex elemento, quod falsum est, cum sit natura corporis supercaelestis. Nec regreditur ad suum fontem, quia ab ipso continue multiplicatur;[29] nec in nihil, quia nihil quod semel est ens cedit in nihil secundum philosophos, quare manet perpetue.

2.2. Praeterea, omne quod corrumpitur corrumpitur in aliquid; si igitur sensitiva et vegetativa corrumpuntur, corrumpuntur in aliud corpus quod est diminutum respectu illorum. Si hoc est verum, tunc vegetativa et sensitiva per agens extra possent educi ex aliquo diminuto, ergo vegetativa et sensitiva canis et etiam corpus per

[24] Averroes, *In Metaphysicam*, XII, com. 44 (ed. Iuntina, VIII, f. 327H): "[…] si aliquae substantiae essent non moventes, essent otiosae."

[25] Cf. Averroes, *In Metaphysicam*, II, com. 5; XI, com. 2; however, it seems to have been used as an adagium; see Thomas Aquinas, *Summa Contra Gentiles*, II, cap. 38, n. 12: "Nam nihil prohibet infinito ex ea parte additionem fieri qua est finitum. Ex hoc autem quod ponitur tempus aeternum, sequitur quod sit infinitum ex parte ante, sed finitum ex parte post: nam praesens est terminus praeteriti." (https://www.corpusthomisticum.org/scg2006.html, accessed 28/03/2021).

[26] Cf. Aristoteles, *De generatione et corruptione*, II.11, 337b 21–25 (AL IX.1, 78–79).

[27] Cf. Aristoteles, *Physica*, III.4, 204a6 (AL VI.1, 114): "Amplius infinitum omne aut secundum appositionem aut secundum divisionem est aut utroque."

[28] ut CO: nec T

[29] nec regreditur ad suum fontem quia ab ipso continue multiplicatur O: *om.* CT

agens extra posset educi ex illo diminuto; tunc canis mortuus iterum viveret, quod falsum est in generatis per propagationem.

[Hic tamen habet dubium de substantia corporea et incorporea quae sunt in eadem potentia <communis> generantis remota igitur primum.][30]

2.3. Praeterea, sicut est in compositione et resolutione secundum rationem, sic est in compositione et resolutione secundum rem, sed ita est in compositione et resolutione secundum rationem, quod, si corrumpantur diversae species, cederent in eandem potentiam communem generantis, quae iterum per rationem ex illa potentia communi possent educi per additionem duarum specialium illi potentiae communi. Sic igitur erit in re, cum igitur sensitiva et vegetativa contineantur sub substantia incorporea et corpus sub substantia corporea, si corrumpantur, cedunt in potentiam communem remotam, quare iterum ex illa potentia remota communi utriusque, scilicet animae et corporis, potest iterum educi utrumque. Et etiam sequitur quod ex corpore possit educi anima et e contrario, quod est impossibile.[31]

2.4. Item, sensitiva non debilitatur, quare nec corrumpitur nec separatur. Primum patet per hoc quod *si senex acciperet oculum*, et cetera.[32]

DICENDUM ad primum quod aliquid potest mori et etiam vivens vita intellectiva potest mori, unde intellectiva potest mori.

Ad primum dicitur quod contingit sumere primum sui esse extra corpus, cum sit permanens non adiunctum successivo, non tamen ultimum sui esse in corpore, quia, licet sit permanens, est tamen adiunctum successivo ut corpori alterabili transmutabili corruptibili et aliis motibus factis in corpore disponentibus ad corruptionem.

Sed contra: Fiat ex nigro album: in illa transmutatione ita est quod contingit sumere ultimum nigredinis et primum albedinis. Et tamen albedo est <aliquid> adiunctum successivo ut corpori alterabili, et cetera, cum sit compositum ex quattuor elementis. Item, albedo fundatur supra principia alterabilia et transmutabilia; cum igitur sit huiusmodi albedinis sumere primum, multo fortius ultimum sui esse in corpore.

Dicitur quod non est simile, quia esse animae in corpore adiungitur successivo in corpore quod disponit finaliter ad eorum separationem. Sic non est ex alia parte, quia, licet albedo sit adiuncta successivo, non tamen alicui quod disponat ad nigredinem vel albedinis corruptionem.

Sed contra: creetur anima; modo est sumere ultimum sui non esse cum sit permanens et primum sui esse cum sit permanens, similiter contingit arguere de intellectiva, et sequitur idem quod prius.

Ad quod potest dici quod contingit sumere ultimum sui non esse et primum sui esse et quod sunt duo instantia; nec inter ea cadit tempus medium, quia non sunt instantia temporis, sed unum instans temporis, ut instans sui esse primum vel adiunctum tempori ultimum sui non esse, est instans aeternitatis. Nec vult Aristoteles

---

[30] It looks like a scribe has incorporated a marginal note here.

[31] quod est impossibile O: ex quo illa potentia communis est istis CT

[32] Aristoteles, *De anima*, I.4, 408b21–22 (Sancti Thomae de Aquino, ed. Leonina Opera Omnia 45, I, ch. x, 47): "si enim accipiat senior oculum iuuenis, uidebit utique sicut et iuuenis."; Averroes, *Commentarium Magnum in De anima*, I, com. 65 (ed. Crawford, 1953, 7).

quod inter talia cadat medium tempus, sed inter duo instantia temporis cadat de necessitate tempus medium. Et si quaeratur, an illa sunt simul an non, dico quod sunt simul secundum convenientiam in ordine existendi, non tamen sunt simul secundum coexistentiam in eodem individuo. Et sic non sunt simul secundum simplicitatem coexistentiae, sunt tamen simul secundum ordinem convenientiae in duratione. Unum enim sequitur ad aliud. Et sic patet qualiter distinguendum est esse simul.

Ad aliud: quod intellectiva habet inclinationem ad corpus non quodcumque, sed humanum, nec quocumque modo, sed in debita complexione et harmonia existens. Unde illa proportio miscibilium in mixto et harmonia debita est essentialis actuali unioni animae cum corpore, habituali tamen unioni animae cum corpore et aptitudinis vivendi est accidentalis; corrupta igitur haec proportione corrumpitur essentiale huic inclinans a parte eius ad quod inclinatur.

Ad aliud quod dicitur quod nulla substantia separata, et cetera, loquitur de intelligentiis separatis quae sunt motores orbium et non de animabus hic inferius.

Ad aliud sequens quod Aristoteles inquantum philosophus concedet infinitas animas separatas actualiter existentes, nec est contra dictum suum in libro *Physicorum*.[33] Non enim loquitur ibi de quacumque appositione numerali, sed de ea quae causatur ex divisione continui; talem enim infinitatem in appositione negat.

Ad ultimum theologice dicitur quod de conditione creaturae est quod nisi per peccatum impediatur, quod vadit de bono ad melius. Cum igitur melius sit esse in se quam in alio, et primo in alio ut possit ibi mereri et demereri, et postea in se cum separatur.

Ad 2

Ad aliud de sensitiva dicendum quod corrumpitur cum corpore, [T f. 85rb] sed per accidens. Corrumpi enim est duplex: aut per contrarium aut per defectum materiae. Secundo modo corrumpitur vegetativa et sensitiva cum corpore. Unde Aristoteles in tertio *Caelo et mundo*:[34] duplex est corruptio flammae: aut a contrario aut a defectu materiae. Et sic terminatur primum problema.

Ad primum in oppositum contra secundum problema respondeo: cum dicit quod tunc, ut videtur, <lux> possit iterum produci ex eodem, scilicet natura elementari, et sic multiplicatio lucis nihil faceret ad generationem, dicatur quod immo irradiatio est agens producens sive annecta oppositio luminosi. Et forte non est necesse quod etsi cedat in perspicuum quod est esse lucis diminutum quod iterum ex eodem producatur ad minus radiositas caelestis, quia, cum sit habitus medii, non est eius absolute, sed per determinatum respectum ad aliquid aliud ut ad luminosum.

Vel dicatur quod luci non debetur corruptio proprie nisi sicut habitum quod corrumpitur per sui privationem. Sed est habitus duplex: unus eductus de potentia materiae ad actum per agens intrinsecum ipsius subiecti, alius per determinatam oppositionem alicuius extrinseci, sicut patet de ydolo in speculo, et illud secundum corrumpitur per subtractionem sui oppositi. Aliter potest dici quod species est sive

[33] See above, n. 24.

[34] Aristoteles, *De caelo*, III.6, 305a10.

similitudo rei cui per se non debetur corruptio cum non sit res, sed solum evanescit a moto corpore obiecto.

Ad primum sustinendo quod lux in medio non est species sola, cum denominet in medio sub nomine proprio (dicitur enim quod illuminat), sic non est de specie albi et nigri (non enim dicitur species albi in medio colorata). Dicendum quod perspicuum est quoddam commune aeri et aquae superiori et inferiori corpori secundum Aristotelem secundo *De anima*.[35] Unde est quoddam esse lucis diminutum et in illud cedit lux in medio; forte non corrumpitur, similiter lux candelae. Unde antiqui dicebant quod lux cedit in substantiam tenuatam elementorum.

Sustinendo tamen quod sit species sola lucis in medio dicendum quod non corrumpitur in aliquid, sed quod evanescit per absentiam eius, cum est species, ut species vel similitudo in speculo evanescit per amotionem obiecti.

Ad aliud de sensitiva: dicunt quidam quod sensitiva nihil aliud est nisi quaedam valde nobilis forma resultans ex nobili harmonia et proportione miscibilium, quorum harmonia corrupta corrumpitur et anima, unde ei non respondet alia potentia quam illi harmoniae. Unde, si aliquod agens educeret illam harmoniam et proportionem ex potentia materiae, educeret in fine actum vitae.

Ad aliud diceret aliquis quod argumentum concludit ponendo quod miscibilia in mixto ita proportionari possunt quod illa proportio et harmonia sub esse nobiliori et sublimato fieret vegetativa et sensitiva; si iterum illa harmonia et proportio conveniens possit per agens extra educi, possit iterum ex ea sensitiva et vegetativa educi. Sed hoc est contra Aristotelem libro *De anima*;[36] dicit enim quod anima non est harmonia et improbat hoc.

Item, natura animae et corporis sive huiusmodi harmoniae sunt ex primis differentibus in genere et distantibus secundum naturam, licet non secundum entitatem similitatis, quia harmonia miscibilium forma corporalis est sive corporea et anima est spiritualis et incorporea, ergo unum non fit aliud sub alia harmonia vel proportione.

Praeterea, forma mixta nec est elementum nec ex elementis, igitur nec anima multo magis, sed illa harmonia est ex elementis; igitur, et cetera. Unde non educitur ex alio tamquam ex illo in quo est in potentia. Quod concedo uno modo.

Diceret aliquis tamen sicut Augustinus forte quod per totas naturas elementares sunt creaturae rationes seminales quae sunt vel dicuntur potentiae activae vel formae diminutae in materia. Unde vegetativa et sensitiva secundum sic ponentes corrumpuntur in suas rationes seminales. Unde forte corpus canis et ratio seminalis vegetativae et sensitivae eius possit iterum sublimari et habilitari per agens extra ut iterum ex eis possit educi canis. Non sequitur resurrectio, quia resurrectio eius ad idem numero reditio, non ad idem specie. Et hic esset reditio ad idem specie, non ad idem numero.

---

[35] Aristoteles, *De anima* non inv.; sed vide *De sensu et sensato*, 3, 439b1–14 ad sensum (Sancti Thomae de Aquino Opera Omnia, ed. Leonina 45, *Sentencia libri De sensu et sensato*, tr. 1, c. 5, p. 33).

[36] Aristoteles, *De anima*, I.4, 408a29 (Sancti Thomae de Aquino, ed. Leonina, Opera Omnia 45, I, ch. ix, 45).

References

Primary Literature

Averroes. *Metaphysica*. In *Aristotelis opera cum Averrois commentariis. Venetiis, Apud Junctas, 1562–1574*, vol. VIII.

Averroes. 1953. *Commentarium Magnum in Aristotelis De Anima Libros*, ed. Stuart Crawford. Cambridge, MA: Medieval Academy of America.

Galfridus de Aspall. 2010. *Questiones in "De Anima"*, ed. Vlatka Čizmić, 3 vols. Munich.

Geoffrey of Aspall. 2012. *Questions on Whether the World in Generable and Perishable. An Edition and Discussion*, ed. Griet Galle. *Documenti e studi sulla tradizione filosofica medievale* 23: 321–355.

———. 2014. *Quaestiones super librum "De somno et vigilia": An Edition*, ed. *Sten Ebbesen. Cahiers de l'Institut de Moyen-Âge Grec et Latin* 83: 257–341.

———. 2017. *Questions on Aristotle's "Physics"*, ed. Silvia Donati and Cecilia Trifogli. Trans. Jennifer Ashworth and Cecilia Trifogli. Oxford: Oxford University Press.

Robertus Grosseteste. 1981. In *Commentarius in Posteriorum Analyticorum Libros*, ed. Pietro Rossi. Florence: Olschki.

Sigerus de Brabantia. 1974. *Ecrits de logique, de morale et de physique*, ed. Bernardo Bazán. Louvain: Publications Universitaries; Paris: Beatrice-Nauwelaerts.

Secondary Literature

Calma, Dragos. 2011. *Études sur le premier siècle de l'averroïsme latin. Approches et textes inédits*. Turnhout: Brepols.

Davies, Rachel. 2019. *Bonaventure, the Body and the Aesthetics of Salvation*. Cambridge: Cambridge University Press.

Donati, Silvia. 2012. Goffredo di Aspall (†1287) e alcuni commenti anonimi ai "Libri naturales" [...]. Parte I. *Documenti e studi sulla tradizione filosofica medievale* 23: 245–320.

———. 2013. Goffredo di Aspall (†1287) e alcuni commenti anonimi ai "Libri naturales" [...]. Parte II. *Documenti e studi sulla tradizione filosofica medievale* 24: 219–418.

Marone, Steven P. 1983. *William of Auvergne and Robert Grosseteste: New Ideas of Truth in Early Thirteenth Century*. Pronceton: Princeton University Press.

Trifogli, Cecilia. 2018. Geoffrey of Aspall on Nature. *Recherches de Theologie et Philosophie Medievales* 85: 47–69.

Chapter 10
Medieval Views on the Subject of Thought and the Intellectual Soul

Cecilia Trifogli

Abstract The speculation of medieval Latin philosophers about the subject of thought and the intellectual soul addresses questions that are of great philosophical importance for an understanding of human nature, such as: What is the subject of thought? Is it the composite of body and the intellectual soul, or is it the intellectual soul alone? If the subject of thought is the intellectual soul alone, does this confer a special status on the intellectual soul compared to the souls of other animals? Does this make the intellectual soul itself somehow independent from the body? And if the intellectual soul is somehow independent from the body, can it still be regarded as the substantial form of a human being? These specific questions of the medieval debate are the subject of this paper, which focuses on Aquinas's replies and solutions. I argue that they constitute his most sophisticated contribution to the Aristotelian theory of human nature and, moreover, were highly influential. However, Aquinas's opinion on this matter did not meet universal consent. Serious objections to Aquinas's view were raised even by one of his closest followers, Giles of Rome, and later on, in the third decade of the fourteenth century, by Thomas Wylton. The critical responses of these two opponents to Aquinas are considered in this paper.

Keywords Subject of thought · Intellectual soul · Thomas Aquinas · Giles of Rome · Thomas Wylton

10.1 Introduction

In Aristotle's philosophy, human beings are conceived of as rational animals. They are animals but animals of a special kind. What is special about them is that they are rational, that is, they are capable of acts of thought (and will). In addition to sensory cognition, which they have in common with all other animals, they have intellectual

C. Trifogli (✉)
University of Oxford, Oxford, UK
e-mail: cecilia.trifogli@philosophy.ox.ac.uk

M. Gensler et al. (eds.), *The Embodied Soul*, Historical-Analytical Studies on Nature, Mind and Action 11, https://doi.org/10.1007/978-3-030-99453-2_10

cognition. In very general and abstract terms, the question I shall address in this paper is how deep the difference between human beings and the other animals is. Are human beings fundamentally unlike the other animals or are they nothing more than the most complex animals? In the medieval Aristotelian tradition, there are two major specific issues relevant to this question: one is the subject of thought and more generally of intellectual acts and the other is the nature of the intellectual soul.

As to the intellectual soul, this paper will focus on Aristotle's considered view, namely, the view that the intellectual soul is the (only) substantial form of a human being.[1] According to this view, human beings, like all the other animals, are hylomorphic composites—composites of matter and substantial form—and the intellectual soul (also called rational soul) is their substantial form. This means that the intellectual soul is what makes a human being the kind of thing it is. And since both acts of sensation and acts of thought define the kind of thing a human being is, the intellectual soul is that in virtue of which a human being is capable of both kinds of act.

The controversial question is whether the intellectual soul performs exactly the same role in acts of sensation and in acts of thought. In the case of sensory acts it is clear that the soul simply provides a principle in virtue of which a human being, that is, a composite of matter and soul, acts: the composite is the subject or agent of a sensory act, whereas the soul is the principle internal to the agent in virtue of which the agent acts. For example, the subject of a visual act are the eyes, which are bodily structures endowed with the visual power (a power of the soul), and not the visual power alone or the soul alone. According to Aristotle, however, there is not an analogous bodily structure in the case of acts of thought: there is not a bodily organ of thought. From this peculiarity of thought some important questions arise. Given that there is not a bodily organ of thought, should the intellectual soul itself be identified with the subject of thought? In another formulation: is the subject of thought the composite of body and the intellectual soul or the intellectual soul alone? And if the subject of thought is the intellectual soul alone, does this confer a special status on the intellectual soul compared to the souls of other animals? Does this make the intellectual soul itself somehow independent from the body? And if the intellectual soul is somehow independent from the body, can it still be regarded as the substantial form of a human being?

These are the specific questions that my paper will address. The main focus will be Aquinas's reply to them.[2] It is arguably his most sophisticated contribution to the Aristotelian theory of human nature and was highly influential. It did not meet, however, with universal consent. Serious objections to Aquinas's view were raised even by one of his closest followers, Giles of Rome, and later on, in the third decade of the fourteenth century, by Thomas Wylton. These are the two opponents to Aquinas I shall consider in this paper.[3]

---

[1] For a study on these topics that takes into account not only Aristotle's view but also Averroes's view and the reactions of some medieval Latin philosophers to Averroes, see McCord Adams, Trifogli (2012).

[2] Important recent studies devoted to Aquinas's view are those of Pasnau (2012) and Bazán in Thomas d'Aquin, *L'Ame et le Corps. Somme de théologie* (2016).

[3] For other medieval discussions about the nature of the intellect and the immortality of the soul, see Dunne, Chap. 9, pp. 163–190, and Lička, Chap. 14, pp. 277–310 in this volume.

10.2 Aquinas's View

An effective way to introduce Aquinas's reply to the questions raised above is by examining his argument for the 'subsistence' of the intellectual soul in his *Summa theologiae*, since it contains the key features of his account of intellectual cognition and of the intellectual soul. The argument is the following:

> (T1) <1> I answer that it is necessary to say that that which is the principle of intellectual operation, which we call the soul of a man, is a principle both incorporeal and subsistent.
>
> <2> For it is clear that in virtue of the intellect a man can know the natures of all bodies. And it is necessary that that which can know some things does not have any of them in its own nature, because that which inheres in it would naturally impede the cognition of other things, just as we observe that a sick man's tongue, being vitiated by a choleric and bitter humour, cannot perceive something sweet, but everything seems bitter to him. Therefore, if the intellectual principle had in itself the nature of a body, it would be unable to know all bodies. But every body has its own determinate nature. Therefore **<C1> it is impossible for the intellectual principle to be a body. It is likewise impossible for it to understand by means of a bodily organ,** since the determinate nature of that bodily organ would impede the cognition of all bodies, just as if a certain determinate colour is not only in the pupil of the eye, but also in a glass vase, the liquid contained in the vase seems to be of that same colour.
>
> <3> Therefore, **<C2> the intellectual principle, which we call the mind or the intellect, has an operation per se, in which the body does not take part.**
>
> <4> But only that which subsists per se can have an operation per se. For nothing can operate except that which is a being in act so that the way in which something operates is the way in which something exists. This is why we do not say that heat heats, but that that which is hot heats. It follows that **<C3> the human soul, which is called the intellect or the mind, is something incorporeal and subsistent.**[4]

Aquinas's argument involves three main claims (printed in bold in text (T1)). The first claim (C1) is that the intellect, understood as the intellectual power (the power for intellectual operations), is incorporeal in the sense that it is not itself a body and more crucially is not a power of a bodily organ, so that there is not an (bodily) organ

---

[4]Thomas de Aquino, *Summa theologiae*, I, q. 75, art. 2, corp.: "<1> Respondeo dicendum quod necesse est dicere id quod est principium intellectualis operationis, quod dicimus animam hominis, esse quoddam principium incorporeum et subsistens. <2> Manifestum est enim quod homo per intellectum cognoscere potest naturas omnium corporum. Quod autem potest cognoscere aliqua, oportet ut nihil eorum habeat in sua natura, quia illud quod inesset ei naturaliter impediret cognitionem aliorum; sicut videmus quod lingua infirmi quae infecta est cholerico et amaro humore, non potest percipere aliquid dulce, sed omnia videntur ei amara. Si igitur principium intellectuale haberet in se naturam alicuius corporis, non posset omnia corpora cognoscere. Omne autem corpus habet aliquam naturam determinatam. **<C1> Impossibile est igitur quod principium intellectuale sit corpus. Et similiter impossibile est quod intelligat per organum corporeum**, quia etiam natura determinata illius organi corporei prohiberet cognitionem omnium corporum; sicut si aliquis determinatus color sit non solum in pupilla, sed etiam in vase vitreo, liquor infusus eiusdem coloris videtur. <3>**<C2> Ipsum igitur intellectuale principium, quod dicitur mens vel intellectus, habet operationem per se, cui non communicat corpus**. <4> Nihil autem potest per se operari, nisi quod per se subsistit. Non enim est operari nisi entis in actu, unde eo modo aliquid operatur, quo est. Propter quod non dicimus quod calor calefacit, sed calidum. Relinquitur igitur **<C3> animam humanam, quae dicitur intellectus vel mens, esse aliquid incorporeum et subsistens.**" (ed. Leonina, electronic version: www.corpusthomisticum.org.; boldface, numbering and English translation are mine).

of thought. The second claim (C2) is that intellectual acts are acts of the intellectual soul alone and not of the composite of body and soul, so that the subject of thought is the intellectual soul alone and not the composite of body and soul. The third claim (C3) is that the intellectual soul is subsistent. Being subsistent means having the mode of being of a substance. The contrast implicit here is that between subsistence and inherence. The intellectual soul is a form but a subsisting form, not an inhering form. It is a form that exists together with matter (as long as the human being of which it is the form exists) but does not depend on matter for its existence.

To get a better grasp of the ideas at work in Aquinas's argument, it is helpful to reformulate the three claims in terms of independence from the body, so that in these simplified formulations (C1) says that the intellect is independent from the body, (C2) says that thought is independent from the body, and (C3) says that the intellectual soul is independent from the body.

The first two claims (C1) and (C2) are the crucial premises of Aquinas's argument for (C3) in text (T1) above. Aquinas first proves (C1), that is, the independence of the intellect from the body (see paragraph <2>). From (C1) he then infers (C2): from the independence of the intellect to the independence of thought (see paragraph <3>). Finally, from (C2) he infers (C3): from the independence of thought to the independence of the intellectual soul (see paragraph <4>). Thus, according to Aquinas's argument, it is because the intellect is independent from the body that also thought is independent from the body, and it is because thought is independent from the body that also the intellectual soul is independent from the body, so that it can survive the death of the body, as Christian theology teaches. This is Aquinas's considered view on the question about the subject of thought and the status of the intellectual soul, a view that he consistently maintains in all his discussions of human nature.

Let us take a closer look at Aquinas's argument. The bulk of it is the proof of claim (C1) about the independence of the intellect. Aquinas's special effort to prove (C1) is possibly due to the fact that it is indeed the crucial premise of his argument for the subsistence of the soul. (C1), however, is universally agreed upon in the Aristotelian tradition. A major reason for this is that Aristotle himself explicitly argues for (C1) in *De anima* III.4.[5] In fact, Aquinas's proof of (C1) in text (T1) is a version of Aristotle's argument. According to some Aristotelian scholars,[6] Aquinas's version is an improvement on Aristotle's original argument, but it is still far from being convincing. The proof is based on an inference 'from content to constitution', that is, from a consideration about the content of thought—the range of things we can think of—to an ontological conclusion about the constitution of the intellect—its independence from the body. The inference is: since we can think of all bodies, then the intellect is not itself a body and is independent from the body to which it belongs so that there is no bodily organ of thought. An inference of this kind presupposes not only certain assumptions about the content, but also about specific cogni-

[5] Aristoteles, *De anima*, III.4, 429a10–27 (trans. Shields, 59).

[6] See Caston (2000, 153–155).

tive conditions and mechanisms that enable us to have cognitive access to that content. The difficulty with Aquinas's inference is that the assumptions about the cognitive conditions and mechanisms are left without an adequate explanation. The explanation provided by Aristotle and followed by Aquinas appeals to an analogy between sensing and thinking, but it is not clear how this analogy actually works. And when one tries to fill in this explanatory gap using the relevant aspects of Aquinas's theory of cognition, typically intentionality, the argument does not seem to work at all.[7] In other contexts, Aquinas offers his own version of the proof of (C1), which is not based on Aristotle's argument of *De anima* III.4. Aquinas's own version also is based on an inference from content to constitution, where the relevant assumption about content in this other version is that we think of universal things: it is because the proper objects of the intellect are universals rather than singulars that the intellect is independent from the body.[8] This version has problems too. Since, however, claim (C1) is universally accepted in the Aristotelian tradition, I will not go into more details in this paper.

Another step of the argument in text (T1) for which Aquinas hints at a proof is the inference from (C2) to (C3) in paragraph <4>, that is, the inference from the independence of thought to the independence of the intellectual soul. Here Aquinas resorts to a very general principle about the correlation between independence in action and independence in being or existence, according to which if A acts independently from B, then A also exists independently from B, and the other way around. In the Aristotelian context, the principle makes the point that the only items that are properly speaking agents are those with the ontological status of a substance, that is, subsisting items. Accidents do not act properly speaking nor do the substantial forms of all material substances other than human beings. Applied to our case, the principle allows the passage from (C2) to (C3): from the claim that the intellectual soul has acts of thought independently from the body to the claim that the intellectual soul also exists or can exist independently from the body. The general principle about the correlation between being and acting invoked by Aquinas is indeed a big one and would need a careful treatment, which Aquinas does not provide. He simply illustrates it with the example of heat and a hot body: the agent of heating is the hot body and not its heat, and the reason for this is that the hot body is subsistent whereas its heat (an accidental form) is not. Aquinas's general principle is not found in Aristotle, nor is it universally agreed upon in the Aristotelian tradition. Ockham would reject it. Indeed, in connection with Aquinas's example, he maintains that it is the heat itself of a hot body and not the hot body that is the proper agent of heating.[9] For the purpose of this paper, however, we can take this principle and the inference from (C2) to (C3) for granted, since the objections to Aquinas's view I intend to focus on do not concern this principle.

---

[7] For more details about this problem, see Trifogli (2021, 143–145).

[8] See, e.g., Thomas de Aquino, *Summa theologiae*, I, q. 75, art. 5, corp. (ed. Leonina, electronic version: www.corpusthomisticum.org)

[9] On this point, see Trifogli (2012, 232).

What is then controversial in Aquinas's view as expressed by his argument for the subsistence of the human soul in text (T1)? First and foremost, it is his inference from (C1) to (C2), the inference from the independence of the intellect from the body to the independence of thought from the body.

10.3 Aristotelian Evidence Against the Independence of Thought from the Body

Anachronistically, Aristotle himself is an authoritative source of objections against Aquinas's inference from (C1) to (C2). As I have pointed out, in *De anima* III.4 Aristotle endorses (C1): the independence of the intellect from the body. However, he seems to reject (C2): the independence of thought from the body. The main textual evidence against (C2) is found in two passages from *De anima* I.[10]

The first of the two passages belongs to the introductory section of the *De anima* where in laying out his plan for the inquiry into the nature of soul Aristotle indicates some major problems that need to be addressed. One of these concerns the affections of the soul and in particular of the human soul.

> (T2) <1> There is also a puzzle about the affections of the soul, concerning whether all are common to what has the soul as well or whether there is something peculiar to the soul itself. This it is necessary to grasp, but not easy.
>
> <2> It seems that in most cases the soul neither is affected nor acts without the body, as, for instance, with being angry or confident or appetitive, or, generally, with perceiving; reasoning, however, would seem most of all to be peculiar to it, but if this is a sort of imagination, or not without imagination, it would not be possible for even this to be without the body.
>
> <3> If, then, some one of the functions or affections of the soul is peculiar to it, it would be possible for the soul to be separated; but if there is nothing peculiar to it, it would not be separable.[11]

The question here is (see paragraph <1>) whether the affections of the soul are all shared by the body so that they are in fact affections of the composite of body and soul or whether there are some that are peculiar to the soul alone and thus independent from the body. In paragraph <2> Aristotle remarks that dependence on the body is clear in many of the soul's affections, such as anger, desire, or sensation.[12] Thinking (reasoning, in Shields's translation) would seem to be the most probable exception. However, Aristotle immediately suggests that in fact thinking is not such an exception. The reason he gives for this is that thinking is either the same as imagination or depends on imagination. The imagination, however, is a sense-power

---

[10] For a more extensive presentation of these two passages and of the conflicting interpretations that Aquinas and Giles of Rome offer of them, see Trifogli (2012, 222–231).

[11] Aristoteles, *De anima*, I.1, 403a3–12 (trans. Shields, 2–3).

[12] For Aristotle's general remarks in support of the dependence of all the affections of the soul on both body and soul, see Aristoteles, *De anima*, I.1, 403a16–b19 (trans. Shields, 3–4).

and so the power of a bodily organ. Therefore, if thinking is the same as or depends on the imagination, then thinking cannot occur without a body. According to Aristotle, then, it is the assumption of the strict connection between thinking and imagination that poses a serious challenge to the separation of thinking. In paragraph <3> Aristotle points out that the existence of an action proper to the soul (rather than to the composite of body and soul) is a necessary and sufficient condition for the separability of the soul from the body. Formulated in our notation, Aristotle's point is that (C2) is a necessary and sufficient condition for (C3), that is, the independence of thinking from the body is required and sufficient for the independence of the human soul from the body. Given Aristotle's reservation about the separability of thinking (the most plausible candidate for being an operation proper to the soul) expressed in paragraph <2>, this seems to lead to the rejection of (C3) that the intellectual soul is separable from the body.

The second passage from the *De anima* that strongly suggests that Aristotle does not endorse independence of thought from the body is the following:

> (T3) Yet saying that the soul is angry would be like saying that the soul weaves or builds. For it is perhaps better not to say that the soul pities or learns or thinks, but that the human being does these things with the soul.[13]

Aristotle here intends to make the point that saying that the soul itself gets angry is a very inappropriate way of speaking, since being angry is no more a property of the soul itself than building or weaving are; therefore, as we do all agree that the soul does not itself build, so we should also agree that the soul does not itself get angry. What we should rather say when we speak in an appropriate way is that the human being, that is, the composite of body and soul, and not the soul alone, has affections and actions like pity, learning, thinking. The soul is that in virtue of which the human being has these affections and actions, but not itself the subject of them. The explicit mention of thinking among the affections/actions that are more appropriately ascribed to the human being than to the human soul provides textual evidence against (C2).

10.4 Aquinas's Defence of the Independence of Thought

How does Aquinas react to this contrary evidence from Aristotle?

Aquinas acknowledges that thought depends on the imagination. It is indeed his considered view that the so-called phantasms, that is, sense-images, which are produced by the imagination, have a crucial role not only in the process of acquiring concepts (intelligible species) but also in the actual use of concepts already acquired, that is, in acts of thought: concepts are obtained by abstraction from phantasms and give rise to actual thought only by being joined to phantasms (the so-called 'turning

[13] Aristoteles, *De anima*, I.4, 408b11–15 (trans. Shields, 14).

to the phantasms').[14] However, Aquinas maintains that the dependence of thought on the imagination is such that it is compatible with the independence of thought from the body. In order to show this, he introduces a distinction between organ-dependence and object-dependence of cognitive acts on the body. This is how he uses this distinction in his commentary on Aristotle's passage (T2):

> (T4) <1> It is important to know, therefore, that there is one kind of operation or state of a soul that needs a body as instrument *and* as object. In this way seeing needs a body (i) as an object, since color (the object of sight) is a body, and also (ii) as an instrument, since vision, even if it is from soul, occurs only through the organ of sight (through the pupil, that is, which serves as an instrument). And in this way seeing belongs not only to soul but also to the organ. There is another kind of operation, however, that needs a body not as its instrument but only as its object. For intellective cognition does not occur through a corporeal organ but needs a corporeal object. For as the Philosopher says in Book III [12.431a14-15], phantasms are related to intellect in the same way that colors are related to sight. But colors are related to sight as objects; therefore phantasms are related to intellect as objects. Therefore, since phantasms do not occur without a body, it follows that intellective cognition does not occur without a body –but in such a way that the body serves as the object, not as the instrument.
>
> <2> Two results follow. One is that having intellectual cognition is an operation that is special to soul and needs body only as its object (as was said). Seeing, however, and other operations and states belong not to soul alone but to the compound. The other result is that what has its operation on its own has its existence and subsistence on its own; what does not have its operation on its own does not have its existence on its own. Hence intellect is a subsistent form; the other powers are forms in matter. And here lay the difficulty of this question: for all soul's states give the appearance of belonging to the compound.[15]

The main idea about the two kinds of dependence that Aquinas presents in this passage is that organ-dependence is much deeper than object-dependence. The reason for this is that the organ-dependence of a cognitive act on a body makes that act

---

[14] See, e.g., Thomas de Aquino, *Summa theologiae*, I, q. 84, art. 6–8; q. 85, art. 1 (ed. Leonina, electronic version: www.corpusthomisticum.org). For more about Aquinas's theory of intellectual cognition and in particular his rejection of innatism, see Pickavé, Chap. 11 in this volume, pp. 213–227.

[15] Thomas de Aquino, *Sentencia libri De anima*, I, c. II, (ed. Gauthier, 9: 50–10: 81): "<1> Sciendum est igitur quod aliqua operatio anime est aut passio que indiget corpore sicut instrumento et sicut obiecto; sicut uidere indiget corpore sicut obiecto quia color, qui est obiectum uisus, est corpus, item sicut instrumento quia uisio, etsi sit ab anima, non est tamen nisi per organum uisus, scilicet per pupillam que est ut instrumentum; et sic uidere non est anime tantum, set etiam organi. Aliqua autem operatio est que indiget corpore, non tamen sicut instrumento, set sicut obiecto tantum; intelligere enim non est per organum corporale, set indiget obiecto corporali: sicut enim Philosophus dicit in III huius, hoc modo fantasmata se habent ad intellectum sicut colores ad uisum, colores autem se habent ad uisum sicut obiecta, fantasmata ergo se habent ad intellectum sicut obiecta; cum ergo fantasmata non sint sine corpore, inde est quod intelligere non est sine corpore, ita tamen quod sit sicut obiectum, non sicut instrumentum. <2> Ex hoc duo sequntur. Vnum est quod intelligere est propria operatio anime et non indiget corpore nisi ut obiecto tantum, ut dictum est; uidere autem et alie operationes et passiones non sunt anime tantum, set coniuncti. Aliud est quod illud quod habet operationem per se, habet esse et subsistenciam per se, et illud quod non habet operationem per se, non habet esse per se; et ideo intellectus est forma subsistens, alie potencie sunt forme in materia. Et in hoc erat difficultas huiusmodi questionis, quia scilicet omnes passiones anime secundum apparenciam uidentur coniuncti." English translation in Thomas Aquinas, *A Commentary on Aristotle's De anima* (1999, 15).

belong to the composite of body and soul, i.e., makes that act dependent on the body. It is not the case instead that the object-dependence of a cognitive act on a body makes that act belong to the composite of body and soul. Object-dependence does not affect the independence of a cognitive act from the body. Accordingly, with Aquinas's examples, what makes seeing an action of the composite is essentially the property of seeing of requiring a bodily organ, and not the bodily nature of its objects (colours). Thinking, on the contrary, is an action of the soul because it does not require a bodily organ, although it depends on a body for its objects. Phantasms, Aquinas claims here, are objects of the intellect just like colours are objects of sight. Therefore, the dependence of thought on phantasms, being simply an object-dependence, does not challenge its independence from the body. One can suppose that, in Aquinas's mind, it is this distinction between two kinds of dependence that explains his inference from (C1) to (C2) in the argument for the subsistence of the soul.

One can still ask, however, whether Aquinas's reply to the contrary evidence provided by Aristotle's passage (T2) based on this distinction is convincing. Far from being so, one obvious problem with it is the analogy between colours and phantasms. Colours are objects of sight in the proper sense of being that which is grasped by sight, the items that are seen. Phantasms, instead, are not objects of the intellect in this proper sense. It is indeed Aquinas's considered view that the objects of the intellect—that which is understood (*id quod intelligitur*)—are the natures of material substances, which are mind-independent items, and not cognitive items (either phantasms or intelligible species).[16] Even more crucially, unlike colours, which are mind-independent items, phantasms are cognitive items, being products of a cognitive power, namely, the imagination. And the fact that phantasms are cognitive items gives rise to a powerful objection against Aquinas's insistence that the dependence of thought on phantasms does not challenge its independence from the body. For it seems to be a compelling requirement for the subject of a cognitive act that it should contain all the cognitive powers necessary for that act, so that if a cognitive act depends on other cognitive acts, then the proper subject of that act should also contain the cognitive powers of the other cognitive acts on which it depends. In the case of thinking, since thinking depends on phantasms and phantasms on the imagination, then the proper subject of thinking should include both the intellect and the imagination, and thus be at least in part a bodily subject. To put it another way, it is true that thinking does not itself have a bodily organ, but it necessarily involves the bodily organ of another cognitive power and thus at least indirectly depends on the body.[17]

Despite Aquinas's authority as an Aristotelian commentator, his reading of the dependence of thought on the imagination is scarcely convincing. The other two philosophers I will consider in this paper, Giles of Rome and Thomas Wylton,

---

[16] See Thomas de Aquino, *Summa theologiae*, I, q. 85, art. 1–2 (ed. Leonina, electronic version: www.corpusthomisticum.org).

[17] On this alternative way of explaining why thinking depends on the body according to Aristotle, see Frede (1992, 106).

depart from it and follow the 'natural' reading of Aristotle's texts, according to which the dependence of thought on the imagination does make thought too an act of the composite of body and soul and not of the soul alone. The list of Aquinas's opponents on this issue is very likely to grow with further historical research. What I found very interesting in Giles's and Wylton's discussions is that they point out that Aquinas's claim (C2) regarding the independence of thought from the body is not simply badly argued but also fundamentally wrong, because it threatens the status of the soul as substantial form of the human being.

10.5 Giles of Rome

Giles's strong disagreement with Aquinas regarding the independence of thinking is very explicit in the section of his *De anima* commentary devoted to Aristotle's passage (T3). Giles's reading of this passage is strikingly different from that of Aquinas. In his commentary, Aquinas quickly dismisses the apparent evidence that this passage provides against his claim about the independence of thought. He maintains that when Aristotle lists thought among the affections of the soul that should be properly ascribed to the human being rather than to the soul alone, he is arguing *ad hominem*, not *ad rem*, that is to say, Aristotle is granting as true the view of those who posit that thinking too has its own bodily organ, a view that Aristotle himself rejects.[18] For Giles, on the contrary, Aristotle is arguing *ad rem*. According to Giles, it is Aristotle's considered view that all the affections of the soul, including thought, belong to the composite and not to the soul alone. Giles devotes two short questions to establishing this point.

The first question asks whether affections like rejoicing and being sad belong to the soul alone.[19] Giles replies:

> (T5) It must be said that, as Aristotle sufficiently indicates in the text, one thing is a principle of acting and another thing is that which acts. It is indeed neither the case that a form acts nor that matter is acted upon; rather, it is the composite that acts in virtue of its form and is acted upon in virtue of its matter. For we have to imagine that just as things are related to being, so they are related to acting. Therefore, the things to which acting by themselves belong are those to which being by themselves belong. It follows from this that the composite acts by itself because the composite has being by itself; a form, instead, does not act by itself but is a principle of acting because it does not have being by itself but is a principle of being. Therefore, hating, rejoicing, being sad and the like, regardless of whether they express an action or a passion, cannot belong to the soul itself, because a thing that does not have being by itself can neither act by itself nor be acted upon by itself. Properly speaking, then, the soul neither rejoices by itself nor is sad by itself, but is a principle in virtue of which the composite rejoices and is sad. Therefore, it has been said correctly that

[18] See Trifogli (2012, 227–228).

[19] Aegidius Romanus, *In libros De anima expositio*, I, cap. 63 (ed. Venice 1500, f. 17rb): "Dubitaret forte aliquis utrum gaudere et tristari et cetera huiusmodi competant ipsi animae secundum se."

it is better to say not that the soul rejoices or is sad but that a man does this in virtue of the soul.[20]

In this reply, Giles makes explicit the metaphysical background of Aristotle's view. In Giles's reading, in his account of change Aristotle distinguishes between (i) the *ratio agendi/patiendi*, that is, the explanatory principle of an action or a passion, that in virtue of which something acts or is acted upon, and (ii) the agent or patient, that is, the subject of an action or a passion, the thing that acts or is acted upon. Given this distinction, Aristotle then posits that only the composite of matter and form, i.e., the material substance, is the agent or patient of a change, whereas its components are principles of change: form is the *ratio agendi*—a principle of the actions of a material substance—, and matter is the *ratio patiendi*—a principle of its being acted upon. In particular, as Giles underlines, it is in the nature of a form that it can only be the principle of an action, but not the subject of an action. The human soul, being the substantial form of the human being, is no exception to this general rule: it can only be the *ratio agendi* of the affections of the human being but not the subject of any of them. As the authoritative Aristotelian scholar Michael Frede puts this point: "Being a form or nature it [i.e., the soul] is a principle of change, something one appeals to to explain change, which hence cannot itself be the kind of thing which is subject to change. Thus it cannot be the kind of thing which has a life of its own."[21]

The second question discussed by Giles is devoted specifically to the case of thought. The question asks whether thought is an exception to the general rule that no affection of the soul belongs to the soul itself.[22] Giles replies:

(T6) **<1> It must be replied that, according to some people, thinking is an operation of the soul in which the soul does not communicate with the body, and such operation, as they say, does not belong to the composite itself.** This is why, as they say, the Philosopher said above that thinking is something different, that is, because it is not an operation of the composite itself, but of the soul itself. For, since the intellect is not an organic power, there is no corporeal organ per se at work in thinking; therefore, such operation seems to belong not to the whole composite, but to the soul alone.

---

[20] Aegidius Romanus, *In libros De anima expositio*, I, cap. 63 (ed. Venice 1500, f. 17rb): "Dicendum quod, ut satis innuit Philosophus in littera, aliud est ratio agendi et aliud est quod agit. Nam nec forma agit nec materia patitur, sed compositum agit per formam et patitur per materiam. Debemus enim imaginari quod sicut res se habet ad esse ita se habet ad agere. Quibus ergo competit per se esse competit per se agere. Compositum ergo quia per se est, per se agit; forma vero quia non per se est, sed est ratio essendi, non per se agit, sed est ratio agendi. Odire ergo, gaudere, tristari et cetera talia, sive dicant actionem sive passionem, non poterunt per se competere animae, quia nihil quod non est per se nec per se agit nec per se patitur. Anima ergo per se loquendo nec per se gaudet nec per se tristatur, sed est ratio quare compositum gaudeat et tristetur. Bene ergo dictum est quod est melius dicere animam nec gaudere nec tristari sed hominem agere haec per animam." (The English translation of the passages from Giles's *De anima* commentary is mine).

[21] Frede (1992, 106).

[22] Aegidius Romanus, *In libros De anima expositio*, I, cap. 63 (ed. Venice 1500, f. 17rb): "Ulterius forte dubitaret aliquis utrum intelligere per se competat animae ita quod possimus concedere quod, quamvis anima non gaudeat nec tristetur, sed homo per animam, nihilominus tamen anima per se intelligit."

<2> **But this cannot stand.** For as long as the soul is joined to the body, it does not have being by itself; and if it does not have being by itself, it does not act by itself. Therefore, thinking too does not belong to the soul itself, but to the composite itself. Even if it were the case that the intellect in no way, i.e., neither per se nor accidentally, used a corporeal organ, however, from the assumption that the soul is the form of the body—an assumption that implies that the soul does not have being by itself, but is a principle of being—it would follow that the soul does not act by itself, but is a principle of acting. Accordingly, thinking itself does not belong to the soul by itself, but just as the soul does not have being by itself, but it is a principle in virtue of which a man has being, so the soul does not think by itself, but it is a principle in virtue of which a man thinks.

<3> **This is the truth that Catholic authors must especially express and confess, namely, that properly speaking we should not say that the soul thinks but that a man thinks in virtue of the soul.** And granted this, the claim that the intellectual soul is the form of the body and is multiplied according to the multitude of bodies must be conceded. What is more, since Averroes himself grants that it is not the soul that learns or discerns, but that a man does this in virtue of the soul, it was necessary for him to say, compelled by this truth, that the intellectual soul is the form of the body, and that there is not only one intellect in number in all men, but that the intellect is multiplied according to the multiplicity of the intellectual souls; and it is necessary for a thing of this kind to be multiplied according to the multiplicity of bodies.[23]

There is little doubt that the opinion Giles refers to in introducing his reply at the beginning of paragraph <1> (printed in bold) is that of Aquinas.[24] This makes it clear that Giles's question specifically devoted to thinking is motivated by the opinion of Aquinas. Giles strongly rejects Aquinas's view. There is something basically wrong with it, as Giles's comment "But this cannot stand" at the beginning of paragraph <2> (printed in bold) suggests. In explaining his comment, Giles points out that ascribing thinking to the soul itself violates the general principle (stated in the

[23] Aegidius Romanus, *In libros De anima expositio*, I, cap. 63 (ed. Venice 1500, f. 17rb): "<1> Dicendum quod, secundum quosdam, intelligere est quaedam operatio animae in qua non communicat cum corpore, et huiusmodi operatio, ut dicunt, non est ipsius coniuncti. Ideo, ut dicunt, Philosophus supra dixit quod intelligere est aliquid alterum huiusmodi, quia non est operatio ipsius coniuncti, sed ipsius animae. Nam cum intellectus non sit virtus organica, in intelligendo per se non laborat organum corporale; quare talis operatio non videtur esse totius coniuncti, sed animae. <2> Sed hoc stare non potest. Nam quamdiu est anima coniuncta corpori, non habet per se esse, et si non competit ei per se esse, non competit ei per se agere. Ipsum etiam intelligere non est ipsius animae, sed est ipsius coniuncti. Dato ergo quod intellectus nullo modo uteretur organo corporali, nec per se nec per accidens, attamen ex quo anima est forma corporis, quo posito non per se est, sed est ratio essendi, sequitur quod non per se agat, sed sit ratio agendi. Quare ipsum intelligere non per se competit animae, sed sicut anima non per se est, sed est ratio quare homo sit, sic non per se intelliget, sed est ratio quare homo intelligat. <3> Hanc autem veritatem specialiter debent exprimere et confiteri catholici tractatores, videlicet quod secundum proprietatem locutionis non dicamus quod anima intelligat, sed quod homo per animam; quo posito oportet concedere animam intellectivam esse formam corporis et multiplicari secundum multitudinem corporum. Immo, cum ipse idem Averroes hoc concedat, quod anima non addiscat nec distinguat, sed homo hoc faciat per animam, ab ipsa veritate coactus oportet eum dicere [quod] animam intellectivam esse formam corporis, et non esse unum intellectum numero in omnibus, sed plurificari secundum plurificationem animarum intellectivarum; huiusmodi autem secundum hoc plurificari oportet secundum plurificationem corporum."

[24] In the analysis of text (T6) I follow closely my paper Trifogli (2012, 234–237).

previous question) that only things that have being by themselves—things that subsist—have actions by themselves. Because it is a form, the intellectual soul does not have being by itself, and therefore does not have actions by itself. Thinking cannot be an action proper to the soul, if the soul is the form of the human being and thus not a subsisting thing.

While Giles's interpretation is closer to Aristotle's text than that of Aquinas, Giles's attack on Aquinas is not fair. It is indeed Aquinas's considered view that the intellectual soul is both a form and something subsisting, i.e., something that has being by itself. Aquinas is well aware that this gives the intellectual soul a special status among the substantial forms of material substances: it is the only subsisting form. Also, Aquinas makes quite some effort to show that the special status of the intellectual soul is compatible with its nature of the form of the human being.[25] But Giles disregards all Aquinas's effort. He simply appeals to the Aristotelian orthodoxy about the nature of a substantial form, without any discussion of Aquinas's arguments for the special status of the intellectual soul.[26]

Giles's additional remarks in paragraph <3> (printed in bold), however, are very interesting. They point to a connection between the issue of the subject of thought and Averroes's view. According to the standard reading of Latin medieval philosophers, Averroes posits that the intellectual soul or the intellect is not the form of a human being but a substance separate from all human beings and one for all human beings. The Catholics who want to reject Averroes's view—Giles remarks—should maintain that the true claim about the subject of thought is that a human being thinks in virtue of the soul and not the claim that the soul itself thinks. The reason for this is that the claim that a human being thinks in virtue of the soul is incompatible with the status of separate substance that Averroes ascribes to the soul and in fact implies that the soul is the form of a human being. On the contrary, the claim that the soul itself thinks is not incompatible with the status of the soul as a separate substance. For Giles, Averroes himself concedes that a human being thinks in virtue of the soul, rather than that the soul itself thinks, and this makes his position inconsistent: if it is a human being that thinks in virtue of the soul, then the soul cannot be a substance separate from a human being, but must be its form.

I think that the target of Giles's polemical remarks in paragraph <3> is again Aquinas. Aquinas, just like Giles and the great majority of medieval Latin philosophers, rejects Averroes's views. Moreover, Aquinas's strongest arguments against Averroes exactly start from the premise 'a human being thinks' (*homo intelligit*) and infer from it that the soul is the form of a human being.[27] The question then that Giles implicitly asks Aquinas, in my reading, is the following: how can Aquinas both (i) maintain that properly speaking the soul thinks rather than that a human

[25] See, e.g., Thomas de Aquino, *Summa theologiae*, I, q. 76, art. 1, ad 1–6 (ed. Leonina, electronic version: www.corpusthomisticum.org).

[26] For a more extensive discussion of Giles's reception of Aquinas's view, see Trifogli (2012, 237–241).

[27] See, for example, text (T7) below. For an outline of Aquinas's critique of Averroes's view and of Aquinas's own position, see McCord Adams, Trifogli (2012, 629–633).

being thinks and (ii) use the claim that a human being thinks as a true premise in his strongest arguments against Averroes? Although Aquinas may have a reply to this question, he does not take enough care to spell it out in details in the relevant contexts. In this respect, Giles's question is legitimate and reveals a lack of transparency in Aquinas. There is indeed an apparent tendency in Aquinas (i) to insist on the truth of the claim that the soul thinks and contrast it sharply with the claim that a human being thinks when arguing for the subsistence of the soul, but (ii) to assume as an undeniable truth that a human being thinks when arguing for the claim that the human soul is the form of a human being.[28]

An excellent example of the kind of tension in Aquinas's account alluded to by Giles is found in the question of the *Summa theologiae* which asks whether the intellectual soul is the form of a human being.[29] In this question Aquinas defends the Aristotelian view that the intellectual soul is the form of a human being against the dualist views of Averroes and Plato. Aquinas presents three main arguments in support of Aristotle's view. In the first and third arguments Aquinas explicitly assumes that the intellectual soul is the principle in virtue of which a human being has its specific vital operations, including thought, so that the claim taken as true here is that a human being thinks and does so in virtue of its soul as principle of the relevant action, namely, thinking.[30] These two arguments then reflect the Aristotelian orthodoxy with respect to the human soul. Actually, as Aquinas remarks, the first argument is the argument of Aristotle himself in *De anima* II.2 for the conclusion that a soul is the form of a living being.[31] Both arguments then are based on the Aristotelian inference: the soul is the principle of action (*ratio agendi*) of a human being, therefore, it is its form.

---

[28] Similar worries about the consistency of Aquinas's view of the human soul are put forwards by Bazán in Thomas d'Aquin, *L'Ame et le Corps. Somme de théologie* (2016, 90–113).

[29] Thomas de Aquino, *Summa theologiae*, I, q. 76, art. 1 (ed. Leonina, electronic version: www.corpusthomisticum.org): "Utrum intellectivum principium uniatur corpori ut forma."

[30] The first argument is: "Illud enim quo primo aliquid operatur est forma eius cui operatio attribuitur, sicut quo primo sanatur corpus est sanitas, et quo primo scit anima est scientia; unde sanitas est forma corporis, et scientia animae. Et huius ratio est, quia nihil agit nisi secundum quod est actu, unde quo aliquid est actu, eo agit. Manifestum est autem quod primum quo corpus vivit est anima. Et cum vita manifestetur secundum diversas operationes in diversis gradibus viventium, id quo primo operamur unumquodque horum operum vitae est anima. Anima enim est primum quo nutrimur, et sentimus, et movemur secundum locum, et similiter quo primo intelligimus. Hoc ergo principium quo primo intelligimus, sive dicatur intellectus sive anima intellectiva, est forma corporis. Et haec est demonstratio Aristotelis in II *De anima*." (ibidem, corp.). The third argument is: "Potest etiam idem manifestari ex ratione speciei humanae. Natura enim uniuscuiusque rei ex eius operatione ostenditur. Propria autem operatio hominis, inquantum est homo, est intelligere. Per hanc enim omnia animalia transcendit. Unde et Aristoteles, in libro *Ethicorum*, in hac operatione, sicut in propria hominis, ultimam felicitatem constituit. Oportet ergo quod homo secundum illud speciem sortiatur, quod est huius operationis principium. Sortitur autem unumquodque speciem per propriam formam. Relinquitur ergo quod intellectivum principium sit propria hominis forma." (ibidem, corp.)

[31] Aristoteles, *De anima*, II.2, 413b11–414a14 (trans. Shields, 25–26).

The second argument is more original. It is worthwhile to quote the relevant parts of it:

(T7) <1> But if someone wants to maintain that the intellectual soul is not the form of the body, he must find a way to explain how it is that this action of understanding is the action of this particular human being. For each one experiences that it is himself who understands. Now an action may be attributed to something in three ways, as is clear from the Philosopher in *Physics* V; for a thing is said to move or act either in virtue of itself as a whole, for instance, as a physician heals; or in virtue of a part of it, as a human being sees by his eyes; or by accident, as when we say that something that is white builds because it is accidental to the builder to be white. Thus, when we say that Socrates or Plato understands, it is clear that this is not attributed to him accidentally; for it is ascribed to him in so far as he is a human being, which is predicated of him essentially. We must therefore say either that Socrates understands in virtue of himself as a whole, as Plato maintained, holding that a human being is an intellectual soul; or we must say that the intellect is a part of Socrates. The first cannot stand, as was shown above, for this reason that it is one and the same human being who experiences both that he understands and that he senses. But one cannot sense without a body; therefore the body must be a part of a human being. It remains therefore that the intellect by which Socrates understands is a part of Socrates, so that in some way it is united to the body of Socrates […].

<2> There remains, therefore, no other way of explaining this than that posited by Aristotle—namely, that this particular human being understands because the intellectual principle is his form. Thus from the very operation of the intellect it is made clear that the intellectual principle is united to the body as its form.[32]

This argument too starts from the premise that the human being thinks, which is taken as an undeniable psychological datum ("each one experiences that it is himself who understands"). Aquinas's explanation of this premise in paragraph <1>, however, is not the Aristotelian one. For in this new explanation the role of the soul is not that of being the principle in virtue of which a human being thinks, but that of being the part of a human being that actually thinks, that is, the part of a human being that is the subject of thought. This is clear from the way in which Aquinas sets the alternative explanations of the fact that a human being thinks, and in particular

[32] Thomas de Aquino, *Summa theologiae*, I, q. 76, art. 1, corp. (ed. Leonina, electronic version: www.corpusthomisticum.org): "<1> Si quis autem velit dicere animam intellectivam non esse corporis formam, oportet quod inveniat modum quo ista actio, quae est intelligere, sit huius hominis actio. Experitur enim unusquisque seipsum esse qui intelligit. Attribuitur autem aliqua actio alicui tripliciter, ut patet per Philosophum, V *Physicorum*. Dicitur enim movere aliquid aut agere vel secundum se totum, sicut medicus sanat; aut secundum partem, sicut homo videt per oculum; aut per accidens, sicut dicitur quod album aedificat, quia accidit aedificatori esse album. Cum igitur dicimus Socratem aut Platonem intelligere, manifestum est quod non attribuitur ei per accidens. Attribuitur enim ei inquantum est homo, quod essentialiter praedicatur de ipso. Aut ergo oportet dicere quod Socrates intelligit secundum se totum, sicut Plato posuit, dicens hominem esse animam intellectivam, aut oportet dicere quod intellectus sit aliqua pars Socratis. Et primum quidem stare non potest, ut supra ostensum est, propter hoc quod ipse idem homo est qui percipit se et intelligere et sentire; sentire autem non est sine corpore; unde oportet corpus aliquam esse hominis partem. Relinquitur ergo quod intellectus quo Socrates intelligit, est aliqua pars Socratis ita quod intellectus aliquo modo corpori Socratis uniatur […]. <2> Relinquitur ergo solus modus quem Aristoteles ponit, quod hic homo intelligit, quia principium intellectivum est forma ipsius. Sic ergo ex ipsa operatione intellectus apparet quod intellectivum principium unitur corpori ut forma." (English translation is mine).

from his rejection of the explanation that a human being thinks as a whole. Aquinas assumes here that the following inference is valid: if a human being thinks as a whole, then the human being is the intellectual soul alone, i.e., the intellectual soul is the whole of a human being. Clearly the inference is based on the implicit premise that that which thinks is the intellectual soul alone and not the composite of body and soul, so that if the whole of a human being thinks, then the whole of a human being must consist of the soul alone, a conclusion that Aquinas rejects. The true explanation for Aquinas, instead, is that a human being thinks because one of its parts, sc. the soul, thinks.[33] Thus, starting from the premise that the human being thinks but assuming the non-Aristotelian premise that the proper subject of thought is the soul, the first conclusion Aquinas reaches is that the intellectual soul is a part of a human being (see end of paragraph <1>). How can this intermediate conclusion lead one to the final conclusion that the intellectual soul is the form of a human being? This is hard to see, and Aquinas fails to show this properly. In the rest of the passage, which I have not reported, he considers accounts alternative to that of Aristotle (basically those of Averroes and Plato) and shows that they are not satisfactory. The conclusion that the intellectual soul is the form of the human being is then reached in paragraph <2> with an indirect argument, based on a process of elimination of the alternative accounts ("There remains, therefore, no other way of explaining this than that posited by Aristotle"). This is a weak argument. I do not want to go into many details, but let me just point out an obvious problem that Aquinas does not address in his argument based on elimination. Consider Aquinas's analogy between the human soul and the eye in paragraph <1>: the human being thinks in virtue of a soul just as it sees in virtue of an eye, so that the soul and the eye here are the parts of the human being that are the proper agents of thinking and seeing respectively. The problem is that the eye is not a form of the human being or something belonging to its soul (it is instead a composite of matter/body and soul). Why then should the soul be the form of the human being? Aquinas does not properly answer this question.

10.6 Thomas Wylton

The early fourteenth-century philosopher and theologian Thomas Wylton is very sceptical about the possibility of finding a convincing answer to this question. He devotes a lengthy disputed question to this issue, the so-called *Quaestio de anima intellectiva*, a question which asks: "Whether it can be established by necessary and evident argument that the intellectual soul is the form of the human body."[34] Wylton's

[33] For this interpretation of Aristotle's claim that the human being thinks, see also Thomas Aquinas, *Summa theologiae*, I, q. 75, art. 2, ad 2 (ed. Leonina, electronic version: www.corpusthomisticum.org)

[34] Critical edition with facing English translation, and Introduction outlining the contents of this question are published in Thomas Wylton, *On the Intellectual Soul*. The passages quoted below

reply is 'yes' and 'no', depending on how we conceive of the intellectual soul. If we conceive of it in the Aristotelian way, then the reply is 'yes',[35] but if we conceive of it in the Catholic way, then the reply is 'no'.[36] Wylton is ready to believe that the Catholic view is the true one, but he believes it by faith alone, since he cannot find any rational argument in its support.[37]

To get a better grasp of Wylton's complex reply, let us see how he defines the intellectual soul and how he draws the distinction between the Aristotelian way and the Catholic way of conceiving of the intellectual soul.

Wylton does not think that the contrast between the Aristotelian and the Catholic views is about the subsistence of the intellectual soul. Unlike Giles, Wylton does not point out any inconsistency in the notion of subsisting form. Moreover, he takes it as agreed by all, including Aristotle, that the intellectual soul is a subsisting form and not an inhering one. So, in clarifying what he means by 'form' when he asks whether we can prove that the intellectual soul is the form of the body, he says:

> (T8) By form, a word which is used equivocally by philosophers, I understand the form that informs the human body or its matter without inhering—in so far as inhering is a property opposed to being by itself or subsisting by nature. For although this is the condition of all other forms which are elicited from the potency of matter, both substantial forms and accidental forms, although in differing ways, still all people—both saints and others—agree on this: that the intellectual soul, which is claimed to be the form of man, is not a form that informs him in this way. In the proposed case, however, I understand by an informing form a form having a natural inclination, an order, and a dependence towards something else, in such a way that from itself and something else one composite is apt by nature to come to be, which by virtue of such a form is specifically distinguished from every other thing.[38]

In this passage, Wylton first draws the distinction between subsisting and inhering forms and classifies the intellectual soul as a subsisting form; he then explains how the intellectual soul as subsisting form 'informs' the human body. The issue at stake here is how being the form of a body is compatible with being subsistent, an issue that Aquinas himself had addressed. In Wylton's explanation the intellectual soul informs the human body not by inhering in it but by having "a natural inclination,

---

and their translations are taken from this work. For an outline of Wylton's highly original defence of Averroes's view in his *Quaestio de anima intellectiva*, see McCord Adams, Trifogli (2012, 637–641).

[35] See text (T10) below.

[36] See text (T11) below.

[37] See the very end of text (T11) below.

[38] Thomas Wylton, *On the Intellectual Soul* (ed. Nielsen-Trifogli, trans. Trimble, Oxford 2010, § 6, 8: 9–18): "Per formam, quae multipliciter dicitur in usu philosophorum, intelligo formam informantem corpus humanum vel eius materiam non-inhaerentem—ut inhaerere est proprietas opposita per se esse vel subsistere per naturam. Nam licet haec sit conditio aliarum formarum omnium quae educuntur de potentia materiae, tam substantialium quam accidentalium, licet aliter et aliter, tamen omnes in hoc conveniunt—et sancti et alii—de intellectiva, quae ponitur forma hominis, quod non est forma sic informans. Sed intelligo in proposito per formam informantem formam habentem inclinationem naturalem, ordinem et dependentiam ad aliud, ita quod ex alio et ipsa natum sit fieri unum compositum, quod per ipsam distinguitur specifice a quocumque alio."

an order, and a dependence" towards the human body. These notions—natural inclination, order, dependence—are those that Aquinas too had used in this connection.[39] The similarity between the two philosophers, however, is only apparent. Wylton's understanding of the natural inclination and dependence of the soul towards the body is in open contrast with Aquinas's view, as is clear from the following passage:

> (T9) I say therefore that on account of the natural inclination and dependence of the material intellect on the body, by which both in its being and in its final operation it depends on an organic body having sensitive powers of this kind, which are necessary preparations with respect to any of its actions, so that to understand does not belong to it but depends essentially on the body as if on that which performs it, therefore from the material intellect and the body one thing comes to be. And with respect to this conception Aristotle says (*On the soul* 1) that if to understand is to imagine or does not come to pass without imagination, it is impossible that it exists outside the body. By means of this he wants to say: to understand is not proper to the intellect, but belongs to the whole conjunct first just as to that which first understands. Consequently, I say that natural dependence of this sort and the mutual order of the principles of man, which is not found in a pile of stones, which are together without any natural commixing or mutual dependence, suffices for the fact that man is truly one being, definable and the subject of demonstration.[40]

Thus, in Wylton's understanding, the relationships of natural inclination and dependence of the intellectual soul towards the body are such that the intellectual soul depends on the body both in its existence and in all of its operations. In particular, thinking too depends on the body, as Wylton makes explicit with reference to Aristotle's passage (T2) above. The intellectual soul's dependence on the body in thinking and existence, however, is exactly the opposite of what Aquinas's claims (C2) and (C3) say.[41]

Furthermore, contrary to Aquinas's view but in agreement with Giles's, Wylton maintains that the assumption that the intellectual soul is the principle and not the agent of thought is crucial for its status as form. For if the intellectual soul is conceived as such a principle, it can then be proved that the intellectual soul is the form of the human body, as Wylton points out in his reply to the Question:

---

[39] See texts mentioned in note 25 above.

[40] Thomas Wylton, *On the Intellectual Soul* (ed. Nielsen-Trifogli, trans. Trimble, Oxford 2010, § 125, 76: 19–31): "Dico ergo quod propter naturalem inclinationem et dependentiam intellectus materialis a corpore, qua et in esse et in operatione sua finali a corpore organico huiusmodi virtutes sensitivas habente dependet, quae sunt praeparationes necessariae respectu cuiuscumque actionis eius, ita quod intelligere non est eius, sed dependet essentialiter ex corpore tamquam ab eo quod operatur, ideo ex his fit unum. Et ad hanc intentionem dicit Aristoteles, 1 *De anima*, quod si intelligere sit imaginari vel non sit sine imaginatione, impossibile est quod sit extra corpus. Per quod vult dicere: intelligere non est proprium intellectui, sed est totius coniuncti primo sicut eius quod primo intelligit. Unde dico quod huiusmodi naturalis dependentia et ordo principiorum hominis ad invicem, quae non invenitur in cumulo lapidum, qui sunt simul sine omni commixtione naturali vel dependentia ad invicem, sufficit ad hoc quod homo sit vere unum ens definibile et subiectum demonstrationis."

[41] Wylton's insistence on the intellectual soul's dependence on the body raises the question of how he understands the special status of the intellectual soul as subsisting form. One suggestion is that he could simply refer to the status of the intellect as non-inhering in matter, that is, to the lack of a bodily organ of thought (Aquinas's claim [C1]).

(T10) Concerning the fourth article, what would seem to me necessary to be said in answer to the question, I say that by understanding by the intellectual soul precisely the form which is our principle of cognizing the abstract quiddities of things, the cognition which everyone experiences in himself that he has when he wants to experience that; also by understanding by the form of the human body the formal principle intrinsic in man by which he is specifically and ultimately distinguished from other species—by not taking into account the special way by which it informs, namely either of something inhering or in another way, but only in general—I say that it can be established evidently—and the Philosopher established this conclusion, *On the Soul* 2, starting from the notion that the soul is that by which we live, sense, move with respect to place and understand—that it is a principle intrinsic to us, not as matter, as he argues by induction there —from this notion he concludes that it is a formal principle. And he establishes this no less concerning the intellectual soul than concerning other things that it is a form in us, since both things are established by the same means.[42]

According to Wylton, the conclusion of Aristotle's argument for the identification of the soul with the form of the living being in *De anima* II.2 is valid for all kinds of souls: the soul of a plant, that of a non-rational animal, and that of a human being. The reason for this is that the argument assumes that the same crucial premise is valid for all kinds of souls, namely, the premise that a soul is the principle of all the vital operations of a living body.

However, if the intellectual soul is conceived as independent from the body in its intellectual operations and in its existence, then not only Aristotle's argument but also all other philosophical arguments would fail to prove the claim that the intellectual soul is the form of the human being, as Wylton points out in the second part of his answer to the Question:

(T11) But if by the intellectual soul were understood not only the form by which we understand the abstract quiddities of things, but in addition to this we were to encompass its other conditions which Catholics attribute to it, namely that it is produced anew simultaneously with the individual whose form it is, and nevertheless that it is eternal into the future, I say, although Adam in the state of innocence and later Solomon in natural cognition in this life had an intellect elevated to the cognizing of many truths about natural things, truths which we do not cognize and cannot establish by natural means, I say regarding them, I do not know whether they established this truth by natural reason or whether they could establish it. Regarding the philosophers, however, I do not know anyone who establishes this by

[42] Thomas Wylton, *On the Intellectual Soul* (ed. Nielsen-Trifogli, trans. Trimble, Oxford 2010, § 177, 106: 16–28): "De quarto articulo, quid mihi videatur dicendum ad quaestionem, dico quod intelligendo per intellectivam praecise formam quae est nobis principium cognoscendi quidditates rerum abstractas, quam cognitionem quilibet experitur in se ipso se habere, cum voluerit; intelligendo etiam per formam corporis humani formale principium in homine intrinsecum quo ab aliis speciebus specifice et ultimate distinguitur—non comprehendendo modum specialem quo informat, scilicet utrum inhaerentis vel alio modo, sed tantum in genere—dico quod evidenter convinci potest—et istam conclusionem convincebat Philosophus, 2 *De anima* ex hoc quod anima est quo vivimus, sentimus, movemur secundum locum et intelligimus—quod est principium intrinsecum nobis non ut materia, ut ipse inducit ibi, ex quo concludit quod est principium formale. Et istud non minus convincit de anima intellectiva quam de aliis quod sit forma in nobis, quia per idem medium convincitur utrumque." In this passage, as in the first argument of Aquinas quoted in note 30, the reference is to Aristotle's argument in Aristoteles, *De anima*, II.2, 413b11–414a14 (trans. Shields, 25–26).

> natural reason or anyone who even posits it. But regarding myself, I say that although I believe without any doubt that this opinion is true, nevertheless I do not know how to establish it by natural reason—let them rejoice who do know how to establish it—but I hold it by faith alone.[43]

Moreover, in Wylton's view it is not only the case that all arguments in support of the Catholic view fail, but it is also not difficult to find arguments to disprove it. A very good one, according to Wylton, is suggested by Aristotle himself in *Metaphysics* VII.[44] In Wylton's report, Aristotle argues that the syllable AB cannot be reduced to its components A and B, but it also contains something else, which should be identified with its form. The principle used by Aristotle here is that A and B can both exist independently of the syllable AB: when A and B are separated one from the other, they both persist but the syllable AB does not. Thus, the existence of A and that of B are not enough for the existence of the syllable AB. A and B are not what make the syllable AB the kind of thing it is. What needs to be added to A and B is the form of this syllable.[45]

It is easy to see how Aristotle's argument can be adapted to the Catholic view of the intellectual soul. Indeed, Wylton's lengthy discussion of it makes it clear that this argument was an important ingredient of the debate about the status of the intellectual soul at his time.[46] In particular, the argument applies perfectly well to a view like that of Aquinas according to which the intellectual soul is the only substantial form of the human being so that the only components of a human being are prime matter and the intellectual soul. According to Aquinas's version of the Catholic view, both the prime matter of an individual human being, say Socrates, and its soul persist when Socrates dies, just like the components A and B of the syllable AB persist when they are separated one from the other so that the syllable AB no longer exists. It follows that the intellectual soul is not the form of Socrates, according to Aristotle's line of argument. Or according to a more general line of argument based on the premise 'when everything that constitutes the entity of a thing remains, the

43 Thomas Wylton, *On the Intellectual Soul* (ed. Nielsen-Trifogli, trans. Trimble, Oxford 2010, § 179, 108: 18–30): "Si autem per intellectivam non solum intelligatur forma per quam intelligimus quidditates rerum abstractas, sed cum hoc comprehendamus conditiones eius alias quas catholici ei attribuunt, scilicet quod sit de novo producta simul cum individuo cuius est forma, et tamen quod sit aeterna ex parte post, dico, cum Adam in statu innocentiae et Salomon post in cognitione naturali viae ad multas veritates rerum naturalium cognoscendas habuerint intellectum elevatum, quas nos non cognoscimus nec ex naturalibus convincere possumus, dico quantum ad illos, nescio si istam veritatem per rationem naturalem convincebant vel convincere potuerunt. Quantum tamen ad philosophos nescio aliquem qui per rationem naturalem convincit nec etiam ponit. Quantum autem ad me, dico quod, licet illam opinionem absque aliqua dubitatione credam veram esse, ipsam tamen per rationem naturalem convincere nescio—gaudeant illi qui convincere eam sciunt—sed sola fide teneo."

44 Aristoteles, *Metaphysics*, VII.17, 1041b12–33 (trans. Bostock, 30–31).

45 Thomas Wylton, *On the Intellectual Soul* (ed. Nielsen-Trifogli, trans. Trimble, Oxford 2010, § 145, 92: 3–17).

46 Thomas Wylton, *On the Intellectual Soul* (ed. Nielsen-Trifogli, trans. Trimble, Oxford 2010, §§ 146–156, 92: 18–98: 15).

thing which is constituted remains',[47] it follows that we should posit a further constitutive principle in Socrates in addition to his matter and his soul, a constitutive principle which would be the most formal component in Socrates, more formal than the intellectual soul itself, a conclusion that not only Aquinas and all the supporters of the unity of the substantial form of a composite substance reject but that is in itself not sound.

In conclusion, according to Wylton, there are serious problems that the Christian doctrine of the independence of the intellectual soul raises for the hylomorphic constitution of the human being.

References

Primary Literature

Aristoteles. 1994. *Metaphysics. Books Z and H. Translation and Commentary by David Bostock*. Oxford, Clarendon Press.

———. 2016. *De anima. Translation, introduction and commentary by Christopher Shields*. Oxford: Clarendon Press.

Giles of Rome. 1500. *In libros De anima expositio*. Venice. Repr. Frankfurt 1982: Minerva.

Thomas Aquinas. 1888. *Summa theologiae*. Rome: Leonine Edition. Electronic version. www.corpusthomisticum.org.

———. 1984. *Sentencia libri De anima*, ed. René-A. Gauthier. Rome–Paris: Leonine–Vrin.

———. 1999. *A commentary on Aristotle's De anima. Trans. Robert Pasnau*. New Haven–London: Yale University Press.

Thomas d'Aquin. 2016. *L'Ame et le Corps. Somme de théologie. Première partie, questions 75 et 76*. Introduction by Bernardo C. Bazán. French translation by Jean-Baptiste Brenet. Paris: Vrin.

Thomas Wylton. 2010. *On the intellectual soul*, ed. Lauge O. Nielsen, Cecilia Trifogli. Eng. trans. Gail Trimble. Oxford: Oxford University Press.

Secondary Literature

Caston, Victor. 2000. Aristotle's argument for why the understanding is not compounded with the body. *Proceedings of the Boston Area Colloquium of Ancient Philosophy* 16: 135–175.

Frede, Michael. 1992. On Aristotle's conception of the soul. In *Essays on Aristotle's De Anima*, ed. Martha C. Nussbaum and Amélie Oksenberg Rorty, 93–107. Oxford: Clarendon Press.

McCord Adams, Marilyn, and Cecilia Trifogli. 2012. Whose thought is it? The soul and the subject of action in some thirteen and fourteenth century Aristotelians. *Philosophy and Phenomenological Research* 85 (3): 624–647.

Pasnau, Robert. 2012. Philosophy of mind and human nature. In *The Oxford handbook of Aquinas*, ed. Brian Davies and Eleonore Stump, 348–368. Oxford: Oxford University Press.

Trifogli, Cecilia. 2012. Giles of Rome against Thomas Aquinas on the subject of thinking and the status of the human soul. *Documenti e studi sulla tradizione filosofica medievale* 23: 221–244.

———. 2021. Thomas Aquinas on cognition as information. In *Information and the History of Philosophy*, ed. Chris Meyns, 137–149. London: Routledge.

[47] Thomas Wylton, *On the Intellectual Soul* (ed. Nielsen-Trifogli, trans. Trimble, Oxford 2010, § 149, 94: 5–9).

Chapter 11
A Medieval Defence of Innatism: The Case of James of Viterbo

Martin Pickavé

Abstract James of Viterbo (ca. 1255–1308) is one of the few later medieval philosophers and theologians who explicitly argued for innate cognition. His defence of innatism is sophisticated and shows great awareness of the anti-innatist arguments of his contemporaries and predecessors, notably Thomas Aquinas. This chapter begins by looking at why Aquinas rejects innatism. For Aquinas, innatism in all of its forms is ultimately based on a mistaken understanding of natural agency and causality, for explaining activities from active principles within the relevant agents and patients seems to undermine the idea that objects in the world can be genuine agents. It then shows how James responds to the concerns raised by Aquinas. According to James, innate active forms—in the intellect, in the will, and in matter—are what make genuine agency possible in the first place rather than undermining it.

Keywords Thomas Aquinas · James of Viterbo · Innatism · Innate perception and cognition · Causation · Seminal reasons · Virtues

11.1 Introduction: Thomas Aquinas on Innatism and Related Mistakes

Innatism—that is, the view that our concepts and knowledge are innate in us rather than acquired through sensory experience—was not very popular among later medieval philosophers and theologians. Innatism seems to be at odds both with the Aristotelian idea, widely endorsed by medieval thinkers, that the human soul is like a blank slate (*De anima*, III 4, 430a1–2), and with the understanding of a human being as a body-soul composite. For if we do not acquire concepts and knowledge through our corporeal sensory powers, why do we need a body in the first place?

M. Pickavé (✉)
University of Toronto, Toronto, ON, USA
e-mail: martin.pickave@utoronto.ca

M. Gensler et al. (eds.), *The Embodied Soul*, Historical-Analytical Studies on Nature, Mind and Action 11, https://doi.org/10.1007/978-3-030-99453-2_11

The history of the late medieval debates concerning innate concepts and knowledge remains to be written.[1] When it is, I have no doubt that Thomas Aquinas will play an important role in it. This will be well deserved, not only because of the eminent position he occupies in the historiography of medieval philosophy and the many arguments against innatism that we can find throughout his works, but also because of his attempt to see innatist views as part of a broader outlook on causation and the soul. In fact, in a couple of his works Aquinas not only contrasts innatism with the completely opposite view, ascribed to Avicenna, according to which rather than being already inside the mind, all intellectual cognition comes from outside by way of emanation from a higher intelligence, but also sees connections between what exponents of these two views at opposite extremes say about (a) the eduction of forms from matter, (b) the acquisition of virtue, and (c) the acquisition of knowledge. In his *Quaestiones disputatae de veritate* Aquinas writes (q. 11, a. 1):

> Some said that all sensible forms are from an extrinsic agent, which is a separate substance or form, which they call a giver of forms (*dator formarum*) or an agent intelligence (*intelligentia agens*). And they said that all lower natural agents are only as it were things that prepare matter for the reception of a form. Similarly, Avicenna says in his *Metaphysics* that our action is not the cause of a good disposition (*habitus*), but that the action prohibits its opposite and thus prepares for the disposition, so that there may occur here a disposition from the substance that perfects the souls of all human beings, which is the agent intelligence or a substance that is similar to it. Likewise, they also say that knowledge (*scientia*) is brought about in us only by a separate agent. Therefore, Avicenna says in book VI of his *Physics* that intelligible forms flow into our mind from the agent intelligence.
>
> Others, however, were of the opposite view, namely, that all these are placed in things and do not have a cause from the outside, but only become manifest through an external action. For some posited that all natural forms exist in actuality (*in actu*) as latent forms in matter and that a natural agent does nothing more than to draw (*extrahere*) them from obscurity into the open. Likewise, some held that all virtuous dispositions (*habitus virtutum*) have been placed in us by nature; and through the exercise of actions impediments (*impedimenta*) are removed, by which the aforementioned dispositions are as it were hidden, just as by filing rust is removed so that the brightness of iron is brought to light. Similarly, some also said that knowledge (*scientia*) of all things is created at the same time as the soul and that the only thing that happens by means of teaching and other such external support is that the soul is brought to remember and to consider things it previously knew. Therefore, they say that learning is nothing but remembering.[2]

The context in which Aquinas presents these two rival views—let us call them the *everything-comes-from-the-top* thesis and the *everything-is-already-inside* thesis, respectively—is highly rhetorical. Against the backdrop of these views at opposite extremes he can go on to present Aristotle's opinion as a middle way (*via media*) which is identical to his own. Who wouldn't prefer to occupy the middle ground? However, not all is rhetoric. Addressing why these two extreme views fail, Aquinas has this to say in a passage that immediately follows upon the previous quotation:

[1] The best treatment currently available is Solère (2018).

[2] All translations are my own. See also Thomas de Aquino, *Quaestiones disputatae de veritate*, q. 10, a. 6 and *Quaestiones disputatae de anima*, q. 15. The references to Avicenna's works are to *Liber de philosophia prima*, IX.5; IX.2; and *De anima*, V.5.

> However, both of these views are without reason (*absque ratione*). For the first view excludes all proximate causes as long as it attributes all effects that occur in lower things only to the first causes. In this it calls into question (*derogatur*) the order of the universe, which is made up of the order and connection of causes, since out of the pre-eminence of its goodness the first cause confers on other things not only that they exist, but also that they are causes. The second view runs into almost the same problem. For since what removes an impediment is a mover only accidentally (*per accidens*), as Aristotle says in *Physics* VIII, if the lower agents do nothing more than to bring something from obscurity into the open by removing impediments by which forms, virtuous dispositions, and knowledge were hidden, then it would follow that all lower agents act only accidentally (*per accidens*).[3]

Despite the obvious differences between the everything-comes-from-the-top and the everything-is-already-inside theses, according to Aquinas they both have in common that they get a basic point about causation and natural agents wrong. For either they deny the existence and efficacy of lower, secondary causes, or they lead to the view that all agents act only accidentally (*per accidens*); but on neither of these views can there be genuine natural agents that are per se causes of their effects. This, for Aquinas, is what ultimately rules them out as serious explanations of the processes they aim to account for—quite a damning verdict.

The direction of Aquinas's attack on the everything-comes-from-the-top and the everything-is-already-inside theses is not the only interesting feature of the passages that I have just quoted. It is fairly easy to put a name to the source of some of the items mentioned under the first view: Avicenna. In fact, Aquinas mentions him twice by name and explicitly attributes to him the idea that virtues and knowledge come from a separate agent or intelligence. The reference to the "giver of forms" (*dator formarum*) in the description of how natural forms come to be in matter is another direct reference to Avicenna. But who are those who maintain that forms, virtues, and knowledge pre-exist in matter and the soul?

No doubt it is plausible that if some philosopher were to hold that forms pre-exist in matter as latent forms and that the role of external natural agents is merely to remove obstacles so that these forms can become manifest, then the same philosopher could use the same model in conceiving of the existence of virtues or knowledge in the soul. This is presumably one of the reasons why Aquinas sees similarities and groups these views together. However, the fact that something is plausible does not mean that it is actually the case. Obviously, the view that knowledge is innate in the soul is a characteristically Platonic one, as is the catchphrase that learning is nothing but remembering.[4] However, nowhere in his works does Aquinas associate Plato with the theory of latent forms. Following Aristotle's remarks in the *Physics*, he usually attributes the latter opinion to Anaxagoras.[5] The view that our virtues are innate is linked to a third figure, namely, John of Damascus, to whom Aquinas ascribes this doctrine (e.g., in the third book of his *Commentary on the Sentences*)

---

[3] See Aristoteles, *Physica*, VIII.8, 255b24.

[4] For authors writing in Latin the association of this catchphrase and Plato goes at least back to Augustine. See, for instance, Augustinus, *De Trinitate*, XII.15.

[5] See, for instance, Thomas de Aquino, *Quaestiones disputatae de virtutibus in communi*, a. 8; see also Aristoteles, *Physica*, I.4, 187a26ff.

and from whom he also takes the example of the polished piece of iron.[6] In other words, all the evidence points to the conclusion that when Aquinas described the everything-is-already-inside view he did not have a single proponent in mind.

However, there was one author in the generation after Aquinas who seems to have defended both that forms pre-exist in matter and that virtues and knowledge pre-exist in the soul—namely, James of Viterbo (ca. 1255–1308), a member of the order of the Hermits of St. Augustine and the successor of Giles of Rome as the Augustinians' regent master of theology at the University of Paris. James's philosophical teaching has recently received more attention and scholars have started to explore his innatism with respect to perception and cognition, his theory of seminal reasons, and his views on innate practical dispositions.[7] But so far as I can see, they have not focused sufficiently on how these three areas of James's teaching hang together. They have also, in my view, overlooked how James's views are shaped by the same worries about agency and causation that are at the forefront of Aquinas's criticism of innatism and views similar to it. All this makes it reasonable to maintain, as I will try to suggest in the following pages, that James developed his own thoroughgoing innatism in reaction to the objections raised by Aquinas.

11.2 James on Innate Perception and Intellectual Cognition

Let us begin with James's defence of innate perception and cognition. You might think that the foremost reason to hold that our soul must already be endowed with ideas or concepts is that our concepts are far richer than what we could ever acquire solely by means of the senses. We perceive colours, smells, sounds, and so on, but how do we get from there to the concept of, say, human being or animal? This is sometimes referred to as the *argument from the poverty of stimulus*. However, this is not the reason why James thinks we ought to hold that our soul contains some sort of innate cognition. Here is how he puts forward his main reason:

> But when it is said that the soul conforms to things in potency (*in potentia*), some assume there to be a purely passive power (*potentia pure passiva*), in the way a mirror is said to be in potency with respect to images, and matter with respect to forms. But this does not seem reasonable. For it seems that in this way it cannot be explained how the actual movement and the informing of the soul itself, according to which it is said to conform to things either by cognizing them or by desiring them, is called an act or operation of the soul (*actio et operatio animae*), and how it is called a vital act and one that remains in the agent [...]. And therefore it seems that one must say that when the soul is said to be potentially conformed to things, such a potency is not a purely passive power, but that it is some sort of incomplete actuality and some sort of start (*inchoatio*) and beginning (*exordium*) and preparation (*praeparatio*) with respect to a further act. Therefore, it can be called a certain aptitude

[6] Thomas de Aquino, *Scriptum super Sententiis*, III, dist. 33, q. 1, a. 2, qc. 1, sol. 1. See also Johannes Damascenus, *De fide orthodoxa*, cap. 59. However, in *Summa theologiae*, I–II, q. 63, a. 1, Aquinas refers to this view as an *opinio Platonicorum*.

[7] On these three topics, see, for instance, Solère (2018), Pickavé and Côté (2018), and Côté (2012).

> (*aptitudo*) and propensity (*idoneitas*) towards a complete act. And such a propensity is connatural to the soul and is in it naturally. As a result, it also always remains in the soul, but sometimes it is imperfect, but at other times it is perfected through an act.[8]

The idea here is fairly simple: there are various activities of the human soul, and perceiving is one of them, as are intellectual understanding and desiring. But how, James asks, can we speak of acts and operations of the soul if we assume that the human soul is purely passive (as some seem to) and that perception and thinking are merely the reception of a form or an impression?

The expression "vital act" (*actus vitalis*) and the reference to the notion of 'immanent act', that is, an act "that remains in the agent", are key here. On a purely passivist understanding of the human soul, perception and similar mental activities are rather like events that happen to the soul, similar to what happens when something gets heated or coloured. We do not attribute the act of heating or colouring, or indeed any activity, to the subject which becomes hot or coloured; rather, we impute such activity to the relevant agent. The passivist account is thus fundamentally at odds with our general conception of the soul as the source of its own (vital) activities, such as perceiving, willing, thinking, etc. The picture of the soul as a purely passive power must therefore be wrong.

This leads James, in the second part of the text quoted above, to the conclusion that the soul must be something more active: in his words, the soul is "some sort of incomplete actuality" (*quaedam actualitas incompleta*). Building on ideas from Anselm of Canterbury and Simplicius's *Commentary on Aristotle's Categories*, James explains the partially active nature of the soul with the idea that the soul has many connatural and innate "aptitudes" (*aptitudines*) or "propensities" (*ideoneitates*) towards its proper acts.[9] And since an incomplete actuality, or aptitude or propensity, is naturally inclined to its complete actuality, it will naturally attain this state of full actualization if nothing prevents it.[10] This may make it sound as if for James an external object and its presence cause a cognition of the object in us only to the extent that they remove an impediment, so that the soul can move itself to a completely actualized cognition of the object. But this is not the case, for the object has a more active role: it "excites" the soul and it is a cause of cognition, even if it

[8] Jacobus de Viterbio, *Quodlibet*, I, q. 7 (ed. Ypma, 1968–1975, 92): "Cum vero dicitur anima conformis rebus in potentia, quidam accipiunt potentiam pure passivam, ad modum quo speculum dicitur potentia ad imagines et materia ad formas; quod quidem non videtur rationabile. Non enim secundum hunc modum videtur posse salvari, quomodo actualis motio et informatio ipsius animae, secundum quam dicitur actu conformis rebus vel cognoscendo vel appetendo, dicatur actio et operatio animae, et quomodo dicatur actio vitalis et in agente manens […]. Et ideo videtur esse dicendum quod, cum anima dicitur esse potentia conformis rebus, huiusmodi potentia non est pure passiva, sed est quaedam actualitas incompleta, et est inchoatio et exordium et praeparatio quaedam, respectu actus ulterioris. Unde potest dici aptitudo quaedam, et idoneitas ad completum actum. Et est huiusmodi idoneitas animae connaturalis et naturaliter indita, ideoque et semper in ipsa manens; sed quandoque imperfecta, quandoque vero perfecta per actus."

[9] See Côté (2009, 2011) and Solère (2018).

[10] Jacobus de Viterbio, *Quodlibet*, I, q. 7 (ed. Ypma, 1968–1975, 95–97).

is not the primary cause.[11] James leaves no doubt that he considers the object to be an efficient cause of the cognitive act.[12]

Note also that James distinguishes between *general* aptitudes, which are the traditional faculties of the soul (intellect, will, senses, etc.), and *special* aptitudes, which are grounded in the common ones and express special determinations of the faculties of the soul.[13] The special aptitudes the soul has with respect to different intelligible and sensible objects are of particular interest here, for as innate dispositions of the underlying faculties of the soul they look like genuine instances of innateness. They also show that James defends innatism not only with respect to intellectual cognition but also with respect to sensory perception.[14]

That James endorses a theory of innate cognition is clear from his answers to the standard objections that medieval authors raise against innatism. According to one objection, James's account of the soul and its aptitudes contradicts Aristotle's view of the soul as a blank slate. For according to the present account, the soul is already, thanks to its aptitudes, in an incomplete actuality with respect to the intelligible objects, and is therefore not completely "blank".[15] In his reply, however, James insists that his account does agree with Aristotle's. To call the soul a blank slate means, according to him, that it is in potency to the *complete* act of cognition with respect to its objects and that no fully actual cognition "has been written" on the soul. But the image does not suggest that the soul is merely in potency with respect to *incomplete* acts of understanding.[16]

This distinction between an incomplete and a complete act of cognition is crucial for understanding why James considers his view to be very different from Plato's theory of recollection. First, he does not combine, as Plato did, a doctrine of innate cognition with the idea of the soul's pre-existence before its union with the body. But more importantly, since for James the soul is endowed only with incomplete acts of cognition, learning is not remembering, for the latter would require that the

---

[11] Jacobus de Viterbio, *Quodlibet*, I, q. 12 (ed. Ypma, 1968–1975, 174–175).

[12] Jacobus de Viterbio, *Quodlibet*, II, q. 6 (ed. Ypma, 1968–1975, 99–100 and 104). See Solère (2018, n. 66) for these and other texts.

[13] Jacobus de Viterbio, *Quodlibet*, I, q. 7 (ed. Ypma, 1968–1975, 93): "Est autem considerandum quod in anima sunt multae et diversae huiusmodi aptitudines, iuxta materiam et diversitatem eorum quibus anima nata est conformari. Et est in huius idoneitatibus ordo quidam, secundum commune et speciale. Quaedam enim huiusmodi aptitudinum sunt generales; et istae sunt illae quae dicuntur animae potentiae ut sensus, intellectus et appetitus. Nam potentia sensitiva non est aliud quam quaedam idoneitas generalis respectu sensibilium potentiarum. Et haec in plures distinguitur secundum numerum sensuum interiorum et exteriorum. Et in quolibet sensu vel sensitiva potentia fundantur speciales idoneitates secundum differentias illius sensibilis quod est obiectum sensus. Similiter autem et potentia intellectiva est idoneitas generalis respectu omnium intelligibilium, idest respectu conformationis actualis ad omnia intelligibilia. Sed super hanc generalem idoneitatem fundantur aliae speciales secundum diversitatem intelligibilium. Et eodem modo dicendum est de appetitu." See also *Quodlibet*, I, q. 12 (ed. Ypma, 1968–1975, 166).

[14] For more on this, see Solère (2018, 191–196).

[15] Jacobus de Viterbio, *Quodlibet*, I, q. 12 (ed. Ypma, 1968–1975, 168).

[16] Jacobus de Viterbio, *Quodlibet*, I, q. 12 (ed. Ypma, 1968–1975, 170).

soul already have had all sorts of complete acts of cognition, but there is no ground for believing that this is so.[17]

The main reason many medieval thinkers dismissed the idea of innate cognition as implausible is that it seems to make the connection between sensory perception and intellectual understanding unintelligible insofar as it abandons the assumption of any direct "transference" of information from the one to the other.[18] James responds to this concern by providing an extensive account of the causal mechanisms involved in sensory and intellectual cognition. True, on James's picture sensory experience merely "excites" the intellect to come to an act of cognition, and sensory experience is not the principal cause of an intellective act. However, this does not mean that the senses and the intellect are only loosely connected. Since there is a natural connection (*naturalis coniunctio*) and a natural order (*ordo naturalis*) between the different powers of the soul, certain acts of the intellect can occur only if the corresponding sensory experience takes place.[19] The reference to a "natural order", or some sort of pre-established harmony, between the acts of the faculties of the human soul can likewise be applied to explain the interaction between sensory objects and the sensory powers, since, as noted above, James also maintains that sensory perception is innate in the form of incomplete sensory acts.

It is obvious that on James's account of the soul, the soul has many innate aptitudes, for the human soul can think of many different objects. This raises the question of just how many innate aptitudes there are. On this, James is surprisingly silent, for he believes that there is no way to find out the answer by natural means in this life.[20] But however this may be, his theory of innate aptitudes in the intellect allows him to address several epistemological problems—for instance, the question of how we are able to have intellectual cognition of a substance even though all our sensory experience is of accidental qualities.[21] This might be a problem for empiricist accounts of cognition, since it is difficult to see what sort of sensory input could ever lead to the concept of a substance in us. In fact, when Richard of Middleton raises the issue of how we know substances in question 42 of his *Quaestiones disputatae*, he mentions the innateness of our concepts of substances as one possible solution.[22] Similarly, if we maintain with James that cognition of substances is innate (as an incomplete actuality) and that sensory cognition of a certain kind does nothing more than to excite the intellect to have the actual cognition of a

[17] Jacobus de Viterbio, *Quodlibet*, I, q. 12 (ed. Ypma, 1968–1975, 170–171). For James's account of learning, see Pickavé (2016, 38–43).

[18] For examples of this line of reasoning in Aquinas, see, for instance, Thomas de Aquino, *Quaestiones disputatae de veritate*, q. 10, a. 6; q. 19, a. 1; *Quaestiones disputatae de anima*, q. 15; *Summa theologiae*, I, q. 84, a. 3.

[19] Jacobus de Viterbio, *Quodlibet*, I, q. 12 (ed. Ypma, 1968–1975, 172–174).

[20] Jacobus de Viterbio, *Quodlibet*, I, q. 7 (ed. Ypma, 1968–1975, 93). However, it makes sense to assume that there is one innate aptitude for each natural kind.

[21] Jacobus de Viterbio, *Quodlibet*, I, q. 13 (ed. Ypma, 1968–1975, 189–190).

[22] Richardus de Mediavilla, *Quaestiones disputatae*, q. 42 (ed. Boureau, 2014, 380–384).

substance—rather than completely cause it—then our ability to entertain concepts of substances looks less mysterious.

As I said earlier, this last issue might appear to be an important motivation to adopt innatism. But James of Viterbo does not explicitly present it as an argument for innatism. Rather, he seems convinced that innate incomplete cognitions and the like are the only way to account for the fact that souls, and by extension living beings, have activities that can be ascribed to them. In other words, innatism is needed to prop up our common-sense ontology of natural agents.

11.3 James of Viterbo on Pre-existing Forms in Matter

Psychological activity is one thing, but changes in material objects are another. Yet there too we see James resorting to the same idea: aptitudes (*aptitudines*), propensities (*idoneitates*), or starts (*inchoationes*) or beginnings (*exordia*), not of acts of cognition, but of forms. To understand why James believes they are necessary, think of normal natural processes such as qualitative changes or the generation of a new natural substance.[23] According to the standard Aristotelian analysis of natural processes, in any such change an external agent brings about a change in a hylomorphic compound (i.e., something composed of matter and form). The end point of any change is a new composite of matter and form: for instance, a new individual animal (as in the case of generation) or a heated object (as in the case of a qualitative change). There seems to be no issue regarding the matter of this new composite: it has been there all along, from the beginning to the end of the process, since it is the substrate in which the change takes place. The problem lies rather in the origin of the new form.

So where does the new form come from? Take, for example, the form of the new animal or the form of heat: it seems that the form cannot come from the external agent, despite the fact that animals bring about other animals and hot objects bring about the heat in heated objects. It does not come from the agent in the sense of the agent giving its own form to the composite that it brings about, for the agent does not lose its form in the exercise of its efficient causality. That it sometimes happens that a heat source like a hot stove cools down when it heats something else is not due to its role as an agent, but rather to the fact that in this scenario it is also a patient; for what is heated and what does the heating affect each other if they are in close spatial proximity.[24] Nor can the form of the external agent inhere in the resulting composite, since one and the same form cannot inhere in two separate chunks of matter. But if, as we have just seen, it is not the agent's form that is transferred from

[23] For the following argument, see Jacobus de Viterbio, *Quodlibet*, II, q. 5 (ed. Ypma, 1968–1975, 75ff.). In this quodlibetal question James in fact gives five arguments for pre-existing forms, but I will focus only on the fifth argument. For more detailed discussion of James's arguments, see Pickavé and Côté (2018), on which the exposition here is based.

[24] Jacobus de Viterbio, *Quodlibet*, III, q. 14 (ed. Ypma, 1968–1975, 189).

the external agent to the final product of the change (i.e., the new composite), then it must be a new form. This means that either the agent produces this new form from nothing, in which case the agent would be engaged in an act of creation, which is impossible, since creatures do not create; or the agent produces this new form from its own substance, which is equally unacceptable. For the latter scenario would require the agent to move itself from potentiality to actuality, because now the agent would undertake an activity all by itself before transmitting this new form to the composite.[25]

For James, this leaves us with only two alternatives: either the forms involved in natural processes are created by God or they already "pre-exist in matter" in potential and imperfect being. The former option, which was adopted by those who posit a "giver of forms" (*dator formarum*), would entail that in every natural process God creates a new form. But this jeopardizes the distinction between natural and supernatural processes, because God now has to intervene in every instance of natural change. Therefore, the only remaining alternative is for forms to come from within the matter from which the composite comes about.[26] On this picture, the role of the natural agent is not to act "by producing something new in matter, but by drawing forth (*extrahendo*) into actuality what was there in potency."[27]

This may make it sound as if James rejects a key idea of medieval Aristotelians, namely, the idea that natural forms are "educed from the potency of matter." Aquinas expresses this idea as follows, here focusing on qualitative change (*Summa contra Gentiles*, III.69):

> For one hot body is not said to heat another body in such a way that the numerically same heat (*idem numero calor*), which is in one heating, crosses over into the heated body. But because of the power of the heat, which is in the heating body, some numerically different heat, which was previously in it only in potency (*qui prius erat in eo in potentia*), becomes actual (*fit actu*) in the heated body. For a natural agent is not something that transfers (*non est transducens*) its own form into another subject, but something that reduces (*reducens*) a subject undergoing a change, from potency to actuality.

There is nothing in this description to which James would object. He does not reject the idea of the eduction of forms. Rather, according to him, for this sort of process to be possible in the first place, the form becoming actual needs already to exist in the subject undergoing the change. It is only because matter is, so to speak, pregnant with pre-existing forms that this account of natural processes is possible. Naturally,

[25] Jacobus de Viterbio, *Quodlibet*, III, q. 14 (ed. Ypma, 1968–1975, 191): "Item forma de subiecto in subiectum migraret vel immitteret aliquid quod non est actu in ipso agente. Hoc autem vel produceretur de nihilo; quod est contra tertiam suppositionem [i.e., that no natural agent acts in the mode of creation], vel produceretur de substantia vel essentia ipsius agentis, quod est contra quartam suppositionem [i.e., that nothing changes itself by moving itself from potentiality to actuality]."

[26] Jacobus de Viterbio, *Quodlibet*, II, q. 5 (ed. Ypma, 1968–1975, 75): "Nam, nisi ponatur quod forma praeexistat in materia modo iam dicto, necessario videtur sequi quod in generatione inducatur forma ab extra, sicut dixerunt ponentes datorem formarum; et ita forma habebit esse per creationem."

[27] Jacobus de Viterbio, *Quodlibet*, II, q. 5 (ed. Ypma, 1968–1975, 85): "Naturale agens non agit producendo aliquid de novo in materia, sed solum extrahendo quod erat in potentia in actum."

forms pre-exist in matter, not in actuality; in this sense James disagrees with Anaxagoras.[28] However, they need to possess a sufficiently robust ontological status. They are clearly not simply nothing before they are fully actualized: they have "potential being" (*esse potentiale*) or "being in potency" (*esse in potentia*).[29] Moreover, James refers to these pre-existing forms as "beginnings of forms" (*inchoationes* or *exordia formarum*), "forms in potency" (*formae in potentia*), and "preparations of forms" (*praeparationes formarum*); and most interestingly, he believes that this is what Augustine meant by "seminal reasons" (*rationes seminales*).[30]

Only if a form pre-exists in potency in matter can it subsequently be educed and brought to actuality. Forms that do not pre-exist in this way cannot be educed naturally. This is why the human soul cannot come about through a natural process, but has to be infused into the body by God through an act of creation.[31] Nevertheless, even if chunks of matter cannot become human beings through a purely natural process, since the human soul does not pre-exist in them in potential being, they can still turn into a large number of other natural things. Thus, it seems as if each chunk of matter contains a large, perhaps even infinite number of forms in their inchoate states. This would not pose much of a problem if existing in potency were nothing more than matter having a bare capacity to receive a form. But clearly for James a form in potential being is something more robust. Note also that he holds that the form in potential being is "the same form in reality" (*eadem forma secundum rem*) as the form in actual being.[32] However, he is not really worried about this consequence of his account. He points out that actuality (*actus*) is what separates and distinguishes things, so forms existing in potential being are not distinct qua being in potency; rather, they are "as if one form and one thing" (*quasi una forma et una res*), although they are distinct insofar as they are related to different actualities.[33]

To conclude this section, let us go back to the relationship between the external agent and the pre-existing form in natural changes. James contrasts them in the following passage:

> In another way something is said to be an active principle by the mode of transmutation. And such a principle needs to be posited in the eduction of forms from potency to actuality, apart from the other principle that is called the beginning of the form. For although this beginning of the form inclines matter so that it becomes a form in actuality, nevertheless this inclination is not sufficient in the absence of something extrinsic, which draws out

[28] Jacobus de Viterbio, *Quodlibet*, II, q. 5 (ed. Ypma, 1968–1975, 65).

[29] See also Jacobus de Viterbio, *Quodlibet*, II, q. 5 (ed. Ypma, 1968–1975, 74): "Esse in potentia medium est inter esse actu et non esse omnino."

[30] Note that *Quodlibet*, II, q. 5, in which James defends his theory of pre-existing forms, asks *Utrum in materia sint rationes seminales?* The answer he gives is affirmative. For more details on how James's views fit the tradition of seminal reasons, see Phelps (1980) and Gossiaux (2021).

[31] Jacobus de Viterbio, *Quodlibet*, II, q. 5 (ed. Ypma, 1968–1975, 77).

[32] Jacobus de Viterbio, *Quodlibet*, II, q. 5 (ed. Ypma, 1968–1975, 71).

[33] Jacobus de Viterbio, *Quodlibet*, II, q. 5 (ed. Ypma, 1968–1975, 86–87).

through a transmutation that which is in potency to actuality. For this is transmutation, namely, a transition from potency to actuality.[34]

This passage illustrates once more that James's pre-existing forms, or beginnings of forms, are more than matter's bare potential for receiving forms; for they are here described as bringing about an inclination and thus as somewhat active. The role of the external agent is not to bring about a form in something that undergoes a transmutation, but merely to bring about the transition from potency to actuality. This raises the question of what accounts for the fact that an external agent has the effect it has. Why does fire heat? Why does a given animal bring about another animal of the same species? Obviously, James cannot resort to the transmission of a form to account for the fact that causes and their immediate effects are often alike. The response James gives to this question is the same he gave to the problem of psychological causal interaction: he insists that there is an order (*ordo*) and an agreement (*convenientia*) between agents and their effects, which again sounds very much like a pre-established harmony.

Whatever one thinks of James's arguments for pre-existing forms in matter, in his own characterization he is not attempting to come up with a revisionary new way of thinking about natural processes. His goal is rather to make sense of everyday causal processes in the natural world, such as generation and qualitative changes. Yet they would not be conceivable if it were not for forms that pre-exist in potential being.

11.4 Innate Virtues

James left a couple of texts on intellectual and moral virtues, but explicit references to the innateness of virtues or dispositions (*habitus*) are extremely rare.[35] Nevertheless, given his account of natural change and the impossibility of new forms being brought about in a subject through a natural process, it seems to follow that virtues must also be to some extent innate. In fact, James explicitly draws the connection between dispositions (such as virtues) and forms:

And from these things it follows that the essence of a disposition [such as a virtue] is naturally placed in the soul according to potential being (*esse potentiale*). And this is clear from the example of the forms which exist in matter, because they are all created together with

[34] Jacobus de Viterbio, *Quodlibet*, II, q. 5 (ed. Ypma, 1968–1975, 89): "Alio modo dicitur aliquid esse principium activum per modum transmutationis. Et tale principium oportet ponere in eductione formarum de potentia ad actum, praeter illud principium quod dicitur inchoativum formae. Licet enim illud inchoativum formae inclinet materiam ad hoc ut fiat in actu, non tamen sufficit illa inclinatio sine agente extrinseco extrahente per transmutationem illud quod est in potentia ad actum. Hoc enim est transmutatio: exitus videlicet de potentia ad actum."

[35] See Osborne (2018) for more on James's teaching on the virtues.

> matter according to potential being, not according to actual being, but they are educed from potency to actuality.[36]

For a medieval Aristotelian, talk of innate virtues sounds much less strange than talk of innate cognition or pre-existing forms. For Aristotle himself mentions natural virtues in book 6 of his *Nicomachean Ethics*. James combines Aristotle's (innate) natural virtues with his own doctrine when he explains that "the virtues which the Philosopher [...] called natural virtues are nothing other than such propensities (*idoneitates*) and aptitudes (*aptitudines*) towards the virtues that are acquired by means of actions (*ad virtutes ex actibus acquisitas*)."[37] Although this makes it sound as if for James full virtues are something new because they are acquired, he sometimes uses more careful language and simply says that these innate propensities and aptitudes are merely perfected by the corresponding acts.[38]

The thing that stands out when James discusses the ontology of virtues is that he uses exactly the same vocabulary of propensities, aptitudes, beginnings, etc. that he also applies to natural forms and in cognitive psychology.[39] This would not be so if he did not see a close connection between pre-existing forms, cognition, and virtues. However, the point of Aristotle's talk of natural virtues is that these natural virtues come in different degrees in different human beings. In this respect, innate virtues are quite different from pre-existing forms and cognition, which seem to exist in the same degree of potential being in all chunks of matter and in all human souls. For James, the difference in natural (innate) virtues stems ultimately from the physiological make-up of the individual human body, which is determined by our environment and especially our parents. This is ultimately one of the reasons why James thinks that stupidity can be inherited.[40]

---

[36] Jacobus de Viterbio, *Quaestio utrum in anima humana* (ed. Côté, 2012, 306). For more about this *quaestio*, see Côté (2012).

[37] Jacobus de Viterbio, *Quodlibet*, I, q. 7 (ed. Ypma, 1968–1975, 93). See Aristoteles, *Ethica Nicomachea*, VI.13, 1144b1–10.

[38] Jacobus de Viterbio, *Quodlibet*, I, q. 7 (ed. Ypma, 1968–1975, 92): "Et haec sunt semina scientiarum et virtutum quae in nobis esse dicuntur, ipsae videlicet idoneitates et aptitudines quas habemus ad scientias et virtutes. Perficiuntur siquidem enim huiusmodi aptitudines per actus. Et quia ex ipsis actibus aggeneratur maior habilitas ad agendum perfectius et promptius, ideo huiusmodi aptitudines perficiuntur etiam per habitus ex actibus acquisitos."

[39] See also Jacobus de Viterbio, *Quodlibet*, IV, q. 23 (ed. Ypma, 1968–1975, 80): "Quaedam vero sunt accidentia animae indita vel innata, ut idoneitates; quaedam sunt aptitudines spirituales ad scientias et virtutes, vel ad horum opposita. Sicut in quibusdam est naturalis quaedam aptitudo ad liberalitatem, in quibusdam ad avaritiam, in aliquibus ad temperantiam [...]."

[40] Jacobus de Viterbio, *Quodlibet*, IV, q. 23 (ed. Ypma, 1968–1975, 81). For more on this particular quodlibetal question, see Loconsole, Chap. 6 in this volume, pp. 91–113.

11.5 Conclusion: Back to Aquinas

In this short contribution, I have presented a philosopher and theologian who defended something like the everything-is-already-inside thesis in the generation after Aquinas. Obviously, there are some differences between the sets of opinions that make up this thesis in Aquinas's description in the *Quaestiones disputatae de veritate* and in James of Viterbo's teachings. James's innatism with respect to cognition is not of the Platonic variety, since it does not entail a theory of recollection; for cognition, both sensory and intellectual, is innate not as complete acts of cognition, but only as incomplete ones. Nor does James defend Anaxagoras's view of latent actual forms in matter. Again, his view is more moderate: his forms exist only in matter and in potential being. And something similar can be said about his views regarding innate virtues. Nevertheless, it is revealing that James defends some version of all three elements of the everything-is-already-inside thesis.

But what about Aquinas's diagnosis of what is wrong with the assumption that everything is already inside—namely, that it makes it incomprehensible how there can be genuine natural agents and that on this account lower agents act only *per accidens*? We have seen over the course of this contribution that James has a rebuttal to many of Aquinas's concerns. His account of innate cognition is driven by the desire to understand how human beings can be cognitive agents and how psychological activities can be attributed to human souls. The same is true for his account of pre-existing forms in matter, for without them, James holds, even the most basic natural processes, such as generation and qualitative change, would not be possible.

I have already hinted at these possible responses to Aquinas, but they seem not to address the second part of Aquinas's critique: the issue of *per accidens* causation. Here Aquinas seems to have a point. For if our cognitive acts exist in our soul as incomplete acts which incline the soul to the corresponding complete act, what role is there to be played by the object? It looks as if the object does not exercise any real causality; all it does is remove, by its mere presence, an impediment to the exercise of our cognitive powers.

James has two responses to this objection. One of these we have already seen: he maintains that the object is indeed a real cause which does something: namely, it excites the soul. He even calls it an efficient cause.[41] Likewise, in his description of natural changes, James insists that the external agent contributes something to the activity, even if the form in potential being is also an active cause of the change in question. In neither case does he speak of the object or the external agent as a mere remover of an impediment. But this response may not be convincing: for it is one thing to describe something in a certain way, but it is another to be entitled to do so.

His second response therefore consists in a clarification of what is meant by calling something a cause *per accidens*. Obviously, there are some causes that are merely incidental and contribute nothing essential to the effect. For instance, if I remove a pillar in a house and this leads the heavy mass supported by the pillar to

[41] See Sect. 11.2, and the texts in footnotes 11 and 12.

fall down, I do not strictly speaking contribute anything essential to the downward movement, which is caused entirely by the weight, but also *per accidens* by the removal of the pillar. However, not all non-essential causes are *per accidens* causes of this type. Another type of non-essential causes are causes "from what follows" (*ex consequenti*). Matter is one of James's examples of a cause *ex consequenti*. Matter is not a per se cause of the activities of, say, the human soul, but since the human soul could not exist without being in matter, and because matter contributes to the ability of the human soul to act, its contributions are not merely coincidental. Causes that dispose or induce something to act are likewise called causes *ex consequenti*, and these show an even more active involvement with the effect they help in producing.[42] It is in this last sense of non-essential cause that James conceives of the causality of objects of cognition and external created agents in natural processes, and presumably also of action in the acquisition of moral virtues. They are not the primary causes, but they are not mere coincidental causes either. This may not be enough to convince Aquinas or his followers, but at least this reply takes out some of the sting associated with the notion of coincidental causation.

I have no argument to demonstrate that James was directly influenced by Aquinas's statements in the *Quaestiones disputatae de veritate* with which I began this short contribution, nor do I wish to claim that this was the case. Aquinas's disputed questions were certainly among his most widely read works but nothing in my argument hangs on James having had direct knowledge of Aquinas's text. However, there can be little doubt that the philosophical issues raised there must have occupied James's mind a fair bit. I hope that this short contribution has shown that James of Viterbo was an original and highly systematic thinker who merits being studied further.[43]

References

Primary Literature

Aristotle. 1894. *Ethica Nicomachea*, ed. L. Bywater. Oxford: Oxford University Press.

———. 1936. *Physica*, ed. William D. Ross. Oxford: Oxford University Press.

———. 1956. *De anima*, ed. William D. Ross. Oxford: Oxford University Press.

Augustine. 1968. *De Trinitate*, ed. William J. Mountain and François Glorie. Turnhout: Brepols.

Avicenna. 1968–1972. *Liber de anima seu sextus de naturalibus*, ed. Simone Van Riet. 2 vols. Louvain: Peeters; Leiden: Brill.

———. 1977–1983. *Liber de philosophia prima sive scientia divina*, ed. Simone Van Riet. 3 vols. Louvain: Peeters; Leiden: Brill.

[42] See Jacobus de Viterbo, *Quodlibet*, II, q. 7 (ed. Ypma, 1968–1975, 108–111). For the importance of this text, see also Dumont (2018, 291–293).

[43] Research for this article was supported by an Insight Grant from the Social Sciences and Humanities Research Council of Canada.

James of Viterbo. 1968–1975. *Disputationes de Quolibet*, ed. Eelcko Ypma. 4 vols. Würzburg: Augustinus-Verlag.

———. 2012. *Quaestio utrum in anima humana sint aliqui habitus naturaliter inditi*. In Antoine Côté, Deux questions inédites de Jacques de Viterbe sur les habitus, 303–308. *Archives d'histoire doctrinale et littéraire du Moyen Âge* 79: 289–311.

John of Damascus. 1955. *De fide orthodoxa*, ed. Eligius M. Buytaert. St. Bonaventure, NY: Franciscan Institute.

Richard of Middleton. 2014. *Quaestiones disputatae, qq. 38–45*, ed. Alain Boureau. Paris: Les Belles Lettres.

Thomas Aquinas. 1888–1906. *Summa theologiae*, ed. Commissio Leonina. 9 vols. Sancti Thomae de Aquino Opera Omnia iussu Leonis XIII P.M. edita, t. 4–12. Rome: Typographia Polyglotta.

———. 1918–1930. *Summa contra Gentiles*, ed. Commissio Leonina. 3 vols. Sancti Thomae de Aquino Opera Omnia iussu Leonis XIII P.M. edita, t. 13–15. Rome: Typographia Polyglotta.

———. 1929–1947. *Scriptum super libros Sententiarum*, ed. Pierre Mandonnet and Marie F. Moos. 4 vols. Paris: Lethielleux.

———. 1965. *Quaestiones de virtutibus in communi*. In *S. Thomae Aquinatis Quaestiones disputatae* 2, ed. P. Bazzi et al., 707–751. Turin–Rome: Marietti.

———. 1970–1976. *Quaestiones disputatae de veritate*, ed. Commissio Leonina. 3 vols. Sancti Thomae de Aquino Opera Omnia iussu Leonis XIII P.M. edita, t. 22. Rome: Editori di San Tommaso.

———. 1996. *Quaestiones disputatae de anima*, ed. Bernardo C. Bazán. Sancti Thomae de Aquino Opera Omnia iussu Leonis XIII P.M. edita, t. 24.1. Roma: Commissio Leonina.

Secondary Literature

Côté, Antoine. 2009. Simplicius and James of Viterbo on Propensities. *Vivarium* 47: 24–53.

———. 2011. Introduction. In *Jacques de Viterbe. L'âme, l'intellect et la volonté: Textes latins introduits, traduits et annotés par Antoine Côté*, 7–59. Paris: Vrin.

———. 2012. Deux questions inédites de Jacques de Viterbe sur les "habitus". *Archives d'histoire doctrinale et littéraire du Moyen Âge* 79: 289–311.

Dumont, Stephen. 2018. James of Viterbo on the Will. In *A Companion to James of Viterbo*, ed. Antoine Côté and Martin Pickavé, 249–305. Leiden: Brill.

Gossiaux, Mark. 2021. James of Viterbo on Seminal Reasons as *inchoationes formarum*. *Vivarium* 59: 52–78.

Osborne, Thomas. 2018. James of Viterbo's Ethics. In *A Companion to James of Viterbo*, ed. Antoine Côté and Martin Pickavé, 306–330. Leiden: Brill.

Phelps, Mary. 1980. The Theory of Seminal Reasons in James of Viterbo. *Augustiniana* 30: 271–283.

Pickavé, Martin. 2016. Some Later Medieval Responses to the Paradox of Learning. In *Schüler und Meister*, ed. Thomas Jeschke and Andreas Speer, 21–44. Berlin: De Gruyter.

Pickavé, Martin, and Antoine Côté. 2018. James of Viterbo's Philosophy of Nature. In *A Companion to James of Viterbo*, ed. Antoine Côté and Martin Pickavé, 127–167. Leiden: Brill.

Solère, Jean-Luc. 2018. James of Viterbo's Innatist Theory of Cognition. In *A Companion to James of Viterbo*, ed. Antoine Côté and Martin Pickavé, 168–217. Leiden: Brill.

Chapter 12
The World of Senses. On the Process of Cognition in Walter Burley

Monika Mansfeld

Abstract "Nothing is in the intellect that was not first in the senses." The Peripatetic axiom, which expresses Aristotle's simple observation that knowledge gathered via sensory experience precedes intellectual cognition, was the starting point for numerous medieval thinkers in their own reflections on the topic. Walter Burley not only discusses it in his commentaries on Aristotle's *De anima* and on the *Parva naturalia* set, but also devotes to it two short independent treatises: *De sensibus* and *De potentiis animae*. Therein, he develops the theory of five external and five internal senses, which are responsible for human cognitive activities, and explains how the information provided by external senses is gathered and reworked by the internal ones until it reaches the intellect. These two treatises offer a sufficient glimpse on Burley's understanding of cognitive processes and on the way in which he made use of the variety of text sources. I argue that he accepts the basic Avicennian doctrinal framework while rejecting some of its features in favour of the Averroist alternative, which, in general, situates his doctrine close to that of Thomas Aquinas.

Keywords Sensory cognition · Internal senses · The five senses · Walter Burley · *De potentiis animae* · *De sensibus*

12.1 Introduction

De sensibus and *De potentiis animae* are two of a dozen of short independent treatises on logic and natural philosophy Walter Burley wrote at an early stage of his academic career. Since they seem to lack perceivable links to William of Ockham of

This work was supported by the Polish National Science Centre (NCN) grant n. UMO-2016/23/B/HS1/00430.

M. Mansfeld (✉)
University of Łódź, Łódź, Poland

M. Gensler et al. (eds.), *The Embodied Soul*, Historical-Analytical Studies on Nature, Mind and Action 11, https://doi.org/10.1007/978-3-030-99453-2_12

any kind, it can be surmised that they were composed before the 1320s.[1] Their form may suggest that they were either written for the general public, e.g., in return for help in receiving a prebend (it is known that Burley held some *in absentia* since 1309),[2] or had been commissioned by one of his protectors. As they remain formally and doctrinally unpolished, they might have not been intended for publication but, in the author's view, were to serve him as a personal reference booklet. The two texts seem to be the works of a mature scholar, which is shown, despite their small size, both in the selection of topics and in their execution. Nonetheless, as Mary Jean Kitchel, the editor of *De potentiis animae*, noted with some embarrassment, it is easy to underestimate the value of Burley's brief essays on psychology and physiology.[3] This is because at first glance they seem to be merely short presentations of ideas derived mostly from Aristotle's *De anima*[4] and other writings, supplemented with opinions taken from works by other authors, such as Themistius, Avicenna, Algazel, Averroes, and Albert the Great.

Kitchel discovered, however, that *De potentiis animae* can be treated as an index of Burley's earliest ideas about the key issues of the philosophical psychology of his times.[5] In my opinion, the same is true of the other treatise, *De sensibus*,[6] which seems to be a sort of appendix to *De potentiis animae* where issues omitted or neglected in the earlier work receive a fair treatment. The information found in the two works is complementary, as if Burley intentionally avoided repeating the same information,[7] and doctrinal inconsistencies between them occur very rarely. Moreover, they share the same style and, for the most part, structure. For these reasons, it can be concluded that the two treatises form a coherent set containing Burley's theory of sense cognition. It deals primarily with human beings, although

[1] The conclusion is based on James Athanasius Weisheipl's proposed dating of Burley's other independent treatises. See Weisheipl (1968, 179). Marta Vittorini, who misidentifies *De sensibus* as one of Burley's logical treatises, claims that it was probably written during Burley's studies in Paris, see Vittorini (2013, 39).

[2] See Martin (1964, 204, 217, 227–228), Vittorini (2013, 20).

[3] See Kitchel (1971, 85).

[4] See Kitchel (1977, 388).

[5] Ibidem.

[6] Herman Shapiro and Frederick Scott published a semi-critical edition of Burley's *De sensibus*. See Gualterus Burlaeus, *De sensibus* (ed. Shapiro, Scott, 1966). Due to its numerous errors, it cannot form the basis for any doctrinal analyses. All quotations in this paper come from my own critical edition of the work. It has been prepared on the basis of all existing manuscripts: Cambridge, Cambridge University Library, MS Hh IV 13, ff. 58v–60v; London, Lambeth Palace Library, MS 70, ff. 116vb–117va; London, Lambeth Palace Library, MS 143, ff. 139vb–140vb; London, British Library, MS Royal 12 B 19, ff. 301v–304r; Oxford, Oriel College, MS 12, ff. 97va–99ra; Vatican, Bibliotheca Vaticana, MS Vat. lat. 2146, ff. 250vb–251vb; Vatican, Bibliotheca Vaticana, MS Vat. lat. 2151, ff. 117rb–119ra; Worcester, Worcester Cathedral Library, MS F. 86, ff. 8v–10r.

[7] This is evident especially in the case of the theory of intellect, which was developed very briefly in *De potentiis animae* and discussed on a surprisingly large scale in *De sensibus*. By contrast, if the text were true to its title, one might expect a comprehensive description of the internal senses in *De sensibus*. Instead, Burley presents only a sketchy description, probably because he believed they had been sufficiently described in *De potentiis animae*.

some of its components, such as the theory of the external senses or of the common sense, apply also to animals.

12.2 The Process of Cognition as a Process of Abstraction

Like many medieval authors, Burley divides human powers of the sensitive part of the soul (*potentia sensitiva*) into two groups of senses: exterior and interior, both sets consisting of five elements.[8] It should be acknowledged that, for him, sensory cognition is a process aimed at transforming the material qualities of the perceived object into the immaterial qualities in our intellect.[9] It comprises several steps, each of which progressively makes the sensory data more abstract and less material.[10] In this respect, Burley follows the Arabic tradition of Avicenna and Averroes,[11] although he slightly modifies the details of how the process of abstraction actually works.

12.2.1 The External Senses

The process of sensory cognition begins when the real qualities of a material object (*agens principale*) move through the external medium (*medium extrinsecum*).[12] These sensible qualities are perceived either with respect to their inherent nature (*sensibilia per se*), or contingently (*sensibilia per accidens*). Burley subdivides the former group into proper (*sensibilia propria*) and common inherent qualities

[8] Gualterus Burlaeus, *De sensibus*, Cambridge, Cambridge University Library, MS Hh IV 13, ff. 58r–v; London, Lambeth Palace Library, MS 70, f. 116vb; Vatican, Bibliotheca Vaticana, MS Vat. lat. 2151, f. 117rb: "Nota quod in homine sunt quinque sensus interiores sicut exteriores [...]."

[9] For more on the process of transferring the material qualities of the perceived object into the immaterial qualities in our intellect in Albert the Great and his predecessors, see Panarelli, Chap. 7 in this volume, pp. 115–132 and Hernández-Rubio, Chap. 8 in this volume, pp. 133–161.

[10] Gualterus Burlaeus, *De sensibus*, Cambridge, Cambridge University Library, MS Hh IV 13, f. 58v; London, Lambeth Palace Library, MS 70, f. 116vb; Vatican, Bibliotheca Vaticana, MS Vat. lat. 2151, ff. 117rb–va: "Et sciendum quod qualitas sensibilis in obiecto habet esse maxime reale et materiale, in medio autem magis spirituale et immateriale, et adhuc in sensu exteriori habet maiorem spiritualitatem. Et sic secundum ordinem sensuum interiorum continue subtiliatur et depuratur, donec veniat ad intellectum, ubi habet omnino esse immateriale."

[11] See Black (2000, 63).

[12] Gualterus Burlaeus, *De potentiis animae*, XVII, 93 (ed. Kitchel, 102): "Themistius aliam rationem ponit, et est quod natura in opere suo semper procedit ab imperfecto ad perfectum. [...] Sic in proportio[ne – *corr. MM*] est ordo determinatus quod a re corporali et materiali ad spiritualem requiritur quod fiat transitus per medium inter utrumque. Unde species in medio spiritualius habet esse quam in obiecto, et materialius quam in organo, et ideo species in medio se habet medio modo inter esse spirituale et materiale."

(*sensibilia communia*).[13] The difference, as he sees it, between proper and common qualities lies in the fact that while the former are perceived by one sense only, e.g., colour by sight, the latter are perceived by more than one sense, e.g., shape is perceived by sight and touch. This does not mean, however, that the perception of common inherent qualities always requires cooperation of several external senses; quite to the contrary: Burley is convinced that a single external sense is sufficient to perceive common qualities alone, but not exclusively this or that particular sense. A good example is provided by the perception of movement: I can see that a ball somebody has kicked is moving, or, if I get hit by it without looking, feel the movement of the ball by touch only, or perceive it by both senses simultaneously. With respect to sensible contingent qualities, Burley gives only the obscure example of *Darius's son*[14] without explaining how relations can be objects of sensual cognition, even if we agree with his claim that they are perceived only secondarily.[15] One can suppose that he is referring to the visual similarity between father and son. The qualities perceived with respect to their inherent natures are the primary objects of our sensory cognition and the crucial role among them is played by proper qualities.

The character of the medium between the object and organs of external senses is relevant for each particular sense. Sight, hearing and smell make use of an external medium (*medium extrinsecum*), which can be either homogeneous, such as pure air or water, or of a mixed character.[16] Taste and touch have so-called connected media (*media coniuncta*), which means that they require direct contact with a body to perceive something. Just like Aristotle,[17] Burley argues that in the cases of touch and

[13] Gualterus Burlaeus, *De potentiis animae*, V, 27 (ed. Kitchel, 90): "Sciendum quod sensibile proprium et sensibile commune sunt per se sensibilia. Unde Philosophus distinguit sensibile primo in duo membra, scilicet in sensibile per se et sensibile per accidens. Et postea distinguit sensibile per se in sensibile commune et sensibile proprium."

[14] Gualterus Burlaeus, *De potentiis animae*, V, 23–26 (ed. Kitchel, 90): "Et sciendum quod triplex est sensibile, scilicet sensibile proprium, sensibile commune, et sensibile secundum accidens. Sensibile proprium est quod non contingit altero sensu sentire et circa quod non contingit sensum errare. Isto modo color est obiectum potentiae visivae, quia color non percipitur ab aliquo sensu nisi a visu et circa colorem non contingit visum errare, sicut postea videbitur. Sensibilia communia sunt quinque: motus, quies, numerus, figura et magnitudo. Et propter hoc dicuntur sensibilia communia, quia percipiuntur a diversis sensibus, sicut patet: magnitudo potest percipi a visu et a tactu, et similiter motus. [...] Sensibile per accidens dicitur ut Darii filius; et propter hoc dicitur sensibile per accidens, quia coniungitur sensibili per se et non percipitur nisi ex hoc quod sensibile per se sentitur."

[15] I suppose that this aspect of Burley's division was problematic for him too, for he does not develop this idea any further, even though he had Aristotle's authority on his side. See Aristoteles, *De anima*, II.6, 418a20–21 (ed. Decorte, Brams, ALDatabase). For more on the division into proper, common, and accidental sensibles in Aristotle, see Knuuttila (2008, 7).

[16] Gualterus Burlaeus, *De sensibus*, Cambridge, Cambridge University Library, MS Hh IV 13, f. 59v; London, Lambeth Palace Library, MS 70, f. 117rb; Vatican, Bibliotheca Vaticana, MS Vat. lat. 2151, f. 118va: "Agens instrumentale seu medium extrinsecum est illud quod intercipitur inter obiectum quod est agens principale et inter medium intrinsecum; et illud aliquando est unum tantum, ut aer tantum vel aqua tantum, et aliquando duo vel tria vel quattuor vel plura."

[17] See Aristoteles, *De anima*, II.7, 419a30–31 (ed. Decorte, Brams, ALDatabase); II.11, 423b20–26 (ed. Decorte, Brams, ALDatabase). The view that in the case of those two contact senses the flesh, rather than being the organ itself, is the medium was also shared by medieval physicians. See Lagerlund (2008, 119).

taste our flesh is the medium, viz. skin for the former and tongue for the latter; the proper organs of these senses are deeper in the body.[18]

Although Burley gives some examples of how things can be perceived through a mixed external medium, almost all of them are restricted to the case that is easiest to explain, i.e., to the sense of sight: seeing through air and water is like seeing through two or more layers of different density, given that they are sufficiently transparent.[19] In his opinion, hearing may also take place with two adjacent media, i.e., water and air, if we treat the connatural air in our ears, although it is actually an inner medium, as one of them.[20] Attempts to explain this phenomenon in the case of smell pose much greater difficulties, of which the author seems to be unaware or unwilling to enter. He says nothing of how multiple layers of medium would possibly allow for smelling anything.[21]

Through the external medium, sensible qualities are transmitted to the interior medium (*medium intrinsecum*), which is one of the external senses. Burley accepts Aristotle's list of them, comprising sight, hearing, smell, taste, and touch. Each external sense has its own interior medium, which is in principle identical with its physical organ. For the sense of sight it is the eye with its humours and membranes, for the sense of touch it is the skin and flesh, for the sense of taste it is the tongue (or at least its upper part), for the sense of smell it is the delicate tissue in the anterior part of the brain, for the sense of hearing[22] it is the connatural air in the ear.[23] Technically, perceiving qualities by external senses does not take place in sense organs: seeing does not take place in the eye only, or tasting in the tongue only, etc. For Burley, it is our brain which is responsible for actions of three of the external senses (seeing, hearing and smelling), while specialised tissues of the body are

---

[18] Gualterus Burlaeus, *De potentiis animae*, XXI, 101 (ed. Kitchel, 104): "Sciendum quod visus, auditus, et olfactus sentiunt per media extrinseca, sed gustus et tactus sentiunt per media coniuncta. Unde medium sensus tactus est caro. […] Medium in gustu est lingua."

[19] Gualterus Burlaeus, *De sensibus*, Cambridge, Cambridge University Library, MS Hh IV 13, f. 59v; London, Lambeth Palace Library, MS 70, ff. 117rb–va; Vatican, Bibliotheca Vaticana, MS Vat. lat. 2151, f. 118va: "Duo: quando res existens in aqua videtur ab oculo existente in aere mediante speculo vel vitro vel aliquo huiusmodi corpore diaphano vel transparente. Per plura tribus, sicut sol et stellae videntur mediante orbe caelesti inferiorum et mediante sphaera ignis et quandoque nubibus densis; et sic possibile est multiplicare media extrinseca ad omnem numerum."

[20] Gualterus Burlaeus, *De potentiis animae*, XIII, 80 (ed. Kitchel, 100): "Unde in aqua potest homo audire si aqua non ingrederetur ad istum aerem. Et quod sit talis aer in aure patet […]."

[21] For more on the sense of smell, see Fedriga, Chap. 2 in this volume, pp. 17–35.

[22] Gualterus Burlaeus, *De sensibus*, Cambridge, Cambridge University Library, MS Hh IV 13, f. 59v; London, Lambeth Palace Library, MS 70, f. 117rb; Vatican, Bibliotheca Vaticana, MS Vat. lat. 2151, f. 118rb: "Intrinsecum est diversum in diversis sensibus: in visu oculi et humores et tunicae oculi; in tactu pellis et caro; in auditu aer connaturalis inclusus in aure; in gustu lingua vel superior pars linguae; in olfactu tela tenuis in anteriori parte cerebri."

[23] Gualterus Burlaeus, *De potentiis animae*, IX, 49 (ed. Kitchel, 94): "Et ideo dicitur quod in aure est aer inclusus qui est connaturalis auri."

responsible for the remaining two (touching and tasting).[24] In the case of touching and tasting, he distances himself from both Aristotle's opinion that these sensations ultimately take place in the heart,[25] and from that of Galen and Avicenna who argue that this role is played by the brain.[26] We may observe, somewhat anachronistically, that his views on that matter resemble the current understanding of the nervous system.[27]

Burley is convinced that the primary objects of our senses are not subject to any illusions. If something is perceived by a particular sense, it is perceived *truly*—there is no doubt here. Burley's opinion on the primary object of sight calls for attention, because it stands out against two much more popular medieval positions on this issue. Some scholars claimed that what the sight primarily perceives is colour (*color* or *coloratum*), because we cannot see anything that does not participate in a colour; others argued that it is light (*lumen* or *lux*) for the same reason. Burley, who was aware of this discord, noticed that colour and light are equally important for our vision and, for that reason, one cannot be a principle of the other. Consequently, he solves the problem by choosing the visible (*visibile*) as a common denominator for both. In his opinion, a perceptible is a member of the set of qualities naturally capable of being perceived, because we can only see what can be seen (*visibile*), we can only touch what can be touched (*tangibile*), etc.; in short, perception entails all and

[24] Gualterus Burlaeus, *De sensibus*, Cambridge, Cambridge University Library, MS Hh IV 13, ff. 59v–60r; London, Lambeth Palace Library, MS 70, f. 117va; Vatican, Bibliotheca Vaticana, MS Vat. lat. 2151, f. 118va: "Unde manifestum est ex istis pro sensibus exterioribus quod visus non est in oculo, sed interius in anteriore parte cerebri in concursu duorum nervorum ad duos oculos protensorum. Nec auditus est in aure, sed interius infra aerem connaturalem colligatum in quadam pellicula in cerebro. Odoratus est in anteriori parte cerebri infra telam praedictam supra visum. Isti tres sensus sunt in cerebro. Gustus non est in carne linguae, sed interius in quodam rete tenui et subtili in ipsa lingua. Tactus vero non est immediate in carne, sed in quodam rete huiusmodi quasi per totum corpus extensum. Isti autem duo sensus angulariter sunt in corpore, alii autem tres anguariter sunt in capite."

[25] See Aristoteles, *De sensu et sensato*, 2, 439a1–2 (ed. Leonina, 27); Aristoteles, *De somno et vigilia*, 2, 456a3–6 (ed. Lulofs, ALDatabase); Aristoteles, *De iuventute et senectute*, 3, 469a10–16 (ed. Hulstaert, ALDatabase).

[26] See Avicenna, *Liber de anima seu sextus de Naturalibus IV–V* (ed. Van Riet, 1968, 176–181); Avicenna, *Canon medicinae*, I, fen. 1, doctr. 6, cap. 1 (ed. Paganinum de Paganinis, 1507, SN).

[27] By doing so, he also distances himself from the views he presented earlier in his commentaries on *De sensu et sensato* and *De somno et vigilia*, where he argued in favour of the heart (or its equivalent in animals which do not have a heart) as the primary organ of touch and taste. See Gualterus Burlaeus, *Commentarium in De sensu et sensato*, 2: "Organum tactus et organum gustus sunt situata iuxta cor, ut per caliditatem cordis reducatur frigiditas terrae ad temperiem." See Gualterus Burlaeus, *Commentarium in De somno et vigilia*, 4: "Generato enim corde in habentibus cor vel proportionali cordi in non habentibus ipsum, quod est prima pars in animali, statim inest sensus tactus, igitur et sensus communis secundum eandem partem, cum adhuc non sit alia pars generata, igitur sensus communis est in eadem parte corporis cum tactu ut in corde."

only that which is perceptible for one or more of the senses.[28] Thus, simple acts of sensation themselves provide sufficient evidence for their authenticity.

Burley, however, finds ample room for error in the interpretation of individual proper sensations; it may be caused by several factors. First, there are distortions of our sensual cognition called indispositions of the object. Burley's example is the perceived size of the Sun as a small spot in the sky, whereas it is actually bigger than the Earth; apparently, for him the object must be the (perceived) size, which is subject to the illusion. Secondly, there are indispositions of the medium itself. We can see that the colour of the Sun varies at different times of the day: it is red in the morning and pale yellow at noon. According to Burley, this is caused by the quality of the air, which is impure and polluted with fumes in the morning, and then gets purified progressively as the day goes by. Thirdly, sensory illusions are due to physiological causes, such as injuries of sensory organs or diseases. For instance, everything tastes bad when we are suffering of fever.[29] Burley does not treat these three factors as mutually exclusive; on the contrary, he seems to allow for all possible combinations of them.

12.2.2 The Internal Senses

From the external senses qualities are transported to the internal ones, where they are subject to other processes. When it comes to enumerating the internal senses, Burley closely follows Thomas Aquinas,[30] for he lists five of them: the common sense, the imagination (also called phantasy), the estimative power, the cogitative power, and the memory. Like in Aquinas, his list seems to be a combination of elements adopted from Avicenna, who enumerated five internal senses (common sense, imagination, phantasy, estimative faculty, and memory), and Averroes, who claimed

[28] Gualterus Burlaeus, *De potentiis animae*, VI, 29 (ed. Kitchel, 91): "Dicendum quod nullus sensus decipitur circa suum primum obiectum. Primum obiectum visus est visibile in sua communitate secundum quod se extendit ad omnia visibilia in totum. Unde primum obiectum actus potentiae est obiectum ei adaequatum sic quod nihil continetur sub isto obiecto quin posset percipi ab illa potentia, et omne quod percipitur a tali potentia continetur sub isto obiecto adaequato. Sic enim visibile est primum obiectum visus, quia omne visibile potest percipi a visu, et nihil potest percipi a visu nisi visibile. Et circa tale obiectum non contingit potentiam errare."

[29] Gualterus Burlaeus, *De potentiis animae*, VI, 29–32 (ed. Kitchel, 91): "Unde visus non iudicat aliquid esse visibile nisi illud sit visibile; circa tamen contentum sub isto obiecto contingit potentiam errare. Et hoc contingit tribus de causis: vel propter indispositionem organi, vel propter indispositionem medii, vel circa indispositiones obiecti. Exemplum primi: sic enim gustus febricitantis decipitur circa sapores, quia enim lingua febricitantis abundat humore cholerico. Ideo febricitans omnia iudicat esse amara. Et hoc accidit propter indispositionem organi gustus. Exemplum secundi: propter enim indispositionem medii visus iudicat solem alterius coloris esse mane et in meridie. Iudicat enim solem esse mane coloris rubei, et hoc quia mane medium est indispositum et vaporosum. Exemplum tertii: circa condiciones obiecti contingit sensum errare. Sic enim visus iudicat solem esse unius pedis tantum cum tamen sol sit maior tota terra."

[30] See Black (2000, 66).

that there are only four of them (common sense, imagination, cogitative power, and memory).[31] Burley's views on the internal sensorium resemble Avicenna's and Averroes's also in another respect, because he strongly emphasises that the actions of all internal senses form a sequence aimed at abstracting the sensory data from matter. For Aquinas, in turn, the internal senses concurrently contribute towards sensible species becoming intelligible.[32] For this reason, Burley ascribes to each internal sense a position in an ascending order. Because the process begins with the common sense, it is regarded as the lowest internal sense. From it the sensory data proceeds to the imagination, which is seen as an internal sense of a somewhat higher position than that of the common sense, and, consequently, the data gets more abstracted in it. Then it is moved to the other internal senses step by step until it reaches the memory—the highest internal sense where the sensory data is most abstracted.[33] Burley seems to be inspired by the empiricist maxim *Nothing is in the intellect that was not first in the senses*[34] so much that he extends it to the internal senses by stating that it can be applied to them in two different ways: first, there is nothing in the internal senses that was not first in the external ones;[35] secondly, there is nothing in the higher internal senses that was not first in the lower ones.[36] Despite the apparent inconsistencies of the positions of Thomas Aquinas on the one hand and of Avicenna and Averroes on the other, Burley tries to reconcile them in his *De potentiis animae* and *De sensibus*—regrettably, as we will see, without much success.

The first of the internal senses, the common sense, collects various perceptions coming from the external senses and distinguishes between them.[37] The author of *De sensibus* does not provide us with any detailed information about it, but we can assume that following Aristotle he must have understood the common sense as a capacity to convert raw sensual perceptions originating in the external senses into global perceptions of objects. The phrase 'distinguishes between', which he uses in his description of the common sense, may be understood in two ways. First, it makes us aware of the difference between sensory data coming from the same external

---

[31] See Black (2000, 59–63).

[32] See Thomas de Aquino, *Quaestiones de anima*, q. 13, resp. (ed. Robb, 190–191).

[33] See Knuuttila (2008, 9).

[34] See *Les auctoritates Aristotelis* (ed. Hammese, 197, 24); Aristoteles, *De sensu et sensato*, 6, 445b16–17 (ed. Leonina, 76).

[35] Gualterus Burlaeus, *De potentiis animae*, XXII, 104 (ed. Kitchel, 104): "Et sciendum quod sensus communis non potest in actus suos nisi per sensus proprios quia, ut dicit Avicenna, sensus communis non percipit nisi ex hoc quod sensus proprii redeunt ad ipsum cum suis lucris."

[36] Gualterus Burlaeus, *De sensibus*, Cambridge, Cambridge University Library, MS Hh IV 13, f. 58v; London, Lambeth Palace Library, MS 70, f. 116vb; Vatican, Bibliotheca Vaticana, MS Vat. lat. 2151, ff. 118rb–va: "Et sciendum quod qualitas sensibilis in obiecto habet esse maxime reale et materiale, in medio autem magis spirituale et immateriale, et adhuc in sensu exteriori habet maiorem spiritualitatem. Et sic secundum ordinem sensuum interiorum continue subtiliatur et depuratur, donec veniat ad intellectum, ubi habet omnino esse immateriale."

[37] Gualterus Burlaeus, *De sensibus*, Cambridge, Cambridge University Library, MS Hh IV 13, f. 58v; London, Lambeth Palace Library, MS 70, f. 116vb; Vatican, Bibliotheca Vaticana, MS Vat. lat. 2151, f. 117rb: "Operatio sive officium sensus communis est distinguere inter sensibilia quinque sensuum exteriorum."

sense (e.g., the difference between sour and sweet tastes), or between sensory data coming from various external senses (e.g., the difference between sensations of sweet and red).[38] Secondly, it distinguishes between the subject and object of perception,[39] implicitly indicating our self-awareness.[40] It is important to note that Burley refutes Averroes's claim that the common sense is a direct source of some sensual cognition and states that it does not perceive any sensual data primarily and without mediation of the external senses.[41] Although it might seem reasonable to apply the same charge to our self-awareness, it actually falls on deaf ears in Burley. The explanation is that, according to him, our self-awareness is always relational and never entirely internal, because it actualizes only when the common sense is triggered by the data coming from the external senses.

The next internal sense to which sensory data is transmitted from the common sense is the imagination, also called phantasy. The author of *De sensibus* follows Averroes's tradition of identifying the phantasy and the imagination as one internal sense,[42] against Avicenna and his followers who distinguished between the two by pointing to their slightly different functions.[43] Burley concedes, however, that the

---

[38] Gualterus Burlaeus, *De potentiis animae*, XXII, 102 (ed. Kitchel, 104): "Sensus communis non solum cognoscit omnia obiecta sensuum particularium, sed etiam cognoscit differentias obiectorum diversorum sensuum particularium."

[39] Gualterus Burlaeus, *De potentiis animae*, XXII, 103 (ed. Kitchel, 104): "Unde per sensum communem percipit se aliquis videre et non per visum."

[40] Feola (2018, 51). Regrettably, Burley does not address the problem of self-awareness in animals. For more information on this issue, see Hernández-Rubio, Chap. 8 in this volume, pp. 133–161 and Evelina Miteva, Chap. 5 in this volume, pp. 73–90, in which she discusses the case of pygmies seen as an intermediary stage between animals and human beings.

[41] Gualterus Burlaeus, *De potentiis animae*, VII, 35, 37 (ed. Kitchel, 91–92): "Nec valet secunda causa [id est quod sensibilia communia per se sentiuntur a sensu communi – *suppl. MM*], quia ab eisdem immutatur communis sensus a quibus immutatur sensus proprius, quia omnes immutationes sensuum propriorum terminantur ad sensum communem. Si igitur sensibilia communia sentirentur a sensu communi, cum per se sentirentur, tunc per se a sensibilibus propriis sentirentur. [...] Nunc autem sensus proprii percipiunt sensibilia communia ratione sui generis ut ratione sensus in commune. Visus enim percipit magnitudinem non inquantum visus, sed inquantum est sensus."

[42] Gualterus Burlaeus, *De potentiis animae*, XXII, 110 (ed. Kitchel, 105): "Commentator ponit tantum quattuor sensus interiores, scilicet sensum communem, phantasticam, aestimativam, et memorativam. Unde imaginativam et phantasticam ponit unum sensum."

[43] Burley describes Avicenna's model in his *De potentiis animae* in detail. He claims that according to Avicenna phantasy and imagination are two separate capacities, the latter of which covers all functions of the former. Gualterus Burlaeus, *De potentiis animae*, XXII, 105–106 (ed. Kitchel, 104): "Post istum sensum est virtus imaginativa, quae [...] etiam retinere potest, quia possumus imaginari res in absentia, quod non esset nisi imaginatio retineret speciem in absentia rei. Etiam potest plus quam sensus communis, quia sensus communis non retinet speciem in absentia rei vel, si retinet, parum retinet. Post imaginationem est virtus phantastica, quae non solum recipit quod facit sensus communis, nec solum retinet quod facit in imaginatio[ne – *corr. MM*], sed componit unam rem cum alia re ut, quia videmus montem et videmus aurum, habemus per phantasiam nostram quod possumus componere montem aureum."

By doing so, Burley joins a large group of Western scholars (including Albert the Great and Thomas Aquinas) who misunderstood Avicenna's position. See Black (2000, 59).

imagination serves two distinct functions: on the one hand, it retains sensory data from the external senses even if the object is no longer present, which is the proper function of the imagination, and, on the other hand, it produces representations of things that have never been perceived by external senses, such as a chimera, a centaur and other non-existing things, which is a characteristic feature of the phantasy.[44]

The third inner sense, the estimative power, judges whether the perceived object can be considered beneficial or harmful, and, as a result, whether it should be approached or not.[45] The example, given in both treatises, that a lamb avoids a wolf and follows its mother, suggests that the estimative power triggers actions of intuitive or perhaps (in modern zoological terminology) instinctive character.[46] In his presentation of it, Burley refers to Avicenna, who claimed that the *vis aestimativa* apprehends not only the sensibles' species, but also receives intentions that have not been sensed yet (*sensationes non sensatas*), such as natural hostility or friendliness, and to Alhazen, who said that it senses the object in a way in which it has not been perceived by the external senses.[47]

Burley does not seem to be concerned with the problematic consequences of this statement: suggesting that the estimative power allows for some unmediated cognition seems to be at odds with (if not completely contradictory to) the main thrust of his teaching that there is nothing in the internal senses that was not first in the external ones. A possible solution here might be that he takes this function of the estimative power to be similar to the abovementioned self-awareness: neither of them is conceived as a fully independent power that can operate without sensory data provided by the external senses. In this case, the sheep flees from a wolf, whose hostility it has never sensed, perhaps even without any prior cognitive process resulting in the conclusion that the wolf is an enemy, but, still, its escape is caused by the perception of some of the wolf's qualities, which it finds alarming. The picture that emerges is one in which the estimative power is imbued with an innate power to recognize an object inasmuch as it is linked to the animal's practical,

[44] Gualterus Burlaeus, *De sensibus*, Cambridge, Cambridge University Library, MS Hh IV 13, f. 58v; London, Lambeth Palace Library, MS 70, f. 116vb; Vatican, Bibliotheca Vaticana, MS Vat. lat. 2151, f. 117rb: "Operatio imaginativae est duplex. Prima, secundum Philosophum secundo *De anima*, est motio facta a sensu exteriori, et haec est fingere rem prius sensatam quasi praesentem, licet sit absens. Secunda operatio eius est fabricare sive fingere individuum sive idolum numquam sensatum, sicut chimaeram et huiusmodi."

[45] Gualterus Burlaeus, *De sensibus*, Cambridge, Cambridge University Library, MS Hh IV 13, f. 58v; London, Lambeth Palace Library, MS 70, f. 116vb; Vatican, Bibliotheca Vaticana, MS Vat. lat. 2151, f. 117rb: "Operatio aestimativae est iudicare inter nocivum et conveniens, sicut agnus per illam virtutem iudicat de lupo ipsum esse nocivum, de virore ipsum esse convenientem; et ideo unum fugit et alium sequitur."

[46] See Phillips (1934, 237).

[47] Gualterus Burlaeus, *De potentiis animae*, XXII, 107 (ed. Kitchel, 104–105): "Quarta virtus est aestimativa, quae non solum recipit species sensibilium, sed etiam intentiones non sensatas. Cuiusmodi sunt amicitia, inamicitia, quae dicunt quasdam intentiones ultra species sensibiles. [...] De hac virtute dicit Avicenna quod est apprehensiva intentionum quae per sensum non apprehenduntur, et ideo recipit species non sensatas. Et Algazel dicit quod virtus aestimativa est virtus apprehendens de sensato quod non est sensatum, sicut ovis inimicitas lupi."

survival-oriented needs and to the release of an appropriate motoric reaction. For that reason, the estimative power in Burley should be understood as primarily interpretive, rather than merely perceptual. It is worth adding that for Burley the human counterpart to the estimative power in animals is called the collative power and is aimed at particular intentions (*virtus collativa intentionum particularium*). According to him, it is responsible for the same type of actions as the estimative power in animals, i.e., deciding whether the object of perception is harmful or desirable for the subject, but it is hard to tell from his description whether it is of an exclusively intuitive character or not.[48]

The cogitative power is the fourth inner sense described by Burley in *De sensibus*. Like Thomas Aquinas, from whom he has probably adopted this idea,[49] he draws a distinction between the estimative and the cogitative power, because the former, he suggests, pertains exclusively to animal subjects, while the latter pertains to human subjects.[50] It is difficult to state, however, whether for Burley the above-mentioned collative power, aimed at particular intentions, which is the human equivalent of the estimative power, is identical with the cogitative one, because he does not provide any information that could help to differentiate between them. If they are one and as such constitute the third, penultimate internal sense in the human cognitive system, which is an equivalent of the estimative power in animals, the total number of internal senses should therefore be reduced from five in a row (for which Burley had argued) to four and it will include the common sense, the imagination, the cogitative power, and the memory (Averroes's set), whereas for animals there is the sequence of the common sense, the imagination, the estimative power, and (for some of them) the memory.

Burley describes the cogitative power as the one that distinguishes between accidents and their subjects and between diverse accidents in the same subject (e.g., between colour, quantity and figure).[51] Although the characteristics he provides are very loosely worded, he seems to mean that the process of comparing different

---

[48] Gualterus Burlaeus, *De potentiis animae*, XXII, 107 (ed. Kitchel, 105): "Unde ista virtus in homine vocatur collativa intentionum particularium; ideo vocatur ratio particularis per quam diiudicatur quid sit conveniens, quid disconveniens. [...] Nam sicut aliquis cognoscit aliquid per intellectum quod non cognoscit per sensum, licet non cognoscat nisi accipiendo a sensu, ita aestimativa cognoscit, licet modo inferiori."

Avicenna claimed that the estimative power in human beings is responsible (together with the practical intellect) for their deeper theoretical understanding of the physical world, cultivating an aesthetic perspective on their surroundings, and transforming their animal appetites into particularly human emotions. Since Burley knew some Avicenna, we can assume that he also considered these to be additional functionalities of the human estimative power. See Black (2000, 61–62).

[49] See Thomas de Aquino, *Summa Theologiae*, I, q. 78, a. 4. (ed. Leonina, 255–257).

[50] Gregorić (2007, 10).

[51] See Gualterus Burlaeus, *De sensibus*, Cambridge, Cambridge University Library, MS Hh IV 13, f. 58v; London, Lambeth Palace Library, MS 70, f. 116vb; Vatican, Bibliotheca Vaticana, MS Vat. lat. 2151, f. 117rb: "Operatio cogitativae est distinguere accidens a suo subiecto et inter diversa accidentia in eodem subiecto, sicut inter colorem, quantitatem et figuram in eodem subiecto, et illa virtus reperitur in solo homine."

features in the cogitative power makes the subject of sensitive cognition become aware of primary substances present in the act of cognition.[52] In this context, Burley's expression 'to distinguish between the subject and its accidents' may mean that by means of the cogitative power a human being becomes aware of the individual as individual (e.g. of a cat as this cat), which is no longer "a bundle of conjoined proper and common sensibles."[53] This mental act, however, is neither identical nor coextensive with the mental act of abstraction found in the active intellect (*intellectus agens*),[54] because it gives us information about an individual but not about the individual's nature or essence.

The fifth and last internal sense is the memory, which Burley calls a treasury of species and intentions.[55] He does not treat reminiscence as something substantially different from memory, because, as in the case of imagination and phantasy, he is convinced that they differ accidentally only.[56] The memory's main function is to recall singular things that have been previously perceived when they are no longer present, in such a way that the subject is aware of the fact that they are no longer present and, moreover, is able to locate their presence in time. Burley's example of a memory concerning an event is such: I remember that I have seen the Pope and I can recall the circumstances of that event.[57] Burley attaches particular importance to

---

[52] See Gualterus Burlaeus, *De potentiis animae*, XXII, 113–115 (ed. Kitchel, 106–107): "Licet omnis sensus sit susceptivus specierum sine materia, et ita in virtute sensitiva videtur esse quodammodo abstractio, tamen in diversis hoc est diversimode. Nam sensus exteriores sic recipiunt species sensibiles [*corr. ex* sensibilis *MM*] sine materia, quia materia formae quae sentitur non est in tali sensu nisi per modum quo cera recipit formam annuli aurei sine auro. Isto modo formae sensibilium recipiuntur in virtute sensitiva. Sed tamen sensus exterior, semper dum habet speciem informantem ipsum, requiret formam cuius est species esse in materia exteriori actualiter. [...] Haec abstractio est imperfecta. [...] Ideo dicitur quod illa imaginatio habet formam sub perfectiori abstractione quam sensus exterior, cum ista tamen abstractione stat alligatio et concomitantia accidentium. [...] Sed virtus aestimativa habet adhuc maiorem abstractionem, quia recipit intentiones non sensatas [...], quae non quantum ex se sunt formae materiales; accepit tamen eis esse in materia. Sed tamen in ista virtute non est totalis et perfecta abstractio formae a materia et ab appendiciis materiae, quia non apprehendet tales formas nisi sub singularitate et sub esse hic-et-nunc, quae sunt condiciones materiae. Sed haec virtus non retinet; ideo requiritur virtus memorativa, et in illa est maior abstractio quam in aestimativa [...]."

[53] Lisska (2016, 243).

[54] See Lisska (2016, 252).

[55] Gualterus Burlaeus, *De potentiis animae*, XXII, 108 (ed. Kitchel, 105): "Quinta virtus est memorativa, quae est thesaurus tam specierum quam intentionum."

[56] Gualterus Burlaeus, *De potentiis animae*, XXII, 121–122 (ed. Kitchel, 108): "Ideo aliud praeteritum est obiectum unius et alterius. Aliter dicitur quod in memoria et reminiscentia sunt eaedem virtutes et solum variantur accidentaliter, quia memoria est ut terminus motus, et reminiscentia sicut motus ad terminum. [...] Rationes factae solum probant quod memoria et reminiscentia variantur accidentaliter."

[57] Gualterus Burlaeus, *De sensibus*, Cambridge, Cambridge University Library, MS Hh IV 13, f. 58v; London, Lambeth Palace Library, MS 70, f. 116vb; Vatican, Bibliotheca Vaticana, MS Vat. lat. 2151, f. 117rb: "Operatio memorativae materialis est apprehendere rem singulariter prius sensatam per modum praeteriti, sicut quod vidi Papam vel Parisium, et ad illam operationem tempus et locus semper concurrunt, sicut quod vidi ipsum in tali loco et in tali tempore."

the act of experiencing the event subjectively. The act is called the direct object of memory (*obiectum proximum*)—in the case of seeing the Pope in Avignon it is the act of seeing him—whereas the object of this act—the Pope—is an object of memory only indirectly (*obiectum remotum*).[58] Of course, not every past event is remembered in such a way, but Burley stresses that remembering in the proper sense has two essential conditions. The first one is that the one who remembers something actually experienced it,[59] which excludes the possibility of remembering anything inferred (e.g., the conclusion that I was born is inferred from the facts that I am alive now and that people come to being through birth), or based on somebody's else's account (e.g., I know that there was World War II because my grandma told me so, or I have read about it in a textbook). Secondly, the subject must be aware of the time that has elapsed since the moment of the experience.[60]

Analysing the senses from the point of view of anatomy and physiology, Burley follows Avicenna's claim that the centre of the human internal sense system is the brain. In it, he distinguishes three coordinated ventricles, each of which is further divided into two parts, namely the anterior and the posterior one. They are responsible for the functioning of similarly coordinated senses: the common sense, located in the anterior part of the first ventricle, the imagination, located in its posterior part, the phantasy, located in the anterior part of the middle ventricle, the estimation, located in its posterior part, the memory, located in the anterior part of the third ventricle, and, finally, the power of locomotion, which is not an internal sense itself, but is nonetheless strictly related to the remaining five[61] and is located in its posterior part.

It can be seen therefrom that according to Burley the process of cognition in the internal sensorium is analysed on two correlated planes. One is psychological and

[58] Gualterus Burlaeus, *De potentiis animae*, XXII, 117 (ed. Kitchel, 107): "Aliud est sciendum quod memoria est respectu duplicis obiecti, scilicet respectu proximi et respectu obiecti remoti. Obiectum proximum memoriae est actus praeteritus ipsius recordantis, sed obiectum remotum est obiectum illius actus. Verbi gratia: recordor quod vidi heri Ioannem hic existentem. Obiectum proximum illius recordationis est actus videndi quem habui heri circa Ioannem, sed obiectum remotum est Ioannes circa quem habui actum."

[59] Gualterus Burlaeus, *De potentiis animae*, XXII, 116 (ed. Kitchel, 107): "Circa virtutem memorativam est dubium an memoria sit solum praeteritorum, et an sit omnium praeteritorum cognitiva. Et est sciendum quod non omnis cognitio praeteriti est memoria. Scio enim mundum fuisse, et me fuisse natum, non tamen recordor. Respectu enim illorum non habeo memoriam; sed ad memoriam requiritur quod cognoscens praeteritum pro tunc circa ipsum habuit hic aliquem actum, et ideo recordatio est cognitio praeteriti inquantum pro tunc circa ipsum habuit recordans aliquem actum."

[60] Gualterus Burlaeus, *De potentiis animae*, XXII, 118 (ed. Kitchel, 107): "Adhuc est sciendum quod ad memoriam requiritur quod recordans percipiat tempus interceptum recordationis et tempus cognitionis rei de qua habetur memoria."

[61] Gualterus Burlaeus, *De potentiis animae*, XXII, 109 (ed. Kitchel, 105): "Unde secundum Avicennam, in cerebro sunt tres ventriculi: primus in prima parte, secundus in media parte, et tertius in parte posteriori. Et quilibet dividitur in partem anteriorem et posteriorem. In prima parte primi ventriculi est sensus communis, et in posteriori eiusdem imaginatio. In anteriori parte medii ventriculi est virtus phantastica, et aestimativa in parte posteriori. In prima parte tertii ventriculi est virtus memorativa, et in parte posteriori virtus motiva secundum locum."

consists of the study of the internal senses' functions, the other is physiological and discusses the processes taking place in those regions of the brain where these senses are anatomically located. This approach allows Burley to explain various difficulties encountered in the process of cognition in a way similar to the one he employs in similar cases with respect to the external senses, for he takes into account both deficiencies of our cognitive faculties themselves, such as interruptions and distractions, and physical injuries of their organ, i.e., the brain. The only difference is that in the internal senses there are no distortions on account of the medium.

12.2.3 The Intellect

Passing through the whole of the human sensorium, sensory data become gradually more and more subtle and less material; nonetheless, even at the stage of the memory, the last of the internal senses, they are still somewhat material. It is only when they reach the intellect that the last remnants of materiality get eliminated. What remains is an immaterial spiritual, i.e., intelligible, quality—the final form of human sensory cognition.[62] Even more precisely, the sensory data from the internal senses reach the intellect gradually, too: first, it is ultimately immaterialized by the active intellect,[63] the faculty of the mind responsible for abstracting intelligible forms from sensory data, then it proceeds to the passive intellect, where it is stored.[64] The

[62] Gualterus Burlaeus, *De sensibus*, Cambridge, Cambridge University Library, MS Hh IV 13, f. 58v; London, Lambeth Palace Library, MS 70, f. 116vb; Vatican, Bibliotheca Vaticana, MS Vat. lat. 2151, ff. 117rb–va: "Et sciendum quod qualitas sensibilis in obiecto habet esse maxime reale et materiale, in medio autem magis spirituale et immateriale, et adhuc in sensu exteriori habet maiorem spiritualitatem. Et sic secundum ordinem sensuum interiorum continue subtiliatur et depuratur, donec veniat ad intellectum, ubi habet omnino esse immateriale."

[63] Gualterus Burlaeus, *De sensibus*, Cambridge, Cambridge University Library, MS Hh IV 13, f. 58v; London, Lambeth Palace Library, MS 70, ff. 116vb–117ra; Vatican, Bibliotheca Vaticana, MS Vat. lat. 2151, f. 117va: "Intellectus agens est intellectus immaterialis qui transfert species rerum sensibilium a materialitate quam habent in omnibus sensibus interioribus ad immaterialitatem quam habent in intellectu possibili. Et haec est causa propter quam ponitur intellectus agens, videlicet quia nullum agens materiale potest sufficienter transferre speciem seu similitudinem suam de materialitate ad immaterialitatem, quia omne quod generatur generatur a suo simili, ut patet secundo *Metaphysicae*, ideo requiritur intellectus agens qui est omnino immaterialis ad istam translationem faciendam."

[64] Gualterus Burlaeus, *De sensibus*, Cambridge, Cambridge University Library, MS Hh IV 13, f. 58v; London, Lambeth Palace Library, MS 70, f. 116vb; Vatican, Bibliotheca Vaticana, MS Vat. lat. 2151, f. 117va: "Et est intellectus possibilis, passibilis seu materialis anima humana rationalis in homine. Et dicitur possibilis, quia est in potentia ad intelligendum et aliquando intelligit actu et aliquando non, et per hoc distinguitur ab intellectu agente qui semper aequaliter intelligit. Idem dicitur passibilis, quia recipit similitudines et cognitiones rerum, et illud recipere est quodammodo pati, sicut est in intellectu agente. Idem dicitur materialis, quia est quasi materia et subiectum specierum et accidentium existentium in ea, et per hoc differt ab intellectu agente qui nihil aliud habet a se in sua substantia."

ultimate result of the whole process is a knowledge of the essence (*essentia*) of the object[65] in the actualized passive intellect (*intellectus adeptus*).

In *De sensibus*, Burley, who follows Averroes in this respect, distinguishes between four types (*modi*) of the intellect: the possible intellect, which is identical with the passible and material one, the active intellect,[66] the speculative intellect (*intellectus adeptus*), seen as an actualized passive intellect, and the imaginative faculty (*virtus imaginativa*),[67] which is the most surprising element of this group. Its presence is justified by the fact that, since the active intellect thinks by means of forms or species, it processes its data conceptually as images, which must be imprinted in the imaginative faculty.[68] For this reason the imaginative faculty ought not to be conceived as a part of the intellect in the proper sense, but as an essential factor in human intellection: a conveyor belt from the material intellect to the active one.[69] Although Burley does not elaborate on this idea either in *De potentiis animae* or in *De sensibus*, he sheds some light on it in his exposition of the *De anima*.[70] There he explains that human intellection takes place when the subject reaches the universal concepts contained in the separated material intellect by means of the sensible images (*forma imaginata*) of the particular material objects connected with

[65] Gualterus Burlaeus, *De potentiis animae*, XXIII, 132 (ed. Kitchel, 110): "Sciendum quod abstrahere est cognoscere vel facere aliquid cognitum absque illo cui tamen est in esse coniunctum. [...] Et ideo intellectus agens dicitur abstrahens, quia per intellectum agentem potest intellectus possibilis cognoscere essentiam rei absque hoc quod cognoscat condiciones materiales quibus essentia est in esse coniuncta; et haec est abstractio intellectus agentis."

[66] The two are seen as immaterial internal parts of human soul. See Gualterus Burlaeus, *De potentiis animae*, XXIII, 124 (ed. Kitchel, 108–109): "Et ideo oportet ponere aliquid aliud [a rebus materialibus – *suppl. MM*] activum, et illud activum non est separatum ab anima nostra sicut quidam antiqui dixerunt quod intelligentia separata est agens cognitionis in animabus nostris."

[67] Gualterus Burlaeus, *De sensibus*, Cambridge, Cambridge University Library, MS Hh IV 13, f. 58v; London, Lambeth Palace Library, MS 70, f. 116vb; Vatican, Bibliotheca Vaticana, MS Vat. lat. 2151, f. 117va: "Intellectus autem secundum Commentatorem tertio *De anima* accipitur quattuor modis, videlicet pro intellectu possibili seu passibili et materiali, qui sunt idem, pro intellectu agente, pro intellectu adepto, pro virtute imaginativa."

[68] Gualterus Burlaeus, *De sensibus* Cambridge, Cambridge University Library, MS Hh IV 13, f. 58v; London, Lambeth Palace Library, MS 70, f. 117ra; Vatican, Bibliotheca Vaticana, MS Vat. lat. 2151, f. 117va: "Intellectus dicitur virtus imaginativa propter similitudinem operationis unius et alterius, quia semper concurrunt in operationibus suis, quia nihil intelligimus sine phantasmate sive imaginativa."

[69] Burley is not precise whether imagination seen as an internal sense and as a type of intellect are the same power or duplicate ones. I get the impression, however, that he treats them as different functions of the same power, of which the first case is primary and the other is auxiliary/instrumental only. This is because Burley focuses on the idea that we think images and, thus, they must be provided by our imagination for intellection to take place.

[70] Burley's exposition of *De anima* was written in Paris between 1315 and 1318. It has not been critically edited yet. There are, however, large passages of it quoted in the articles devoted to the problem of whether Burley presented views characteristic of Latin Averroists or not. See Maier (1955, 477–499), Kuksewicz (1981, 341–377).

the abovementioned concepts. These images, received by individual human souls, are in turn processed by the imaginative faculty.[71]

In *De potentiis animae* Burley claims, contrary to the presentation in *De sensibus*, that actually there are only two types of intellect: the passive one and the active one. All other kinds traditionally distinguished by philosophers, such as the speculative one (*intellectus adeptus*), the effected one (*intellectus in effectu*), the potential one (*intellectus in potentia*), or the assimilated one (*assimilatus intellectus*), can be reduced to either of the two. This does not mean, however, that he treats the latter group as pure synonyms, for they describe various functions or states of either of the two intellects. For instance, he sees the assimilated intellect as the one that has perfect knowledge of the prime cause and the potential intellect as the potency to speculate or acquire knowledge. The point he wants to make is that none of them actually constitutes a separate intellectual potency in the way the active and passive ones do.[72]

In my opinion, the disagreement between Burley's views on the intellectual powers presented in *De sensibus* and in *De potentiis animae* is not as substantial as it may seem, because in each case he is analysing them from a different point of view. In the former work, he is paying attention to the constitutive elements of intellectual cognition understood as a process. Describing them step by step, he must take into account all the factors that contribute to acquiring knowledge, from the moment when the active intellect acts upon the passive one with the images provided by the imaginative faculty, to the point when knowledge is actualized in the *intellectus adeptus*. In the latter text, his interest lies in the activity or passivity of the intellect's functions, regardless of their content. This is why he does not even make a distinction between the potential passive intellect and the actualized one (*intellectus adeptus*) and ignores the role of the imaginative faculty.

---

[71] Gualterus Burlaeus, *Commentarium in II De anima Aristotelis*, MS Vatican, Vat. lat. 2151, f. 22vb: "Ad primum istorum respondet Commentator tertio *huius* commento quinto dicens quod non dicimur actu intelligentes nisi per hoc quod intellecta in actu copulantur nobiscum; et intellecta in actu, scilicet intellecta speculativa, copulantur nobiscum per copulationem formarum imaginatarum in virtute imaginativa. Unde non sum actu intelligens nisi per intentionem in intellectu materiali copulatam in actu, et ista intentio est mihi copulata in actu per formam imaginatam existentem in virtute imaginativa in ea moventem intellectum materialem."

Although there are always concepts of all things in the passive intellect, a man comes into contact with them only when his sensible images join the intellect, and only then he does perceive a given object. Gualterus Burlaeus, *Commentarium in II De anima Aristotelis* (MS Vat. lat. 2151, f. 23rb): "Ad aliud dicendum quod intelligere est in potestate nostra, quia intellectus materialis semper est dispositus ad recipiendum per speciem imaginatam in imaginativa nostra, supposita dispositione ex parte imaginativae nostrae, et quia dispositio ex parte imaginativae nostrae est in potestate nostra, ideo actus intelligendi est in potestate nostra."

[72] Gualterus Burlaeus, *De potentiis animae*, XXIV, 137–138 (ed. Kitchel, 111–112): "Aliis nominibus nominatur intellectus a philosophis; dicitur enim intellectus formalis et intellectus adeptus, intellectus in effectu et intellectus in potentia, et intellectus assimilatus. Sed ista non sunt diversae potentiae intellectivae. [...] Intellectus adeptus dicitur intellectus possibilis cum adeptus fuerit cognitionem multarum rerum, ut cum intellectus possibilis sit bene habituatus. Intellectus in effectu dicitur intellectus actualiter considerans. Intellectus in potentia dicitur intellectus secundum quod est in potentia ad considerandum vel ad scientiam inquirendam [*corr. ex* inquirendum *MM*]. Intellectus assimilatus dicitur intellectus possibilis habens perfectam cognitionem primae causae, quia ad eius similitudinem creatus est intellectus; et ideo cum eius cognitionem habeat, assimilatur ei qui est origo omnis cognitionis."

12.3 Active and Passive Elements of the Human Sensorium

Another important aspect of Burley's theory of sensual cognition is the distinction between the active and passive elements of the human sensorium. Burley is aware of the fact that, since the external senses are passive faculties (*motum-non movens*) acted upon by the object (*movens-non motum*), there always has to be some medium (*movens-et-motum*) between them, which makes the sensation possible. The medium is moved by objects and, in turn, moves the sense.[73] This is because no immediate action is possible between accidents of different genera;[74] in this case they are accidents from two opposed *genera generalissima*, i.e., action and passion. Although Burley follows Aristotle to a large extent in this respect, he does not accept his ideas uncritically. The central point of the controversy is Aristotle's claim expressed in the second book of the *De anima*, where he characterised the external senses as completely passive faculties. It resulted from his debate with Plato and other philosophers about the nature of vision: contrary to Plato, who believed in extramission, i.e., emissions of rays from the eyes to the object, he argued that vision occurs through intromission, i.e., emissions of likenesses from the object to the eyes. Extrapolating from this idea, Aristotle and his followers thought that it is not only sight which works this way, but also the other external senses; consequently, they are passive faculties only. While dealing with this problem, Burley tacitly accepts Aristotle's conclusion that if we discover how sight actually works and determine whether it is an active or a passive faculty, we will be able to apply this finding to all the other external senses. And, indeed, he does not deal with any other of them either in *De potentiis animae* or in *De sensibus*, for he restricts himself to analysing the views on sight and vision.

Burley rejects the opinion that seeing involves exclusively a representation of the object, provided that the medium and the external sense are disposed to receive it. From this point of view, as Burley puts it, we cannot speak of any action or passion of either the object alone or of the sense alone.[75] Seeing things reduced to representations would be like reflections in a mirror. Burley rejects this option for several reasons. First, it is against our experience: if I stare at something shiny, e.g. the Sun,

---

[73] Gualterus Burlaeus, *De potentiis animae*, XVII, 90 (ed. Kitchel, 102): "Sensibile est movens-non-motum; sensus est motum-non-movens. Sed inter movens-non-motum et motum-non-movens oportet ponere movens-et-motum medium, et illud dicitur esse medium in sensatione. Medium enim movetur ab obiecto et medium movet sensum, et sic medium est movens-et-motum; sensibile non est motum ab aliquo nec sensus est movens."

[74] Gualterus Burlaeus, *De potentiis animae*, XVII, 91 (ed. Kitchel, 102): "Ratio quare movens-non-motum non potest immediate agere in motum-non-movens est quia illa quae agunt immediate et patiuntur ad invicem debent esse eiusdem generis."

[75] Gualterus Burlaeus, *De sensibus*, Cambridge, Cambridge University Library, MS Hh IV 13, f. 59r; London, Lambeth Palace Library, MS 70, f. 117ra; Vatican, Bibliotheca Vaticana, MS Vat. lat. 2151, ff. 117vb–118ra: "Una opinio est quae ponit visionem fieri neque per actionem obiecti neque sensus, nec per passionem, sed per solam repraesentiam obiecti supposita dispositione medii et sensus."

or brightly coloured, e.g. a clump of trees in spring, and suddenly turn my eyes to look at something else, the second object will appear as shiny or bright as the first one in the first moment of my looking at it.[76] Secondly, the organ can be destroyed by exposing it to some qualities of the object that exceed its physical capacities.[77] Moreover, if the sensible quality reached only the surface of the eye, just like it reaches the mirror, its species would not be taken over by any internal sense, because it would disappear in the same moment when the reflection was gone. As a result, we would not remember it.[78] Accordingly, Burley notes that there must be something in the sense that can be acted upon by an object.

An even more important question is, however, whether vision can take place if the object acts upon the sense and the sense receives the sensation passively only,[79] which is, as we remember, the position of Aristotle himself. To use a modern simile, according to this view, our senses act like the radio, i.e., radio waves (sensible forms) are emitted by the objects and the receiver (our sensory system) immediately receives them. Burley's answer to this question is negative. He sees the weakness of Aristotle's theory in ignoring the fact that, using the above example, the radio receiver (our external sensorium) is not always on. When we are asleep, it is obviously off; the same happens when we get distracted or tired—in those cases, there is, at best, a lot of interference.[80] According to Burley, there must be an active element in our external sensorium mobilising the senses to perceive and focusing their

[76] Gualterus Burlaeus, *De sensibus*, Cambridge, Cambridge University Library, MS Hh IV 13, ff. 59r–v; London, Lambeth Palace Library, MS 70, f. 117rb; Vatican, Bibliotheca Vaticana, MS Vat. lat. 2151, f. 118ra: "Item, secundum Alhacen primo *Perspectivae*: si quis respexerit viridarium intensum diu et stabiliter et divertat se ad obiectum alterius coloris, illud secundum obiectum apparebit viride et tandem apparet proprii coloris, quod non esset nisi oculus pateretur ab obiecto. Item, si quis diu respiciat solem et claudat oculos vel divertat se ad locum tenebrosum, apparet sibi per aliquod tempus quod videat solem splendidum sicut prius, postea apparebit minus splendidus donec appareat purpureus, et adhuc apparebit remitti quousque apparet subniger et tandem apparebit totaliter obscuratus, quod non esset nisi oculus pateretur a sole."

[77] Gualterus Burlaeus, *De sensibus*, Cambridge, Cambridge University Library, MS Hh IV 13, f. 59r; London, Lambeth Palace Library, MS 70, f. 117rb; Vatican, Bibliotheca Vaticana, MS Vat. lat. 2151, f. 118ra: "Contra primam arguitur sic secundum Philosophum tertio *De anima*: excellens sensibile corrumpit sensum, ergo agit in sensum."

[78] Gualterus Burlaeus, *De sensibus*, Cambridge, Cambridge University Library, MS Hh IV 13, f. 59r; London, Lambeth Palace Library, MS 70, f. 117rb; Vatican, Bibliotheca Vaticana, MS Vat. lat. 2151, f. 118ra: "Praeterea, secundum hoc nihil causaretur per visum in sensus interiores et per consequens nec in memoriam, ergo memoria non reciperet aliquam speciem vel similitudinem obiecti visibilis, ergo nec memoraretur illius in eius absentia."

[79] Gualterus Burlaeus, *De sensibus*, Cambridge, Cambridge University Library, MS Hh IV 13, f. 59r; London, Lambeth Palace Library, MS 70, f. 117ra; Vatican, Bibliotheca Vaticana, MS Vat. lat. 2151, f. 118ra: "Alia opinio fuit quod visio fit per actionem obiecti et per passionem visus tantum."

[80] Gualterus Burlaeus, *De sensibus*, Cambridge, Cambridge University Library, MS Hh IV 13, f. 59v; London, Lambeth Palace Library, MS 70, f. 117rb; Vatican, Bibliotheca Vaticana, MS Vat. lat. 2151, f. 118ra–rb: "Item, si ibi esset tantum actio obiecti, cum obiectum aequaliter agat sive sensus advertat sive distrahatur, sequitur quod sensus aequaliter sentiret advertens et distractus, quod est contra experimentum in vigilantibus et in dormientibus, nam distracti et dormientes recipiunt speciem obiecti, et tamen non iudicant de obiectis."

attention on receiving sensory data of a given kind, which is the proper operation of each sense, such as seeing for the sense of sight or touching for the sense of touch.[81] Although this solution assumes some sort of immanent activity of the sense itself, it cannot be concluded that Burley accepts some form of extramission (transient action).[82] By distinguishing between immanent and transient actions and assigning only the former to the senses, Burley tries to reconcile Aristotle's assertion that senses are passive faculties[83] with his own belief that attention, i.e., being active, is necessary for sensory cognition.

Unlike the external senses, the internal ones are—according to Burley—unquestionably active. Their activities range from apprehending (common sense) and maintaining (phantasy) the sensible forms to apprehending (estimative faculty) and maintaining (memory) intentions that have not been sensed (*intentiones non sensatae*).[84]

---

[81] Gualterus Burlaeus, *De sensibus*, Cambridge, Cambridge University Library, MS Hh IV 13, f. 59v; London, Lambeth Palace Library, MS 70, f. 117rb; Vatican, Bibliotheca Vaticana, MS Vat. lat. 2151, f. 118ra–rb: "Contra secundam opinionem arguitur sic: visio est propria operatio visus et ista causatur a visu, quia quaelibet forma naturalis habet suam operationem propriam, ergo visus non tantum patitur, sed aliquo modo agit. [...] Praeterea, iudicare est quodammodo agere, sed visus iudicat de visibili, ergo agit."

[82] Gualterus Burlaeus, *De sensibus*, Cambridge, Cambridge University Library, MS Hh IV 13, ff. 59r–v; London, Lambeth Palace Library, MS 70, f. 117ra–rb; Vatican, Bibliotheca Vaticana, MS Vat. lat. 2151, f. 118ra–rb: "Tertia ponit quod visio fit per extramissionem radiorum visualium ab oculo per totum medium usque ad obiectum et sic, ut videtur, communiter ponunt perspectivi. [...] Contra opinionem tertiam arguitur sic: si illa opinio esset vera, visus non pateretur a visibili, quod est prius improbatum. [...] Quarta opinio est quod visio fiat per actionem obiecti et etiam per actionem sensus, ita quod fiat quaedam mixtio temperata ex radiis visualibus et ex specie obiecti. [...] Contra quartam opinionem potest argui sic: ita est in sensatione visus sicut in sensatione aliorum sensuum, sed in sensatione aliorum sensuum non oportet quod sensus emittat aliquos radios nec aliquod tale miscibile cum specie sensibili, ut patet in tactu, gustu et olfactu et auditu, ergo non est ita in visu, sicut ponit ista opinio. Praeterea, etsi ita esset sicut ponit ista opinio, hoc tamen non sufficeret ad sensationem, nam ibi requiritur actio immanens ipsius sensus, videlicet ipsa visio seu actus sentiendi, sicut est ostensum contra opinionem secundam, quia ipsa est in sensu visus et non extra, sicut est de voluntate et intellectu qui habent operationes suas immanentes, id est interius manentes in virtute agente illam actionem et non transeuntes extra."

[83] Gualterus Burlaeus, *De sensibus*, Cambridge, Cambridge University Library, MS Hh IV 13, f. 59v; London, Lambeth Palace Library, MS 70, f. 117rb; Vatican, Bibliotheca Vaticana, MS Vat. lat. 2151, f. 118rb: "Ad illud dicendum quod duplex est actio, scilicet immanens et transiens. Ideo Philosophus negat a sensu actionem transeuntem in obiectum secundum quod mathematici posuerunt de visu. Et loquendo de passione opposita tali actioni, videlicet actioni transeunti, sensus est passivus tantum ab obiecto et non activus in obiectum. Sed loquendo primo modo de actione, scilicet actione immanente, non negat Philosophus actionem a sensu."

[84] Gualterus Burlaeus, *De potentiis animae*, XXII, 112 (ed. Kitchel, 106): "Hoc est quod vult Avicenna, *Sexto Naturalium*, qui dicit quod virtutes interiores aut sunt virtutes quarum actus sunt circa species formarum sensibilium, aut quarum actus sunt circa intentiones non sensatas. Si primo modo, aut ut recipiens et sic est sensus communis, aut ut conservans et sic est virtus phantastica; si secundo modo, hoc potest esse dupliciter, aut ut recipiens et sic est aestimativa, aut ut conservans et sic est memorativa."

The final element of the human cognitive system is the intellect. Its active power manifests itself in the active intellect, whereas its passive power is manifested in the possible intellect.[85] According to Burley, both are equally necessary for human cognition and in a way complementary. From the observation that sometimes we understand something and sometimes we do not he concludes that there must be an intellectual faculty responsible for receiving an act of understanding, which is the possible intellect.[86] In accordance with it, the passive faculty would not work without its active counterpart either: there must also be something responsible for actualising the knowledge. It cannot, however, be the material object we perceive, because it is only potentially intelligible, nor can it be separated from our soul, or we would not be able to understand when we want it; as a result we must admit the active intellect as an internal faculty of our soul.[87] In fact, Burley sees one of its operations as a sort of intellectual illumination, which enables the possible intellect to receive species from the internal senses when they have finally been stripped of all of their materiality.[88] In my opinion, an illumination like that, seen as an immanent operation of the intellect, is similar to attention, the active internal element of the external senses. The other, transient operation of the active intellect is abstraction, aimed at providing the possible intellect with the essence of what has been perceived.[89] As regards the possible intellect, Burley follows the tradition and

[85] Gualterus Burlaeus, *De potentiis animae*, XXIII, 123 (ed. Kitchel, 108): "Sequitur de potentia intellectiva quae dividitur in duas potentias, scilicet in potentiam activam et potentiam passivam. Intellectus agens est potentia activa, et intellectus possibilis est potentia passiva."

[86] Gualterus Burlaeus, *De potentiis animae*, XXIII, 124 (ed. Kitchel, 108): "Quod in parte intellectiva sit ponere potentiam passivam manifestum est, quia quilibet in se ipso experitur et aliquando intelligit et aliquando non intelligit. Sed in eo quod aliquando est in potentia est aliquando in actu oportet ponere potestatem passivam quae sit receptiva talis actus."

[87] Gualterus Burlaeus, *De potentiis animae*, XXIII, 124 (ed. Kitchel, 108–109): "Sed cuilibet potentiae passivae correspondet potentia activa, aliter frustra poneretur potentia passiva. Huic igitur potentiae passivae correspondet aliquod activum, reducens ipsum de potentia ad actum. Illud activum quod reducit intellectum possibilem de potentia ad actum non est aliqua res materialis, quia res materiales sunt solum in potentia intelligibiles. [...] Sicut oportet dicere activum esse aliquid ipsius animae, cuius probatio est, nam in potestate nostra est intelligere cum volumus postquam habuerimus species intelligibiles, sicut quilibet in se ipso experitur, sed si illud activum ducens intellectum nostrum de potentia ad actum esset separatum ab anima nostra, non esset in potestate nostra intelligere cum volumus."

[88] Gualterus Burlaeus, *De potentiis animae*, XXIII, 126 (ed. Kitchel, 109–110): "Et ideo sicut ad hoc quod coloratum gignat speciem suam in visum requiritur lumen solis vel alicuius corporis lucidi, sic ad hoc quod species existens in virtute phantastica gignat speciem in intellectu possibili requiritur quoddam lumen spirituale, et illud est lumen intellectus agentis. [...] Et ideo actio intellectus agentis non est nisi illuminatio."

[89] Gualterus Burlaeus, *De potentiis animae*, XXIII, 132 (ed. Kitchel, 110): "Et ideo intellectus agens dicitur abstrahens, quia per intellectum agentem potest intellectus possibilis cognoscere essentiam rei absque hoc quod cognoscat condiciones materiales quibus essentia est in esse coniuncta; et haec est abstractio intellectus agentis."

compares it to a blank slate which is initially empty but progressively receives intellectual species, sc. universal concepts.[90]

12.4 Object of Cognition

Having traced the steps of human sensory cognition, the last important question remains to be answered: what do we actually perceive? What is the nature of the sensible? For a realist that Burley is, the answer it clear: the proper object of our sensory cognition are abstracted accidents, such as light, colour, sound and flavour.[91] Concrete accidents, in turn, such as bright, red, loud and sour, are a combination of abstracts and their substrates and are perceived indirectly through their abstracts.

Burley explains the relation between perceiving abstract and concrete accidents using the example of light: I can see brightness of itself because it is the proper object of sight; but when I can see the Sun, which is bright, I perceive it because there is a brightness in it, hence secondarily only. Burley develops the example further in a way that reveals its ontological premises: on the one hand, there can be light as such without a substrate and I will still be able perceive it; on the other hand, a subject without its brightness will not be perceived at all.[92]

[90] Gualterus Burlaeus, *De potentiis animae*, XXIII, 133, 136 (ed. Kitchel, 111): "De intellectu possibili dicit Philosophus quod in principio est sicut tabula rasa in qua nihil est depictum. [...] Eodem modo intellectus possibilis in principio est in potentia ad recipiendum species omnium rerum intelligibilium absque abiectione alicuius ab intellectu possibili. [...] Sed intellectus cognoscit universale propriate, et ibi est appropriatio a parte universalis. Intellectus enim cognoscit universale quod universale non cognoscitur nisi ab intellectu."

[91] Gualterus Burlaeus, *De sensibus*, Cambridge, Cambridge University Library, MS Hh IV 13, f. 60r; London, Lambeth Palace Library, MS 70, f. 117va; Vatican, Bibliotheca Vaticana, MS Vat. lat. 2151, f. 118va–vb: "Ex his manifestum est quod omnia quae sentiuntur per se sunt abstracta, ut lux, color, sonus, odor, calor, frigus, et sic de aliis."

[92] Gualterus Burlaeus, *De sensibus*, Cambridge, Cambridge University Library, MS Hh IV 13, f. 60r; London, Lambeth Palace Library, MS 70, f. 117va; Vatican, Bibliotheca Vaticana, MS Vat. lat. 2151, f. 118vb: "Concreta autem non sentiuntur per se, sed tantum ad sensationem suorum abstractorum quae important, ut lucidum quod est congregatum ex duobus, videlicet ex subiecto et accidente, non sentitur per se seu per suum subiectum, sed tantum per lucem, quia, posito per imaginationem quod lux existeret per se sine subiecto, videretur sicut nunc, sed si subiectum lucis, ut sol vel luna vel ignis, esset per se sine luce vel colore, nullo modo videretur, quia non per se, quia per se subiectum non est activum in visu, nec per accidens, quia non habet aliquod accidens visibile in eo, et ita est de omni qualitate visibili et de omni substantia."

12.5 Conclusion

All things considered, Burley's views on human cognition from his *De potentiis animae* and *De sensibus* do not present a complete picture of his opinion on the topic. These two treatises, however, offer a sufficient glimpse into his understanding of its processes and on the way in which he made use of the variety of text sources. Burley chose a literary format different from his commentaries on Aristotle's natural philosophy that allowed him to present a unified account or a theory of sensation and perception which takes into account all available aspects taken from psychology, physiology and anatomy. Doctrinally, he accepts the basic Avicennian framework while rejecting some of its features in favour of the Averroist alternative. This, in general, situates his doctrine close to that of Thomas Aquinas. Although all elements of his discussion of the human cognitive system have antecedents in the texts of his predecessors, their arrangement and development appear to be his own.

References

Primary Literature

Aristoteles. 1943. *De somno et vigilia*, ed. Hendrik Joan Drossaart Lulofs. Leiden: Burgersdijk & Niermans.

———. 1985. *De sensu et sensato*, ed. Comissio Leonina. Roma.

———. 1999. *De iuventute et senectute*, ed. Kristien Hulstaert. Aristoteles Latinus XVI.1.2. Turnhout: Brepols.

———. 2003. *De anima*, ed. Jos Decorte, and Jozef Brams. Turnhout: Brepols.

Avicenna. 1507. *Canon medicinae*, ed. Paganinum de Paganinis. Venetiis.

———. 1968. *Liber de anima seu sextus de Naturalibus IV–V*, ed. Simone Van Riet. Louvain–Leiden: Éditions Orientalistes.

Gualterus Burlaeus. *Commentarium in II De anima Aristotelis*. Vatican: Bibliotheca Vaticana, MS Vat. lat. 2151, ff. 1ra–88rb.

Gualterus Burlaeus. 1966. *De sensibus*, ed. Herman Shapiro, and Frederick Scott. In Walter Burley's "De Sensibus". *Mitteilungen des Grabmann-Instituts der Universität München* 13: 3–10.

———. 1971. *De potentiis animae*, ed. Mary Jean Kitchel. In The "De potentiis animae" of Walter Burley. *Mediaeval Studies* 33: 85–113.

Gualterus Burlaeus. *De sensibus*. Cambridge: Cambridge University Library, MS Hh IV 13, ff. 58v–60v.

Gualterus Burlaeus. *De sensibus*. London: Lambeth Palace Library, MS 70, ff. 116vb–117va.

Gualterus Burlaeus. *De sensibus*. London: Lambeth Palace Library, MS 143, ff. 139vb–140vb.

Gualterus Burlaeus. *De sensibus*. London: British Library, MS Royal 12 B 19, ff. 301v–304r.

Gualterus Burlaeus. *De sensibus*. Oxford: Oriel College, MS 12, ff. 97va–99ra.

Gualterus Burlaeus. *De sensibus*. Vatican: Bibliotheca Vaticana, MS Vat. lat. 2146, ff. 250vb–251vb.

Gualterus Burlaeus. *De sensibus*. Vatican: Bibliotheca Vaticana, MS Vat. lat. 2151, ff. 117rb–119ra.

Gualterus Burlaeus. *De sensibus*. Worcester: Worcester Cathedral Library, MS F. 86, ff. 8v–10r.

Gualterus Burlaeus. (Forthcoming-a.) *Commentarium in De sensu et sensato*, ed. Monika Mansfeld. In *Walter Burley's Commentaries on "Parva Naturalia"*.

———. (Forthcoming-b.) *Commentarium in De somno et vigilia*, ed. Marek Gensler. In *Walter Burley's Commentaries on "Parva Naturalia"*.

Les auctoritates Aristotelis, ed. Jacqueline Hamesse. 1974. Louvain–Paris: Publications universitaires.

Thomas de Aquino. 1889. *Summa theologiae*, ed. Comissio Leonina. Roma.

———. 1968. *Quaestiones de anima*, ed. James H. Robb. Toronto: Pontifical Institute of Mediaeval Studies.

Secondary Literature

Black, Deborah L. 2000. Imagination and Estimation: Arabic Paradigms and Western Transformations. *Topoi* 19: 59–75.

Feola, Giuseppe. 2018. The Unity of Sense-Power in the De anima and Parva naturalia. In *The Parva naturalia in Greek, Arabic and Latin Aristotelianism Supplementing the Science of the Soul*, ed. Börje Bydén and Filip Radovic, 51–63. Dordrecht: Springer.

Gregorić, Pavel. 2007. *Aristotle on the Common Sense*. Oxford: Oxford University Press.

Lagerlund, Henrik. 2008. Pietro d'Abbano and the Anatomy of Perception. In *Theories of Perception in Medieval and Early Modern Philosophy*, ed. Simo Knuuttila and Pekka A. Kärkkäinen, 117–130. Dordrecht: Springer.

Kitchel, Mary Jean. 1977. Walter Burley's Doctrine of the Soul: Another View. *Mediaeval Studies* 39: 387–401.

Knuuttila, Simo. 2008. Aristotle's Theory of Perception and Medieval Aristotelianism. In *Theories of Perception in Medieval and Early Modern Philosophy*, ed. Simo Knuuttila and Pekka A. Kärkkäinen, 1–22. Dordrecht: Springer.

Kuksewicz, Zdzisław. 1981. The Problem of Walter Burley's Averroism. In *Studi sul XIV secolo in memoria di Anneliese Maier*, ed. Alfonso Maierù and Agostino Paravicini Bagliani, 341–377. Firenze: Edizioni di Storia e Letteratura.

Lisska, Anthony J. 2016. *Aquinas's Theory of Perception: An Analytic Reconstruction*. Oxford: Oxford University Press.

Maier, Anneliese. 1955. Ein unbeachteter "Averroist" des 14. Jahrhunderts: Walter Burley. In *Medioevo et Rinascimento. Studi in onore di Bruno Nardi*, ed. Bruno Nardi, vol. 2, 477–499. Firenze: Giulio Cesare Sansoni.

Martin, Connor. 1964. Walter Burley. In *Oxford Studies Presented to Daniel Callus*, ed. Oxford Historical Society, 194–230. Oxford: Clarendon Press.

Phillips, Richard Percival. 1934. *The Modern Thomistic Philosophy: An Explanation for Students*, vol. 1: *The Philosophy of Nature*. London: Newman Bookshop.

Vittorini, Marta. 2013. Life and Works. In *A Companion to Walter Burley*, ed. Alessandro D. Conti, 17–38. Leiden–Boston: Brill.

Weisheipl, James Athanasius. 1968. Ockham and Some Mertonians. *Mediaeval Studies* 30: 163–213.

Chapter 13
"Is Touch One Sense or Several?" A Late Medieval Scientific Question

Chiara Beneduce

Abstract "Is touch one sense or is it several senses?" Aristotle (d. 322 BC) addressed this question in his *De anima*. Late medieval authors, both in commentaries on Aristotle and in works of theoretical medicine, discussed the problem. In my paper, I focus on the doctrines concerning touch—and in particular on the question whether touch is one sense or several senses—advanced by the French philosopher John Buridan (d. ca. 1361) and the Italian physician Tommaso del Garbo (d. ca. 1370). I thereby demonstrate the relevance of this question to understanding the reception of Aristotelian psychology and the relationship between natural philosophy and medicine in the late Middle Ages.

Keywords Sense of touch · John Buridan · Tommaso del Garbo · Natural philosophy · Medicine

13.1 Introduction: The Issues Presented by Touch

The complexity of the sense of touch is currently at the center of debates in contemporary scientific and philosophical analysis, which involves posing questions such as "What constitutes the sense of touch?", "Is touch multisensory?".[1] Yet the issue has been interrogated by philosophical inquiry since its beginnings. Aristotle

[1] See Fulkerson (2016).

C. Beneduce (✉)
Radboud University, Nijmegen, The Netherlands
e-mail: c.beneduce@ftr.ru.nl

M. Gensler et al. (eds.), *The Embodied Soul*, Historical-Analytical Studies on Nature, Mind and Action 11, https://doi.org/10.1007/978-3-030-99453-2_13

(d. 322 BC) was the first to recognize the importance of touch among the senses.[2] Nonetheless, he also stressed its uniquely problematic status, questioning both what constituted the proper organ of touch (its *sensorium*) and whether touch should be considered one sense or several.[3] Aristotle never gave a clear answer to these questions. In the case of the organ of touch, his position is not consistent across texts. In *De anima* (II.11, 423b18–27), Aristotle locates the organ of touch close to the heart and states that flesh is the medium of touch.[4] In *De sensu et sensato* (439a2–3), he confirms that the organ of touch (and taste) is located close to the heart.[5] In *De partibus animalium*, we find two different views: in 647a19 and 653b25, Aristotle says that flesh is the organ of touch but, in 656b34, he returns to the more predominant position that the organ of touch is located "internally".[6] His response to the question of the unicity or plurality of touch, bearing on the more general issue of the number of the senses, is even hazier. In the second book of *De anima* (II.11, 422b20), Aristotle identifies the problem; he does not, however, offer a solution of any kind.[7] His position is sketched some lines later, at the beginning of the third book (*De anima*, III.1, 424b22–425a13), where he states that there are only the five senses (sight, hearing, taste, smell, and touch) and adds somewhat tautologically, referring specifically to touch, that everything tangible can be perceived by touch.[8]

The issue of the organ and medium of touch became particularly compelling in the Latin Middle Ages, since the Aristotelian position was not only unclear in itself but also divergent from the medical tradition: Avicenna (d. 1037) clearly identified

[2] With Aristotle touch acquired a clearly recognized function, distinct from the other senses. Before Aristotle, in pre-Socratic philosophy and in Plato, touch was mostly associated with taste or seen as an instrument through which the soul or the other senses acquired sensorial stimuli. Aristotle instead stressed the role of touch as an independent and important sense when, in the second book of the *De anima*, he stated both that touch is the only sense that all animals possess and that without touch it is impossible to have any other sense. See Aristoteles, *De anima*, II.2, 413b (ed. Barnes, 1984, 658). Aristotle restated this concept in the third book of the *De anima*, where he also stressed that without the sense of touch animals die. See Aristoteles, *De anima*, III.13, 435b4–5, 18–19 (ed. Barnes, 1984, 692). This renders touch a fundamental sense to the extent that all animals must possess it in order to be in contact with the external world and accomplish their proper functions. Aristotle moreover emphasized human touch as particularly efficient and eminent over the same sense in other animals: human touch is more delicate and sensitive, confirming human primacy in the hierarchy of organisms, especially as far as cognitive processes are concerned. See Aristoteles, *De anima*, II.9, 421a20–23 (ed. Barnes, 1984, 670). Touch enables human beings to reach a higher level of knowledge than other animals. It is true that Aristotle stressed the primacy of sight over the other senses in e.g., Aristoteles, *Metaphysica*, IV.6, 1011a32–34 (ed. Barnes, 1984, 1596); *De anima*, III.3, 429a3–4 (ed. Barnes, 1984, 682); *Metaphysica*, I.1, 980a22–28 (ed. Barnes, 1984, 1552). However, he nonetheless granted touch a special status in the sensory system, especially as far as natural philosophy is concerned. On this, see Pogliano (2015, 17) and Goldner (2018). On Aristotle and the sense of touch, see especially Freeland (1992).

[3] Aristoteles, *De anima*, II.11, 422b17–23 (ed. Barnes, 1984, 672).

[4] Aristoteles, *De anima*, II.11, 423b18–27 (ed. Barnes, 1984, 674).

[5] Aristoteles, *De sensu et sensato*, 439a2–3 (ed. Barnes, 1984, 697).

[6] Aristoteles, *De partibus animalium*, 647a19, 653b25, and 656b34 (ed. Barnes, 1984, 1007, 1018, 1023).

[7] Aristoteles, *De anima*, II.11, 422b20 (ed. Barnes, 1984, 672).

[8] Aristoteles, *De anima*, III.1, 424b22–425a13 (ed. Barnes, 1984, 675).

the nerves as the organs for touch and denied the need of a medium for touch to function.[9] The issue therefore became as controversial as the problem of localizing the common sense (in the heart or the brain?), or the roles of the male and the female in reproduction. In other words, "the organ (and medium) of touch" belongs to the set of medieval topics constituting the so-called "controversy between philosophers and physicians" debating the biological themes on which the Aristotelian tradition diverged from the medical tradition.[10] The controversial character of the issue and its place at the crossroads of the natural philosophical and the medical stances has attracted significant recent scholarly attention.[11] On the contrary, the late medieval concern with the unicity or plurality of touch has been largely neglected. In this paper, I take a closer look at the issue of the unicity or plurality of touch in the late Middle Ages.

The first thing to note is that the issue of the unicity or plurality of touch widely attracted the attention of medieval scientists and frequently appeared in late medieval texts conceived within the university *milieu*, including both commentaries on Aristotle and works of theoretical medicine.[12] The problem was addressed by Averroes (d. 1198),[13] Thomas Aquinas (d. 1274),[14] and John Duns Scotus (d. 1308),[15] the question also appears in the Anonymous Bernardini's *Quaestiones super librum de anima* (ca. 1250–1260),[16] in the Anonymous Gauthier's *Lectura in librum de*

[9] See Avicenna, *Liber de anima seu sextus de naturalibus*, pars II, cap. 3 (ed. Van Riet, 1972, 138).

[10] On the so-called "controversy between philosophers and physicians," see Miteva (2018); Chandelier (2020); Martorelli Vico (2000); de Asúa (1997, 1999, 2013), Jacquart and Micheau (1990, 167–203); Jacquart (1993, 284–287); Siraisi (1980, 1981, 186–202; 1990, 80–82), Ottosson (1984, 219–239).

[11] See especially the studies by Hasse (2000, 2001, 2010), Lagerlund (2008), and a recent article by Beneduce (2019, 211–212), which also contains a list of late medieval questions on the organ and medium of touch. See also Knuuttila and Kärkkäinen (2013, 76–78), who incidentally state: "The question of whether touch is a single sense was popular throughout the middle ages." Knuuttila and Kärkkäinen (2013, 77).

[12] As some of the following examples show, the topic appeared also in theological works.

[13] Averroes, *Commentarium magnum in Aristotelis de anima libros*, II, com. 106 (ed. Crawford, 1953, 293–294): "De tangibili autem et de tactu idem est sermo. Tactus enim si non fuerit unum genus, sed plus, necesse est ut tangibilia etiam sint plura uno. Sed est dubium utrum sint unum aut plura." Averroes, *Commentarium magnum in Aristotelis de anima libros*, II, com. 128 (ed. Crawford, 1953, 323–325): "Quoniam autem non est alius sensus praeter istos quinque, scilicet visus et auditus et olfactus et gustus et tactus […]."

[14] Thomas de Aquino, *Summa theologiae*, I, q. 78, art. III ad tertium (Romae, 1889, 253–255): "Utrum convenienter distinguantur quinque sensus exteriores." Thomas de Aquino, *In Aristotelis librum de anima commentarium*, II, lect. 22 (ed. Pirotta, 1959, 129–132): "Inquirit num sensus tactus sit unus, vel plures: et solvit quod plures sunt […]."

[15] Johannes Duns Scotus, *Quaestiones super secundum et tertium de anima*, q. 1 (ed. Noone et al., 2006, 1–12): "Utrum sensus tactus sit unus vel plures."

[16] Anonymus Bernardini, *Quaestiones super librum de anima*, q. 84 (ed. Bernardini, 2009, 258–264): "Quaeritur de tactu, et primo utrum sit sensus unus. Et primo utrum differentiae tangibiles reducantur ad unam naturam per quam sentiantur." Anonymus Bernardini, *Quaestiones super librum de anima*, q. 85b (ed. Bernardini, 2009, 258–264): "Quaeritur tunc utrum sensus tactus sit sensus unus." Anonymus Bernardini, *Quaestiones super librum de anima*, q. 85a (ed. Bernardini, 2009, 258–264): "Quaeritur tunc utrum unus sensus sit unius contrarietatis."

anima (ca. 1245–1250),[17] in the Anonymous Bazán's *Quaestiones de anima* (1246–1247),[18] in Peter John Olivi's (d. 1298) *Quaestiones in secundum librum Sententiarum*,[19] in Peter of Spain (thirteenth century)'s *Quaestiones super libro de animalibus Aristotelis*,[20] in Giles of Rome's (d. 1316) *Super libros de anima*,[21] in Peter of Abano's (d. 1316) *Conciliator*,[22] in Turisanus's (d. ca. 1320) *Plusquam commentum*,[23] in Adam Burley's (1327/8) *Quaestiones de anima*,[24] in John of Jandun's (d. 1328) *Super libros Aristotelis de anima*,[25] in Thaddeus of Parma's (first half of the fourteenth century) *Quaestiones de anima*,[26] in John Buridan's (d. ca. 1361) *Quaestiones de anima*,[27] in Tommaso del Garbo's (d. ca. 1370) *Summa medicinalis*,[28] in Marsilius of Inghen's (d. 1396) *Quaestiones de anima*,[29] in Blasius

---

[17] Anonymus Gauthier, *Lectura in librum de anima*, II, lect. 20, q. 1 (ed. Gauthier, 1985, 391–393): "Dubitatur primo utrum tactus sit unus sensus."

[18] Anonymus Bazán, *Quaestiones de anima*, lect. 21 (ed. Bazán, 1971, 256–270): "De eo autem quod potest tangi et tactu eadem est ratio: si enim tactus non est unus sensus, sed plures, necessarium est et quae possunt tangi sensibilia plura esse. Habet autem dubitationem utrum plures sit aut unus."

[19] Petrus Johannis Olivi, *Quaestiones in secundum librum Sententiarum*, q. 61 (ed. Jansen, 1922–1926, vol. 2, 574–587): "Quantum ad secundum, quod scilicet tactus in plures tactivas potentias dividatur." In a 2008 article, Mikko Yrjönsuuri analysed this question by Olivi, pointing out Olivi's attention to the problem of bodily self-awareness in connection with his theory of the sense of touch. See Yrjönsuuri (2008). On the link between the issue of the unicity or plurality of touch and the problem of proprioception, see *infra*, footnote 92.

[20] Petrus Hispanus, *Quaestiones super libro de animalibus Aristotelis*, XIV, q. 4 (ed. Navarro Sánchez, 2015, 94): "Utrum tactus sit unus sensus vel sint plures sensus."

[21] Aegidius Romanus, *Super libros de anima*, II (Venetiis, 1500, repr. 1982, f. 49rb–49va): "Dubitaret forte aliquis utrum tactus sit sensus unus."

[22] Petrus de Abano, *Conciliator*, diff. 42 (Venetiis, 1565, f. 64va–b): "Utrum caro sit organum tactus, necne" which contains the following sub-question: "Adhuc id dependet ex praemoto secundo *De anima*, ac tactus sit unus sensus, aut plures."

[23] Turisanus, *Plusquam commentum*, 2.31 (Siraisi, 1981, 335): "Utrum sensus tactus sit plures sensus."

[24] Adamus Burlaeus, *Quaestiones de anima*, q. 13 (ed. Synan, 1997, 72–75): "Utrum sensus tactus sit unus formaliter."

[25] Johannes de Janduno, *Super libros Aristotelis de anima subtilissimae quaestiones*, II, q. 32 (Venetiis, 1552, f. 45vb): "[A]n sensus sint tantum quinque." Johannes de Janduno, *Super libros Aristotelis de anima subtilissimae quaestiones*, II, q. 27 (Venetiis, 1552, f. 49rb): "[A]n tactus sit unus sensus."

[26] Thaddaeus de Parma, *Quaestiones de anima*, II, q. 24 (Vanni Rovighi, 1951, XXI–XXXIX): "Utrum tactus sit unus sensus."

[27] Johannes Buridanus, *Quaestiones de anima (III red.)*, II, q. 19 (ed. Sobol, forthcoming, 1): "Utrum tactus sit unus sensus vel plures." Johannes Buridanus, *Quaestiones de anima (III red.)*, II, q. 20 (ed. Sobol, forthcoming, 1): "Utrum sint tantum quinque sensus exteriores." For more detailed information on this edition, see *infra*, footnote 55.

[28] Thomas de Garbo, *Summa medicinalis*, q. 82 (Venetiis, 1531, f. 79ra): "[I]n isto quaesito singulariter tria declarabimus: [...] Secundo de eius unitate et pluralitate."

[29] Marsilius de Inghen, *Quaestiones de anima*, II, q. 20 (Marshall, 1983, 85): "Utrum sint tantum quinque sensus exteriores."

of Parma's (d. 1416) *Quaestiones de anima*,[30] in Jacques Despars's (d. 1458) *Commentarium in Avicennae Canonem*,[31] in John Erycles's (fifteenth century) *Quaestiones de anima*,[32] and in Offredus Apollinaris Cremonensis's (fifteenth century) *Quaestiones de anima*.[33]

It should not be surprising that the question on the unicity or plurality of touch occurs so frequently in late medieval texts. Medieval philosophers and physicians could not overlook the evidently thorny points of Aristotle's theory of touch. The first of these is the problem of touch as a contact sense. When we perceive through touch, we are in contact with the sensible object (*De anima*, III.13, 435a17–18) or, in other words, the contact with the object qualifies our sensation as tactile (*De anima*, III.1, 424b27–28).[34] This way, the sense of taste could be regarded as merely a type of touch: we taste by touching the food with our tongue, *directly and by contact*. Yet, we distinguish taste from touch and give them different names. If other perceptions occur through direct contact with sensible objects, we cannot exclude that these perceptions are tactile too. One could rightly ask whether there is, therefore, only one sense of touch encompassing all contact-based perceptions, or whether there are many senses of touch, all functioning by contact but not reducible to *one* sense of touch. In other words, it is legitimate to wonder whether all contact-based perceptions are to be labelled as "touch", or whether we can ascribe another name to some of them, i.e., refer them to an additional sense.[35] Second, there is the problem of the proper object of touch. According to Aristotle, each and every sense must refer to a proper object of sensation: sight to colour, hearing to sound, smell to

[30] Blasius de Parma, *Quaestiones de anima*, II, q. 24 (Federici Vescovini, 1974, 16): "Utrum quinque sint sensus exteriores et non plures nec pauciores." Blasius de Parma, *Quaestiones de anima*, II, q. 21 (Federici Vescovini, 1974, 16): "Utrum sensus tactus sit unus et non plures."

[31] Iacobus de Partibus, *Commentarium in Avicennae Canonem*, lib. I, fen. 1, doct. 6, cap. 5 (Jacquart, 1998, 535): "Utrum sint tantum quinque sensus exteriores seu virtutes comprehensivae manifeste."

[32] Johannes Erycles, *Quaestiones de anima*, II, q. 20 (Marshall, 1983, 135): "Utrum sint tantum quinque sensus exteriores." Johannes Erycles, *Quaestiones de anima*, II, q. 19 (Marshall, 1983, 135): "Utrum tactus sit unus sensus an plures."

[33] Offredus Apollinaris Cremonensis, *Quaestiones de anima*, II, q. 29 (manuscript Paris, Bibliothèque nationale de France 6447, f. 67va): "Utrum sensus gustus a tactu realiter distinguatur." Offredus Apollinaris Cremonensis, *Quaestiones de anima*, II, q. 30 (manuscript Paris, Bibliothèque nationale de France 6447, f. 69rb): "Utrum sensus tactus sit sensus unus."; Offredus Apollinaris Cremonensis, *Quaestiones de anima*, II, q. 36 (manuscript Paris, Bibliothèque nationale de France 6447, f. 83va): "Utrum tantum quinque sint sensus exteriores, scilicet visus, auditus, odoratus, gustus et tactus." This list of occurrences of the issue of the unicity or plurality of touch in late medieval texts does not aim to be complete. The observation that the question of the unicity or plurality of touch was widely addressed in the late Middle Ages comes also from the Conimbricenses' *Commentarium in tres libros de anima Aristotelis*, II, q. 2 (Coloniae, 1609, repr. 2006, f. 326–330): "[S]it ne unus tactus, an plures." The text lists several authors treating the problem.

[34] Aristoteles, *De anima*, III.13, 435a17–18 and III.1, 424b27–28 (ed. Barnes, 1984, 692, 675).

[35] Johansen (1997, 178–188); Goldner (2018).

odour, taste to flavour. We cannot, however, easily identify a proper and single object of touch. This could imply that touch is a plural, or multiple, sense.[36] Third, there is the problem of the pairing of contraries. The senses, Aristotle claims, each refer to one pair of contrary qualities: e.g., sight to black and white, hearing to high and low, and taste to sweet and bitter.[37] Again, touch is an anomaly. Since touch refers to multiple pairs of contraries (hot/cold, wet/dry, hard/soft), it could be made up of a plurality of senses.[38]

In what follows I will take up the work of two fourteenth-century authors as case studies on the debate over the unicity or plurality of touch in the late Middle Ages: John Buridan (d. ca. 1361), Master of Arts at the University of Paris, and his commentary on Aristotle's *De anima*, and Tommaso del Garbo (d. ca. 1370), the Italian physician who authored the *Summa medicinalis*,[39] a work of theoretical medicine. Undoubtedly, other authors could have been taken as case studies for this paper as well and a complete research on the topic of the unicity or plurality of touch should analyse as many texts as possible. Yet, since this research is just at its inception, I found it worth starting with John Buridan and Tommaso del Garbo. Both present an extended treatment of the problem and offer a unique occasion to analyse two almost contemporary theories of the unicity or plurality of touch, elaborated within two different textual traditions (a commentary on Aristotle and a medical work) and shaped in two different environments (France and Italy).

Based on my analysis, the conclusions of this paper will run as follows. Studying the issue of the unicity or plurality of touch, I argue, can facilitate a better understanding of the late medieval reception of Aristotle's psychology and the relationship between natural philosophy and medicine in the late Middle Ages.

---

[36] Aristoteles, *De anima*, III.1, 425a20 (ed. Barnes, 1984, 676); II.6, 418a10–15 (ed. Barnes, 1984, 665); II.7, 418a26 (ed. Barnes, 1984, 665); II.8, 419b4–5 (ed. Barnes, 1984, 667); II.11, 422b30 (ed. Barnes, 1984, 672).

[37] Aristoteles, *De anima*, II.11, 422b23–30 (ed. Barnes, 1984, 672).

[38] Aristoteles, *De anima*, II.11, 422b23–30 (ed. Barnes, 1984, 672). For other medieval reflections on touch and taste within this volume, see Fedriga, Chap. 2 in this volume, pp. 17–35.

[39] For a profile of Tommaso del Garbo, see Park (1985, 202–211); Cappellini (1950); Guido (1970). For recent research on Tommaso del Garbo, see Chandelier (2017). Boughan (2006, 132–219) presents some aspects related to the doctrine of the soul in Tommaso's *Summa medicinalis*. For a general presentation of John Buridan's work as a Master of Arts in Paris, see Zupko (2003).

13.2 Two Late Medieval Reflections on the Unicity or Plurality of the Sense of Touch

13.2.1 Tommaso del Garbo on the Unicity or Plurality of the Sense of Touch

Tommaso del Garbo discussed the sense of touch in his work of theoretical medicine *Summa medicinalis*. He devoted question 82 entitled "Utrum tactus ad sui sensationem requirat medium quo mediante fiat sensatio tactus" (ff. 79ra–82rb) to problems related to touch.[40] The question is divided into three sub-questions dealing respectively with the necessity of the sense of touch for animals, touch's unicity or plurality, and its organ.[41] Thus, one of the problems Tommaso finds especially intriguing is the issue of the unicity or plurality of touch (ff. 80ra–80va). He immediately notes the aporetical character of the problem raised by Aristotle in the second book of the *De anima*.[42] Tommaso goes on to make the following distinction: with regard to the organ of sensation touch is only one sense,[43] yet it is plural with respect to the qualities disposing the organ to be affected by the tangibles.[44] Tommaso specifies that touch is only one sense in terms of the organ because there is no evident distinction in the sense-organ of touch throughout the body, based on the different tangible objects.[45] He ascribes the position of the singularity of touch to some further unspecified "doctors saying that materially and according to (its ?) subject, the sense of touch is one."[46] Tommaso then dedicates much of the subsequent part of his text to outlining the plurality of touch as far as the qualities are concerned.[47] For instance,

---

[40] Thomas de Garbo, *Summa medicinalis,* q. 82 (Venetiis, 1531).

[41] Thomas de Garbo, *Summa medicinalis,* q. 82 (Venetiis, 1531, f. 79ra): "[I]n isto quaesito singulariter tria declarabimus: primus est quis sensus est tactus et quam in corpore habet necessitatem. Secundo de eius unitate et pluralitate. Tertio de principali quod quaeritur: in quo apparebit de organo eius."

[42] Thomas de Garbo, *Summa medicinalis*, q. 82 (Venetiis, 1531, f. 80ra): "[...] Philosophus secundo *De anima* videtur illud sub dubitatione inquirere ad utramque partem arguendo."

[43] Thomas de Garbo, *Summa medicinalis*, q. 82 (Venetiis, 1531, f. 80ra): "Et breviter dicendum quod sensus tactus quantum ad organum est unus sensus."

[44] Thomas de Garbo, *Summa medicinalis*, q. 82 (Venetiis, 1531, f. 80ra): "Est autem tactus plures quantum ad qualitates quibus perficitur organum quae sunt causa ut organum possit moveri a sensibilibus tangibilibus omnibus [...]."

[45] Thomas de Garbo, *Summa medicinalis,* q. 82 (Venetiis, 1531, f. 80ra): "[...] quia non apparet manifesta distinctio organi sensus tactus in universo corpore respectu omnium tangibilium [...]."

[46] Thomas de Garbo, *Summa medicinalis,* q. 82 (Venetiis, 1531, f. 80ra): "[…] doctores qui dicunt quod materialiter et secundum subiectum sensus tactus est unus." On this position, see *infra*, footnote 58.

[47] It takes part of f. 80ra and all f. 80rb.

in order to sense hot and cold we need a different quality in the flesh and in the nerves than when we sense moist and dry.[48] Tactile qualities are of many different contrary types and different genera; this is not the case with the other sensibles. In the case of sight, for example, light and colour are not of different genera. Light enters the essence of the colour and makes it possible for us to see the colour. On the contrary, tactile qualities do not contribute to the sensation of one another.[49]

In the second part of his discourse, Tommaso introduces a different viewpoint. He says that both Aristotle and Averroes seem to have supported the position that touch is only one sense in terms of its internal medium (*medium intrinsecum*), i.e., the flesh; while it is plural as far as the instrument of touch (*instrumentum tactus*) is concerned. By the latter concept Tommaso means the tactile nerves, which are the organ(s) of touch. He explains that there are as many tactile instruments as distinct contrary qualities (*contrarietates*) that can be perceived by touch.[50] In other words, different tactile contrary qualities correspond to different nerves (different "instruments of touch") that can perceive them. Yet, Tommaso rejects this different viewpoint: it is not necessary for each contrary quality to have a proper instrument (or organ). The multiplicity of touch lies not in the organ but in the qualities. To confirm touch's plurality, it is sufficient to assume a difference in qualities within the same instrument, i.e.,

---

[48] Thomas de Garbo, *Summa medicinalis,* q. 82 (Venetiis, 1531, f. 80ra): "Est ergo intentio Philosophi, quando vult quod tactus sit plures, hoc modo: quia, ad sensationem unius contrarietatis in tactu, verbi gratia, ad sensationem caliditatis et frigiditatis, sit aliqua qualitas requisita in carne vel nervo et principians vel qualitercumque aliter requisita et qua remota non posset fieri in membro species caliditatis et frigiditatis et per consequens nec sensatio posset fieri in membro. Et similiter quia in eadem parte potest fieri sensatio humiditatis et siccitatis quae est alia contrarietas sequitur alia qualitas distincta ab illa quae requiritur ad caliditatem et frigiditatem et sic de aliis contrarietatibus."

[49] Thomas de Garbo, *Summa medicinalis,* q. 82 (Venetiis, 1531, f. 80ra): "[...] et quia tales qualitates tangibiles sunt plurium contrarietatum et diversorum generum quod non est in aliis sensibilibus. Ideo illo modo tactus est plures sensus quod in aliis non contingit." Thomas de Garbo, *Summa medicinalis,* q. 82 (Venetiis, 1531, f. 80rb): "Nec valet si diceremus in visu quod visus habet pro obiecto colorem et lucem [...] color et lux non differunt sicut diversa genere [*satis* genera ? *CB*] quia lux est de esse coloris et eius essentiam ingreditur [...]. Similiter etiam non valet quia lux est causa in videndo colores et est sicut universalis causa necessario concurrens ad omnem visionem. [...] In tactu autem non est sic, quia sensibilia tactus existentia diversorum generum non sunt talia quod unum sit causa concurrens necessario ad sensationem alterius et praecipue hoc habere veritatem in primis duabus contrarietatibus primarum qualitatum nec etiam unum istorum est alterum constituens et illius essentiam ingrediens sicut in luce et colore."

[50] Thomas de Garbo, *Summa medicinalis,* q. 82 (Venetiis, 1531, f. 80rb): "Et circa hoc attentione dignum quod Philosophus et Averroes, si quis bene considerat dicta eorum, videntur velle quod tactus ex parte qua medium intrinsecum requisitum ad ipsum, quod est caro, est unum sit sensus unus. Sed ex parte instrumenti tactus quod est intus, quod est multiplex, est sensus plures. Et videntur in superficie verborum velle quod tot sunt praedicta instrumenta quot sunt distinctae contrarietates perceptibiles per tactum et ideo contrarietas inter calidum et frigidum perciperetur per unum nervum et illa quae est inter humidum et siccum perciperetur per alium et sic de aliis."

the same organ, which is necessary to perceive sensible objects of different genera.[51]

Here, Tommaso's position is ultimately neat: touch is a plural sense as far as this difference in qualities is concerned, while it can be considered only one sense as far as the organ is concerned. His reasoning is less clear than his position. He first bases his explanation on the distinction between the organ and the qualities (which I call the "first distinction"); he then introduces yet another distinction between the medium and instruments—or organs—of touch (which I call the "second distinction"). If in the "first distinction" the plurality of the sense of touch lies in the qualities and not in the organ, in the "second distinction" plurality is affirmed for the organ (the different nerves). However, Tommaso finally rejects that "second distinction". He takes touch to be one sense as far as both the organ and the medium are concerned,[52] and as a plural sense as far as the qualities are concerned. It should also be noted that, if in the "second distinction" Tommaso patently identifies the instrument(s) or organ(s) of touch as the nerves,[53] in the "first distinction" he does not clarify to what organ he is referring (the heart?, the nerves?). We learn, however, from the subsequent part of his discussion on touch that he clearly takes the nerves to be the organ of touch.[54]

---

[51] Thomas de Garbo, *Summa medicinalis,* q. 82 (Venetiis, 1531, f. 80rb–va): "Sed pro firmo talis distinctio nervorum nullatenus est necessaria quia cum idem nervus sit commixtus ex calido et frigido, sicco vel humido et sic de aliis qualitatibus per eundem nervum necessario habent omnes illae contrarietates percipi [...]. Sufficit ergo quod illa talis diversitas instrumenti quae facit tactum esse plures sensus sit in qualitatibus diversis in instrumento existentibus necessario requisitis ad sensationem diversorum sensibilium diversorum generum per modum dictum supra et hoc erit tam in carne quam in nervo. Et hoc expresse asserit Avicenna capitulo de tactu cum dicit quod si tactus et gustus essent simul dispersi per totum, sicut sunt in lingua, tunc dicerentur unus sensus." Tommaso mentions Avicenna's chapter on touch (in the *Liber de anima seu sextus de naturalibus*) to support this last proof. The passage from Avicenna is Avicenna, *Liber de anima seu sextus de naturalibus*, pars II, cap. 3 (ed. Van Riet, 1972, 141): "Istae enim sunt primae actiones sensus, et oportet ut unumquodque genus earum habeat virtutem propriam; sed quia istae virtutes diffusae sunt per omnia membra aequaliter, ideo putaverunt eas esse unam virtutem, veluti si tactus et gustus essent diffusi per totum corpus sicut sunt diffusi in lingua, putaretur quod eorum principium una virtus esset, sed quia diversi sunt extra linguam, cognita est eorum diversitas."

[52] See *supra*, footnote 51: "[…] et hoc erit tam in carne quam in nervo [...]."

[53] See *supra*, footnotes 50 and 51.

[54] Thomas de Garbo, *Summa medicinalis,* q. 82 (Venetiis, 1531, f. 82ra): "Credimus tamen veriorem esse opinionem quae ponit nervi esse proprium et principalius organum tactus." Thomas de Garbo, *Summa medicinalis,* q. 82 (Venetiis, 1531, f. 82rb): "[…] qui pura eorum [*sc.* Averrois et Aristotelis *CB*] intentio visa est esse quod nervus sit organum et caro medium." Tommaso curiously ascribes the position that the nerves are the organ of touch to Aristotle. This strange attribution of the position to Aristotle is also found in the Anonymus Bazán's *Quaestiones de anima*, II, q. 31 (ed. Bazàn, 1971, 450–451).

13.2.2 John Buridan on the Unicity or Plurality of the Sense of Touch

Buridan addresses the problem of the unicity or plurality of touch in two questions of his commentary on the second book of *De anima* (third redaction): question 19 ("Utrum tactus sit unus sensus vel plures") is thematically devoted to the issue, while question 20 ("Utrum sint tantum quinque sensus exteriores") directly refers only to the general problem of the number of senses, despite in fact being an extensive further reflection on the unicity or plurality of the sense of touch.[55]

To answer the question of whether touch is one sense or many (question 19) Buridan introduces what he calls the "common opinion" (*respondetur communiter*): touch is one sense "materially" (*materialiter*), but it is plural "formally" (*formaliter*). More specifically, he explains, in the sense of touch, one and the same organ perceives tangible objects, while the qualitative dispositions (*dispositiones qualitativae*), namely the qualities disposing the organ to perceive those different tangible objects, are different, i.e., plural.[56] This "common opinion" reported by Buridan reminds of Tommaso's position above. Tommaso himself referred the first part of this idea, i.e., that touch is one sense "materially and according to (its ?) subject," to a previous tradition (to some "doctors").[57] It should be acknowledged here that this "common opinion" can actually be found

[55] At the moment, the only available edition of the second book of Buridan's *De anima* is Johannes Buridanus, *Quaestiones de anima (III red.)*, II (ed. Sobol, 1984). However, the edition I am referring to in this article is the new, forthcoming, edition of John Buridan's *De anima*. See Zupko (2018) announcing the edition. I am grateful to Peter Sobol who shared with me questions 19 and 20 of the second book as prepared for the new, forthcoming, edition. Indication of pages are referred to the files containing the two questions as I received them by Sobol on March 2, 2021. Buridan devotes other parts of his commentary on the *De anima* to the sense of touch, as well as some parts of his commentary on the *Parva naturalia*. In *De anima*, he devotes to the topic especially q. 17 ("Quaeritur circa capitulum de olfactu utrum odor multiplicatur realiter per medium vel spiritualiter seu intentionaliter"), a question on smell with some references to touch, q. 21 ("Utrum sensibile positum supra sensum facit sensationem, id est, sentitur"), and q. 25 ("Utrum in organis exterioribus sensuum subiective fiat actualis sensatio vel solum receptio specierum sensibilium et non sensatio nisi in corde"). In *Parva naturalia*, Buridan talks about touch especially in q. 2 ("Utrum omni animali necessarii sint tactus et gustus") and q. 8 ("Utrum organum tactus et gustus sint terrae a dominio") of the *De sensu et sensato*. The edition of Buridan's *Parva naturalia* is Johannes Buridanus, *Quaestiones super Parva naturalia* (ed. Stanek, 2015). On some aspects of Buridan on the sense of touch, see Beneduce (2019) and Sobol (2017). For a list of questions on the sense of touch in the *Parva naturalia* tradition (*De sensu et sensato*), see the catalogue by Ebbesen, Thomsen Thörnqvist, Decaix, 2015.

[56] Johannes Buridanus, *Quaestiones de anima (III red.)*, II, q. 19 (ed. Sobol, forthcoming, 3): "Et respondetur communiter quod bene est unus sensus materialiter sed non unus formaliter, quia idem est organum secundum materiam et quantitatem, sed secundum aliam dispositionem qualitativam est perceptivus calidi et frigidi, et secundum aliam humidi et sicci."

[57] See *supra*, footnote 46.

in other late medieval texts preceding theirs.[58] Buridan ultimately agrees with the "common opinion". Yet, he wants to inquire further on the topic of the unicity or plurality of touch. In his own words, he wants to "better articulate the truth."[59] First, he reformulates the "common opinion" and explains: if by the word "sense" we mean the soul or the part of the soul that perceives, then we can say that we perceive tangible objects with only one sense; if by "sense" we mean the composite of body and soul, then again we must intend touch as only one sense. Yet, if we refer to the composite of body and soul *plus* the qualitative dispositions through which the composite of body and soul perceives sensible objects, then the sense of touch must be conceived as a plural sense.[60] Second, Buridan discloses what these dispositions are and why are they implied in the discourse on the plurality of touch. The sensible species of contrary qualities (e.g., hot and cold, wet and dry) are received in the organ thanks to different qualitative dispositions (also called by Buridan "soul's dispositions").[61] According to Aristotle, Buridan says, touch cannot be completely denuded of the contrary qualities it perceives; consequently, the organ of touch is constituted by the middle mean of those qualities and denuded of their extremes. It is because of the middle proportion of hot and cold that touch is perceptive of the qualities of hot and cold, and

[58] For example Thomas de Aquino, *In Aristotelis librum de anima commentarium*, II, lect. 22 (ed. Pirotta, 1959, 131): "Unde formaliter loquendo, et secundum rationem, sensus tactus non est unus sensus, sed plures; subiecto autem est unus." Anonymus Bazán, *Quaestiones de anima*, II, q. 30 (ed. Bazàn, 1971, 449): "Sed considerando tactum formaliter, sic sunt plures sensus." Johannes Duns Scotus, *Quaestiones super secundum et tertium de anima*, q. 1 (ed. Noone et al., 2006, 5): "Respondeo quod sensus tactus sunt formaliter duo, non tamen ita diversi vel divisi sicut alii ab invicem." Johannes de Janduno, *Super libros Aristotelis de anima subtilissimae quaestiones*, II, q. 27 (Venetiis, 1552, f. 45vb): "De quaestione communiter dicuntur duo. Primum quod tactus non est unus formaliter et essentialiter, immo plures sensus. Secundo, quod ipse est unus materialiter." Adamus Burlaeus, *Quaestiones de anima*, q. 13 (ed. Synan, 1997, 73): "Ad quaestionem dicendum est quod sensus tactus non est unus formaliter, sed solum materialiter."

[59] Johannes Buridanus, *Quaestiones de anima (III red.)*, II, q. 19 (ed. Sobol, forthcoming, 4): "Sed magis dearticulando veritatem pono tertiam conclusionem [...]."

[60] Johannes Buridanus, *Quaestiones de anima (III red.)*, II, q. 19 (ed. Sobol, forthcoming, 4): "[...] quod si hoc nomen *sensus* poneretur supponere pro composito substantiali ex anima et corpore, adhuc eodem sensu sentimus praedicta sensibilia, et hoc satis est manifestum. Sed quinta conclusio est quod si *sensum* capiamus pro aggregato ex huiusmodi composito substantiali et qualitativa dispositione secundum quam organum est receptivum speciei sensibilis et secundum quam huiusmodi compositum substantiale est cognitivum ipsius sensibilis, tunc non eodem sensu cognoscimus dicta sensibilia, sed alio sensu calidum et frigidum, alio humidum et siccum, et alio dulce et amarum."

[61] Johannes Buridanus, *Quaestiones de anima (III red.)*, II, q. 19 (ed. Sobol, forthcoming, 4): "Sed etiam secundum alias dispositiones qualitativas recipiuntur species calidi et frigidi in organo sensus, et secundum alias species humidi et sicci, vel etiam secundum aliam dispositionem qualitativam organi anima iudicat de calido et frigido, et secundum aliam de humido et sicco." Johannes Buridanus, *Quaestiones de anima (III red.)*, II, q. 19 (ed. Sobol, forthcoming, 4): "Nam per totum corpus inveniuntur qualitativae dispositiones ad cognoscendum calidum et frigidum, et humidum et siccum, sed aliae sunt dispositiones speciales ad cognoscendum dulce et amarum quae solum inveniuntur in lingua vel prope."

it is because of the middle proportion of wet and dry that touch is perceptive of the qualities of wet and dry.[62] In following Buridan's text, qualitative dispositions are thus explained in terms of the middle proportions of qualities that can be perceived by the organ of touch. In other words, qualitative dispositions are nothing other than the middle proportions of tactile qualities in the touch organ, as Buridan himself incidentally states: "[...] that disposition or middle proportion [...]."[63]

Having clarified the qualitative or soul's dispositions, Buridan re-states his view: touch is a plural sense as long as by "touch" we understand both the aggregate of the substantial composite (body and soul) *and* qualitative dispositions. For, it is on account of the different dispositions that the power of touch is perceptive of both hot and cold and wet and dry and nothing prevents the different qualitative dispositions from being in the same subject at the same time.[64] In other words, nothing prevents the dispositions from being in the same organ, which makes touch only one sense as far as the organ is concerned. In the final part of the question (*ad rationes*), Buridan also underlines why touch is unique compared to other senses. Like Tommaso, Buridan compares touch to sight: colours and light work through the same qualitative disposition or, in other words, the eye is not receptive to the species of light and colours on account of different dispositions. The fact that we have two eyes does not

---

[62] Johannes Buridanus, *Quaestiones de anima (III red.)*, II, q. 19 (ed. Sobol, forthcoming, 4–5): "Tactus enim, prout vult Aristoteles, cum non possit esse denudatus totaliter a calido et frigido, humido et sicco, debet consistere, quantum ad organum, in media proportione huiusmodi qualitatum, ita quod sit denudatus ab excellentiis earum ut possit percipere huiusmodi excellentias. Oportet enim sensum esse denudatum a sensibili quod debet apprehendere vel cuius debet speciem recipere, secundum determinationem Aristotelis et Commentatoris. Nunc igitur videmus quod sensus tactus seu eius organum non est denudatum ab excellentiis calidi et frigidi ea ratione qua est in media proportione humidi et sicci, sed ea ratione qua est in media proportione calidi et frigidi. Ideo non secundum mediam proportionem humidi et sicci tactus est cognoscitivus calidi et frigidi excedentium, nec est perceptivus specierum suarum, sed secundum mediam proportionem calidi et frigidi. Et e converso, pari ratione, non secundum mediam proportionem calidi et frigidi tactus est cognoscitivus humidi et sicci excedentium, sed secundum mediam proportionem humidi et sicci. Sed constat quod *alia est dispositio qualitativa quae est proportio media calidi et frigidi, et alia quae est proportio media humidi et sicci*, licet contingat eas esse in eodem subiecto simul. Igitur sequitur propositum, scilicet quod secundum aliam et aliam dispositionem qualitativam potentia tactiva est perceptiva calidi et frigidi ex una parte et humidi et sicci ex alia parte." Italics are my own.

[63] Johannes Buridanus, *Quaestiones de anima (III red.)*, II, q. 19 (ed. Sobol, forthcoming, 6): "Sed propter solutiones rationum, notandum est primo quod secundum eamdem dispositionem qualitativam organi nos percipimus calidum quod in caliditate excedit *illam dispositionem seu mediam proportionem*, et frigidum quod etiam in frigiditate excedit illam mediam proportionem." Italics are my own. See also the italicized text quoted in the next footnote.

[64] Johannes Buridanus, *Quaestiones de anima (III red.)*, II, q. 19 (ed. Sobol, forthcoming, 5): "Sed constat quod alia *est dispositio qualitativa quae est proportio media calidi et frigidi, et alia quae est proportio media humidi et sicci*, licet contingat eas esse in eodem subiecto simul. Igitur sequitur propositum, scilicet quod secundum aliam et aliam dispositionem qualitativam potentia tactiva est perceptiva calidi et frigidi ex una parte et humidi et sicci ex alia parte. Ideo sic est duplex sensus tactus, capiendo *tactum* pro aggregato ex composito substantiali et dispositione qualitativa, etc., sicut a principio proponebatur."

change the theory: eyes make perceptive judgements by means of the same qualitative disposition (i.e., transparency).[65]

At the beginning of question 20, Buridan harmonizes the position stated in question 19—namely that touch is a plural sense with regard to the qualitative dispositions in the touch organ—with the Aristotelian statement that there are only five senses. Buridan tries to avoid contradicting Aristotle by explaining that, in his enumeration of the five senses, Aristotle takes the sense of touch to be single in accordance with the common way of speaking (*secundum modum loquendi vulgarem*). This is because people commonly distinguish among the different senses on the basis of their specific organ of perception, and not on the basis of the different qualitative dispositions.[66] Yet, Buridan continues, "Aristotle does not mean to say that in reality there are not six senses in the above-described way, but only that there would be only five if those two kinds of touch were considered only one, as people commonly take it."[67] That is to say, if the qualities are not taken into account and touch is considered only with regard to its sense-organ—as it is in the most common way of numbering the senses—then it is easy to acknowledge that there are no more than five senses. This premise allows Buridan not to contradict the Aristotelian position that there are only five senses, without having to reject the idea that touch is plural as far as qualitative dispositions are concerned. Therefore, in question 20, Buridan can state that "we must posit the conclusion with Aristotle that we should count only five senses"[68] and prove, following Aristotle, that we do not need more than five senses. This holds true "as far as the organs are concerned,"[69] "as far as the mediums

---

[65] Johannes Buridanus, *Quaestiones de anima (III red.)*, II, q. 19 (ed. Sobol, forthcoming, 6–7): "Ad tertiam dicendum est quod colores quodammodo participant naturam lucis, licet obumbrate propter opacitatem. Ideo in eodem subiecto secundum eamdem dispositionem qualitativam illius subiecti faciunt suas species, scilicet ratione diaphaneitatis. Ideo non oportet organum visus, secundum aliam et aliam dispositionem qualitativam ipsius, esse receptivum speciei lucis et speciei coloris. Et non est ita de calido et humido et de eorum speciebus. Quod autem dicitur quod habemus duos oculos, apparet quod illi in iudicando de visibilibus sunt eiusdem rationis tali modo, quia de quibus unus oculus iudicat de eis alius iudicat, et secundum similem qualitativam dispositionem, puta diaphaneitatem, quamvis sit secundum quantitatem et situm diversitas."

[66] Johannes Buridanus, *Quaestiones de anima (III red.)*, II, q. 20 (ed. Sobol, forthcoming, 4): "Unde notandum est sicut dicit Commentator quod, licet sit alius sensus calidi et frigidi et alius humidi et sicci secundum praedicta in praecedenti quaestione, tamen in hac enumeratione sensuum quinque Aristoteles accipit illos duos sensus tangendi tamquam unum, secundum modum loquendi vulgarem. Vulgus enim non distinguit inter eos quia in nullo membro invenitur unus sine alio."

[67] Johannes Buridanus, *Quaestiones de anima (III red.)*, II, q. 20 (ed. Sobol, forthcoming, 4): "Sed tamen non est intentio Aristotelis quin praedicto modo sint sex secundum rei veritatem, sed est intentio eius quod solum essent quinque si illi duo tactus non essent nisi unus, sicut ponit vulgus." And shortly after he says again, Johannes Buridanus, *Quaestiones de anima (III red.)*, II, q. 20 (ed. Sobol, forthcoming, 4): "Ponenda est igitur prima conclusio cum Aristotele quod praedicto modo, solum sunt ponendi quinque sensus exteriores *si non esset nisi unus tactus*, [...]." Italics are my own.

[68] Johannes Buridanus, *Quaestiones de anima (III red.)*, II, q. 20 (ed. Sobol, forthcoming, 4): "Ponenda est igitur prima conclusio cum Aristotele quod praedicto modo, solum sunt ponendi quinque sensus exteriores [...]."

[69] Johannes Buridanus, *Quaestiones de anima (III red.)*, II, q. 20 (ed. Sobol, forthcoming, 5): "Quod ergo nullus deficit nobis probat Aristoteles bene ex parte mediorum, obiectorum, et organorum."

are concerned,"[70] and "as far as the sensibles are concerned"[71] (including proper sensibles[72] and common sensibles[73]).

Nevertheless, Buridan dedicates an extended part of question 20 to interrogating the idea that touch is only one sense. He addresses a particular issue, i.e., whether the sensation felt during the passage of sperm through the genitals in ejaculation should be considered to be produced by the sense of touch or by another, additional, sense.[74] This issue makes up a large part of Buridan's question and is analysed in some detail. Buridan's final conclusion is that there is no need for an additional sense besides touch to explain genital pleasure during ejaculation. Yet, it is worth considering *how* Buridan questions the Aristotelian position on the number of senses in the case of ejaculation. Buridan firstly notes that if we distinguish touch from taste, we should also distinguish touch from genital sensation.[75] In the same way, Buridan says, the sensations we feel in our stomachs can also be ascribed to a

[70] Johannes Buridanus, *Quaestiones de anima (III red.)*, II, q. 20 (ed. Sobol, forthcoming, 6): "Secundo etiam ostenditur quod non deficit nobis sensus aliquis ex parte mediorum [...]."

[71] Johannes Buridanus, *Quaestiones de anima (III red.)*, II, q. 20 (ed. Sobol, forthcoming, 6): "Tertio ostenditur quod non deficiat nobis sensus ex parte sensibilium communium nec ex parte propriorum."

[72] Johannes Buridanus, *Quaestiones de anima (III red.)*, II, q. 20 (ed. Sobol, forthcoming, 6): "Primo probatur quod non ex parte propriorum [...]." Buridan raises three objections to this conclusion, Johannes Buridanus, *Quaestiones de anima (III red.)*, II, q. 20 (ed. Sobol, forthcoming, 7): "Sed contra hanc deductionem potest esse triplex cavillatio." The first is related to the sense of touch, Johannes Buridanus, *Quaestiones de anima (III red.)*, II, q. 20 (ed. Sobol, forthcoming, 7): "Prima est quia falsum est quod tactus noster percipiat calidum et frigidum et omne medium ipsorum, quia dicit Aristoteles quod non percipit similiter calidum et similiter frigidum." His answer is Johannes Buridanus, *Quaestiones de anima (III red.)*, II, q. 20 (ed. Sobol, forthcoming, 7): "Sed respondetur ad huiusmodi cavillationes. Ad primam dicitur quod si per manum meam non sentio sibi similiter calidum, tamen illud sentire possum per pedem quem habeo frigidiorem, et illud etiam calidum et in eodem gradu senties per manum quando manus erit frigidior. Unde dolens caput habens frontem calidam et manum frigidiorem experitur quod, ponendo manum ad frontem, sentit caliditatem frontis et per frontem frigiditatem manus."

[73] Johannes Buridanus, *Quaestiones de anima (III red.)*, II, q. 20 (ed. Sobol, forthcoming, 8): "Demum etiam probatur quod non deficit sensus nobis ex parte sensibilium communium, [...]."

[74] Johannes Buridanus, *Quaestiones de anima (III red.)*, II, q. 20 (ed. Sobol, forthcoming, 2): "Item arguitur aliquantulum difficile quia membra genitalia per suas proprias naturas et complexiones sentiunt appropriatas qualitates quas alia membra non sentiunt nec alia organa. Igitur in eis est alius sensus et alterius rationis praeter illos qui in aliis membris inveniuntur. Consequentia patet quia aliter tu non posses habere rationem quod esset alius sensus tactus et gustus. Sed antecedens manifestum est quia membra genitalia sentiunt qualitates spermatum secundum quas delectantur, quas non sentirent alia membra, ut manus aut pes, quod patet quia non sic in illis delectantur." Note that the connection between touch and the topic of sexual pleasure was present in medieval (and Renaissance) thinking. See for example Coucke (2009) and Cadden (1993, 156–158), who address the topic of touch and sexual pleasure in Peter of Abano. On touch and sexual pleasure, see also *infra*, footnote 92. Note that Buridan touches upon the topic of pleasure derived from touch in his commentary to Aristotle's *Ethics*. On this, see Grellard (2017).

[75] Johannes Buridanus, *Quaestiones de anima (III red.)*, II, q. 20 (ed. Sobol, forthcoming, 9): "Sed tamen sic dicere non solvit dubitationem, quia sic gustus etiam est quidam tactus. Et ideo vel tu non distingues gustum contra tactum, et sic erunt solum quattuor sensus, vel, si distinguis gustum contra tactum, ita tu distinguis etiam istum sensum qui appropriate est in genitalibus contra tactum, et sic erunt sex sensus. Et hoc totum est contra Aristotelem."

sense different than the sense of touch (we do not feel hungry or thirsty in the hand, but specifically in the stomach).[76] However, Buridan adds, the sensation felt in the genitals during ejaculation is even more complicated due to the fact that it is linked to intense pleasure and there is no pleasure without a cognitive process.[77] We cannot solve the issue by claiming that this sensation is felt (or processed as a sensitive cognition) by the common sense or fantasy, because they are in any case activated by an external sense;[78] nor is ejaculation felt in the same way we feel the common sensibles or sensibles by accident.[79] Thus, the big problem (*magna dubitatio*) is what is felt as a proper sensible during ejaculation.[80] The proper sensibles of the sensation of ejaculation are not the qualities of hot, humid, or dry: these are all sensations that we can ascribe to the sperm *after* ejaculation.[81] The quality which is the proper sensible of the sense in the genitals has no name, Buridan says. It can just be defined as the quality whose perception is followed by intense pleasure in the genital member.[82]

---

[76] Johannes Buridanus, *Quaestiones de anima (III red.)*, II, q. 20 (ed. Sobol, forthcoming, 9): "Et apparet quod etiam simile argumentum fieret de stomacho, per quem percipimus nos esurire et sitire, quod non percipiemus per manum vel per pedem, in quibus est tamen sensus tactus."

[77] Johannes Buridanus, *Quaestiones de anima (III red.)*, II, q. 20 (ed. Sobol, forthcoming, 9): "Et dubitatio magis augetur quia in spermatizando est delectatio etiam intensa et non est delectatio vel tristitia nisi mediante cognitione, non intellectiva in proposito, quia huiusmodi delectatio convenit brutis."

[78] Johannes Buridanus, *Quaestiones de anima (III red.)*, II, q. 20 (ed. Sobol, forthcoming, 9): "Igitur est cognitio sensitiva, et non solum secundum sensum communem vel phantasiam quia sensus communis et phantasia non moventur ad actus suos nisi mediante alio sensu prius facto in actu, scilicet mediante sensu quem vocamus exteriorem." Buridan makes this remark *en-passant*. However, its importance should not be underestimated. The issue of how pleasure and pain are felt in the body was an important one in medieval philosophy. While Olivi, for example, referred the feeling of pleasure and pain to the common sense, Peter of Abano linked pleasure and pain to the sense of touch. On this, see Yrjönsuuri (2008). At least in this part of *De anima* and with reference to sexual pleasure, it seems that Buridan rejects the idea that the feeling of pleasure could be *simply* related to the common sense. His discourse in q. 20 clearly connects the feeling of sexual pleasure to touch: an additional sense cannot be identified because the pleasure linked to ejaculation is felt by the sense of touch.

[79] Johannes Buridanus, *Quaestiones de anima (III red.)*, II, q. 20 (ed. Sobol, forthcoming, 10): "Et non sentimus ipsum per modum sensibilis communis nec per modum sensibilis per accidens solum, quia sic nihil sentimus nisi concomitanter cum sensibili per se et proprie."

[80] Johannes Buridanus, *Quaestiones de anima (III red.)*, II, q. 20 (ed. Sobol, forthcoming, 10): "Tunc ergo est magna dubitatio quid per modum sensibilis proprii sentimus in emissione spermatis."

[81] Johannes Buridanus, *Quaestiones de anima (III red.)*, II, q. 20 (ed. Sobol, forthcoming, 10): "Si enim respondeatur quod sentimus et iudicamus aut calidum vel frigidum, humidum vel siccum, hoc apparet falsum; quamvis enim sperma sit calidum, tamen non iudicamus ipsum esse calidum ita nec frigidum nec humidum nec siccum, nisi forte cum iam sit extra emissum."

[82] Johannes Buridanus, *Quaestiones de anima (III red.)*, II, q. 20 (ed. Sobol, forthcoming, 11): "Et si aliquis quaereret, dicens 'assigna ergo mihi istam qualitatem quae est proprium sensibile isto sensu, ex quo ista nec est caliditas nec frigiditas nec humiditas nec siccitas,' ad hoc respondetur quod sibi non est nomen impositum, sed circumlocutive dicitur quod est ista qualitas ad cuius perceptionem sequitur talis intensa delectatio in illis membris, quam qualitatem et sensationem eius natura ingeniavit ut sequeretur ista delectatio, propter quam animal inclinaretur ad exercendum istam operationem ut inde proveniret sibi similis generatio, finaliter intenta a natura."

Buridan further asks what the sensation followed by that pleasure is, why that pleasure occurs in the genitals and not in the hand, and what—ultimately—the genital members sense.[83] He finally answers the question of what the genital members sense by stating that what is felt in the genital members during ejaculation is "[...] the excitement of spirits and an inflaming windiness, and the flow of sperm, and other coincident external frictions."[84] This statement pairs with a definition of touch Buridan provided some lines before—one quite different from the discussion of touch previously encountered in question 19 and question 20. The nature of touch, Buridan says here, is the feeling of movement (which more precisely means that in the touch sensation we do not feel the contrary qualities of hot/cold or wet/dry, but rather just movement). Accordingly, Buridan writes, we can feel a fly just by perceiving its movement; we do not need to feel the hotness of its body. The same is true for perception of the wind: we can perceive it without perceiving its coldness or hotness, but just by sensing its movement on our body.[85] In this way, in accordance with Aristotle's teachings, Buridan excludes the necessity of adding another sense to explain sensation in the genitals. For, since what is perceived in the genitals is a movement and since all tactile sensations are based on the feeling of movement, we do not need to postulate an additional sense to explain sensation in ejaculation.

Buridan also finally turns to the issue of why the intense pleasure that takes place in the genitals does not occur in the hand or the foot.[86] In order to answer this question he refers to Aristotle's *Rhetorica*:[87] a sensible motion proper (*conveniens*) to nature causes pleasure, while a sensible motion not proper (*disconveniens*) to nature

---

[83] Johannes Buridanus, *Quaestiones de anima (III red.)*, II, q. 20 (ed. Sobol, forthcoming, 12): "Tota ergo dubitatio est cum sperma ibi non sentiantur secundum quod calidum et frigidum aut siccum, secundum quid ergo sentitur ad cuius sensationem illa delectatio consequitur? Et quare tanta delectatio consequitur quae tamen non consequitur in manu sentiente? Quidquid ista membra sentiunt?"

[84] Johannes Buridanus, *Quaestiones de anima (III red.)*, II, q. 20 (ed. Sobol, forthcoming, 13): "Tunc ergo ad propositum diceretur quod in virga virili sentitur commotio spirituum et ventositatum inflammantium et fluxus spermatis et exterius aliae confricationes ad hoc concurrentes. Nec propter hoc oportet dicere quod ibi sit sensus alius nisi tactus." Describing ejaculation in terms of *ventositas* is typical of medieval theories on generation, see Jacquart and Thomasset (1985, 110) and Laqueur (1990, 45). One should not forget that Buridan himself authored a text on human generation. See Johannes Buridanus, *Quaestiones de secretis mulierum* (ed. Beneduce and Bakker, 2019).

[85] Johannes Buridanus, *Quaestiones de anima (III red.)*, II, q. 20 (ed. Sobol, forthcoming, 12): "Ad hoc respondetur quod tactus, sive in manu aut in pede aut in aliis membris, bene percipit aliqua absque hoc quod iudicet de eorum caliditate, frigiditate, humiditate, vel siccitate. [...] Et muscam scilicet sentimus super manum nostram movere, licet non iudicemus an sit calida. Et aerem vel ventum contra nos motum percipimus et sentimus, licet sit nobis similis calidus." Johannes Buridanus, *Quaestiones de anima (III red.)*, II, q. 20 (ed. Sobol, forthcoming, 12): "[...] Et haec est natura tactus, scilicet sentire tales motus ex eoquod medium est sibi coniunctum vel medio nobis connaturali."

[86] Johannes Buridanus, *Quaestiones de anima (III red.)*, II, q. 20 (ed. Sobol, forthcoming, 13): "Sed ratio quaerebat ultra: quare ex sensatione huiusmodi motuum in illis membris consequitur delectatio sic intensa et non sequitur ex sensatione talium in aliis membris, ut in manu vel in pede."

[87] Aristoteles, *Rhetorica*, I.11, 1369b33–35 (ed. Barnes, 1984, 2180).

causes pain.[88] Pleasure is therefore felt in an organ when a sensation is proper (*conveniens*) to its nature. This explains why the pleasure occurring in the genitals would not be felt in the hand or the foot. We do not need to assume the existence of a sense other than touch, we just have to recognize that the sensation of pleasure accompanying ejaculation is proper to the nature of the genitals and not to that of other corporeal parts. For those other bodily parts do not serve human reproduction, which gives the most agreeable pleasure, since it is connected to the most natural function in living beings.[89]

13.3 Conclusion

The question whether touch is one sense or several senses seems to have been largely addressed in the late Middle Ages. This paper focuses on two fourteenth-century examples of its treatment to bring about the following observations. The

---

[88] Johannes Buridanus, *Quaestiones de anima (III red.)*, II, q. 20 (ed. Sobol, forthcoming, 13): "Et pro solutione huius dubitationis oportet notare quod determinat Aristoteles primo *Rhetoricae*, scilicet quod motus sensibilis in existentem naturam, id est conveniens naturae, est causa delectationis, et apprehensionem eius consequitur naturaliter delectatio. Et sic e converso, motus sensibilis disconveniens naturae est causa tristitiae et eius apprehensionem sequitur naturaliter tristitia."

[89] Johannes Buridanus, *Quaestiones de anima (III red.)*, II, q. 20 (ed. Sobol, forthcoming, 13–14): "In quo ergo membro et quando motus est conveniens naturae, si sentitur, consurgit delectatio. In quo autem membro vel quando motus qui sentitur non est conveniens naturae, non consurgit ex huiusmodi sensatione delectatio, quod apparet quia si homo famescit et sitit, descensus cibi et potus per os in stomachum est valde delectabilis, quia est conveniens naturae. Sed cum saturati sumus, tunc non est nobis talis motus delectabilis, sed abominabilis, quia non est conveniens naturae. Nunc ergo, cum generare sibi simile sit in viventibus naturalissimum operum, ut dicitur secundo huius, sequitur quod coitus et emissio spermatis et alii motus ad hoc ordinati et in membris in quibus sunt ad hoc ordinati, et quando natura hoc exigit, sunt valde delectabiles. Et non essent tales motus delectabiles in aliis membris non ad hoc per naturam ordinatis."

In conclusion to this section devoted to Buridan, a remark should be added. According to Buridan, the soul's sensitive power is distributed throughout the body, exercising its particular powers only wherever there is the right mediating qualitative disposition to do so. On this, see Zupko (2008) and (2003, 165–167). The case of touch as presented in this paper offers an occasion to elaborating somewhat on this position of Buridan. Buridan agrees with the "common" view that touch is a plural sense as far as the qualitative dispositions are concerned or, to put it differently, that we have plural possible qualitative dispositions enabling us to exercise tactile sensation. This can explain why sensation perceived in the genitals during ejaculation is touch (and not an additional, different, type of sense). For also the genitals can possess a right mediating qualitative diposition to let touch exercise its functions, even if different, in principle, by other qualitative dispositions enabling body to exercise other tactile sensations in other corporeal parts. Moreover, the tactile sensation felt in the genitals is felt only there, and not in other corporeal parts, such as hands or foots, because, as Buridan explains, the sensation of pleasure felt in the genitals is proper to the genitals and not to the other corporeal parts, meaning perhaps that the other corporeal parts do not have the same qualitative disposition as the genitals.

general ideas articulated by Tommaso del Garbo's and John Buridan's texts on the unicity or plurality of touch overlap. They both ultimately consider touch a plural sense as far as the qualities disposing the organ to perceive the tangibles are concerned, while they identify touch as only one sense as far as the organ of perception is concerned—in line with what they consider to be the common or traditional way of addressing the issue. The presence of a "common opinion" on the issue of the unicity or plurality of touch suggests that this question had a certain relevance for medieval scientific thought. In other words, the topic of the unicity or plurality of touch attracted the attention of medieval scientists to the point that a "common opinion" on it emerged. The texts by Tommaso and Buridan also show that (at least some) late medieval scientists were able to reflect widely on the discourse on the unicity or plurality of touch. Both texts show that the "common opinion" could be used just as a starting point to be further elaborated, which supports the idea that the issue of the unicity or plurality of touch attracted a genuine interest among medieval scientists. Moreover, the texts by Tommaso and Buridan show that the issue of the unicity or plurality of touch stimulated medieval authors in putting Aristotle into serious question. In their discourses, they both remained ultimately faithful to Aristotle's position on the number of senses, but they offered a highly personalized reading of the topic of the unicity or plurality of touch and questioned the Aristotelian view in details. Buridan's attempt to propose a position extending the outline of the "common opinion" is evident from how he builds his argument ("better articulating the truth," he says). Moreover, while confronting the issue of the unicity or plurality of touch, he continues to put Aristotle's position in question and open new lines of reasoning. This emerges especially from the discourse on ejaculation, where he repeatedly casts doubt on the unicity of the sense of touch. Tommaso's discourse is less rich than Buridan's one. Nonetheless, we also see evidence of elaboration of the topic in his text, which develops the "common opinion" in further detail and particularly explores the plurality of touch as far as tactile qualities are concerned.

Interestingly, the physiological discourse on ejaculation is present in Buridan and absent in Tommaso. This markedly physiological aspect of Buridan's discourse on the unicity or plurality of touch is somehow striking considering that, unlike Tommaso, he was not trained in medicine and worked in an institutional environment—the Parisian University—where the faculties of Medicine and Arts were much more separate than in Italy.[90] This element suggests that discussions on the unicity or plurality of touch are revealing of information on the relationship between

[90] Note that Buridan's use of medical elements in his discussion of the sense of touch is evident in his entire treatment of the sense of touch. While talking about the issue of the proper organ of touch, Buridan supports Avicenna's idea of the nerves as the organ of touch, see Beneduce (2019). The same idea of the nerves as the organ of touch was endorsed by Tommaso del Garbo, but Tommaso tried to reconcile this view with Aristotle by saying that Aristotle himself supported this position (see *supra*, footnote 54). This reconciliation attempt is not present in Buridan, who endorses a straight medical position on the point. On Buridan's use of medical thought in his natural philosophy, see Beneduce (2017).

natural philosophy and medicine in the late Middle Ages. Indeed, this issue is a useful vantage point for addressing that relationship, to the extent that the nature of touch was treated both in natural philosophical and medical discussions on sensation, in commentaries on Aristotle as well as in works of theoretical medicine, and in different geographical and institutional environments, too. For this reason, if the topic of the organ of touch has attracted scholarly attention as theoretically rich for exploring the interaction between natural philosophy and medicine in the Middle Ages, the same should be true for the question of the unicity or plurality of touch.[91]

In short, the high number of questions about the unicity or plurality of touch, both in natural philosophical and in medical works; the presence of a "common opinion" on the topic; examples of detailed elaborations of the issue extending the "common opinion" and putting Aristotle's view of the external senses into question; the different ways of treating the topic in different institutional environments—all of this suggests that the issue of the unicity or plurality of touch is relevant for understanding the reception of Aristotelian psychology and the relationship between natural philosophy and medicine in the late Middle Ages. Therefore, I strongly invite a more systematic study on how the issue was addressed and negotiated in late medieval science.[92]

[91] Moreover, as Buridan's and Tommaso's texts testify, the questions of touch's proper organ and its unicity or plurality were strongly intertwined. The two issues cannot be taken separately if we want to reconstruct their respective treatment in late medieval science. This also suggests that questions on touch should be studied as a unified corpus of issues, as claimed in Beneduce (2019).

[92] The research undertaken in this paper is developed within the project "The Sense of Touch. From Anomaly to Paradigm" (NWO-Veni grant project number VI.Veni.191F.005). I especially plan to devote further study to the specific topic of medieval and Renaissance views on the unicity or plurality of the sense of touch in relation to the topic of sexual pleasure. The set of problems around *the exact number of the senses, what the objects of touch are, whether touch is one or several senses, whether (sexual) pleasure is among the objects of touch,* and *whether or not sexual pleasure is a "sixth sense"* are especially relevant to understanding the problem of muscular sensation as distinct from the sensation of touch (and, more broadly, of proprioception). To be more detailed: the sixteenth-century debate between Girolamo Cardano (d. 1576) and Julius Caesar Scaliger (d. 1558) on the objects of touch and Scaliger's related discourse on sexual pleasure as the "sixth sense" have been considered the pre-history of the recognition of muscular sensation as distinct from touch sensation. On this, see Hamilton (1863, 867); Smith (2011); Scheerer (1987); Brett (first 1921, repr. 2002, 148–149). Similar questions about touch and sexual pleasure as a possible "sixth sense" are also found in other Renaissance authors, such as Girolamo Fracastoro (d. 1553) and Giulio Cesare Vanini (d. 1619); see Pogliano (2015, 38–39, 48–49) and Eschenbaum (2015). The analysis undertaken in this article, with the case of Buridan in particular, has shown that interesting antecedents of that Renaissance debate on the number of the senses, objects of touch, unicity or plurality of touch, and sexual pleasure as a possible "sixth sense" had already been present in scientific scholastic texts on sensation much earlier (see also the case of Olivi, mentioned *supra*, footnote 19, connected, in Yrjönsuuri's article, to the issue of proprioception). *Supra*, footnotes 74 and 78 moreover indicate additional aspects of how the topic of the sense of touch was related to pleasure in medieval philosophy and science.

References

Primary Literature

Adamus Burlaeus. 1997. *Quaestiones de anima*. In *Questions on the "De anima" of Aristotle by magister Adam Burley and Dominus Walter Burley*, ed. Edward A. Synan. Leiden: Brill.

Anonymus Bazán. 1971. *Quaestiones de anima*. In *Trois commentaires anonymes sur le "Traité de l'âme" d'Aristote*, ed. Maurice Giele, Fernand van Steenberghen, and Bernardo C. Bazàn. Louvain–Paris: Publications universitaire.

Anonymus Bernardini. 2009. *Quaestiones super librum de anima*. In *Anonymi magistri artium "Quaestiones super librum de anima"*, ed. Paola Bernardini. Firenze: SISMEL Edizioni del Galluzzo.

Anonymus Gauthier. 1985. *Lectura in librum de anima*. In *Anonymi magistri artium (c. 1245–1250) Lectura in librum de anima a quodam discipulo reportata*, ed. René A. Gauthier. Grottaferrata: Collegium S. Bonaventurae.

Aristotle. 1984. *The Complete Works of Aristotle. The Revised Oxford Translation*, ed. Jonathan Barnes, 2 vols. Princeton: Princeton University Press.

Averroes. 1953. *Commentarium magnum in Aristotelis de anima libros*. In *Averrois Cordubensis Commentarium magnum in Aristotelis De anima libros*, ed. Stuart Crawford. Cambridge, MA: The Medieval Academy of America.

Avicenna. 1972. *Liber de anima seu sextus de naturalibus, II*, In *Avicenna latinus. Liber de anima seu sextus de naturalibus, I–II–III*, ed. Simone Van Riet. Louvain: Peeters; Leiden: Brill.

Conimbricenses. 1609. *Commentarium in tres libros de anima Aristotelis*. Coloniae. Repr. 2006, Hildeshaim–Zürich–New York: Georg Olms Verlag.

Aegidius Romanus. 1500. *Super libros de anima*. Venetiis. Repr. 1982, Frankfurt am Main: Minerva.

Johannes Buridanus. 1984. *Quaestiones de anima (III red.), II*. In *John Buridan on the Soul and Sensation. An Edition of Book II of his Commentary on Aristotle's Book on the Soul with an Introduction and a Translation of Question 18 on Sensible Species*, ed. Peter Sobol. Dissertation. Bloomington: Indiana University.

———. 2019. *Quaestiones de secretis mulierum*. In *John Buridan's "Quaestiones de secretis mulierum". Edition and Introduction*, ed. Chiara Beneduce, Paul J.J.M. Bakker. *Vivarium* 57: 127–181.

———. 2015. *Quaestiones super Parva naturalia*. In *Jana Burydana "Quaestiones super Parva naturalia Aristotelis". Edycja krytyczna i analiza historyczno-filozoficzna*, ed. Maciej Stanek. Dissertation. Katowice: Uniwersytet Śląski.

Johannes Duns Scotus. 2006. *Quaestiones super secundum et tertium de anima*. In *B. Ioannis Duns Scoti Quaestiones super secundum et tertium De anima*, ed. Timothy B. Noone et al. Washington, DC: The Catholic University of America Press; St. Bonaventure, NY: Franciscan Institute Publications St. Bonaventure University.

Johannes de Janduno. 1552. *Super libros Aristotelis de anima subtilissimae quaestiones*. Venetiis.

Offredus Apollinaris Cremonensis. *Quaestiones de anima*. MS Paris, Bibliothèque nationale de France 6447.

Petrus de Abano. 1565. *Conciliator controversiarum, quae inter philosophos et medicos versantur*. Venetiis.

Petrus Johannis Olivi. 1922–1926. *Quaestiones in secundum librum Sententiarum*. In *Quaestiones in secundum librum Sententiarum*, vol. 2, ed. Bernhard Jansen. Quaracchi–Florence: Collegium S. Bonaventurae.

Petrus Hispanus. 2015. *Quaestiones super libro de animalibus Aristotelis*, ed. Francisca Navarro Sánchez. Farnham: Ashgate.

Thomas de Aquino. 1889. *Summa theologiae, I*, In *Thomae Aquinatis doctoris angelici opera omnia iussu impensaque Leonis XIII P. M. edita, tomus quintus, pars prima Summa Theologiae, a quaestione L ad quaestionem CXIX*. Roma: Typographia Polyglotta.

———. 1959. In *Aristotelis librum de anima commentarium*, In *Sancti Thomae Aquinatis In Aristotelis librum de anima commentarium*, ed. Angelo M. Pirotta. Torino–Roma: Marietti.

Thomas de Garbo. 1531. *Summa medicinalis*. Venetiis.

Secondary Literature

Beneduce, Chiara. 2017. *Natural Philosophy and Medicine in John Buridan. With an Edition of Buridan's "Quaestiones de secretis mulierum"*. Dissertation. Nijmegen: Radboud University.

———. 2019. La fisiologia del tatto nel XIV secolo. Il caso di Giovanni Buridano. In *Rappresentazioni della natura nel Medioevo*, ed. Onorato Grassi and Giovanni Catapano, 207–220. Firenze: SISMEL Edizioni del Galluzzo.

Boughan, Kurt Martin. 2006. *Beyond Diet, Drugs, and Surgery: Italian Scholastic Medical Theorists on the Animal Soul, 1270–1400*. Dissertation. Iowa City: University of Iowa.

Brett, George Sidney. 1921. *A History of Psychology. Mediaeval & Early Modern Period*, vol. 2. Repr. 2002, London: Routledge.

Cadden, Joan. 1993. *Meanings of Sex Difference in the Middle Ages. Medicine, Science and Culture*. Cambridge: Cambridge University Press.

Cappellini, Icilio. 1950. Date importanti per la biografia di Maestro Tommaso del Garbo e per gli inizi dell'insegnamento medico nello Studio Fiorentino desunte dai codici del Fondo Vaticano latino. *Rivista di storia delle scienze mediche e naturali* 41: 212–218.

Chandelier, Joël. 2017. *Avicenne et la médecine en Italie. Le "Canon" dans les universités (1200–1350)*. Paris: Honoré Champion.

———. 2020. Medicine and Philosophy. In *Encyclopedia of Medieval Philosophy. Philosophy between 500 and 1500*, ed. Henrik Lagerlund, 1138–1146. Dordrecht: Springer.

Coucke, Gijs. 2009. *Non adeo est honesta ut delectabilis*. Sexual Pleasure in Medieval Medicine: The Case of Petrus de Abano. In *Piacere e dolore. Materiali per una storia delle passioni nel Medioevo*, ed. Carla Casagrande and Silvana Vecchio, 117–148. Firenze: SISMEL Edizioni del Galluzzo.

de Asúa, Miguel. 1997. Pedro de Abano y la controversia entre médicos y filósofos. *Patristica et Mediaevalia* 18: 49–66.

———. 1999. Medicine and Philosophy in Peter of Spain's Commentary on "De animalibus". In *Aristotle's Animals in the Middle Ages and Renaissance*, ed. Carlos Steel, Guy Guldentops, and Pieter Beullens, 189–211. Leuven: Leuven University Press.

———. 2013. War and Peace. Medicine and Natural Philosophy in Albert the Great. In *A Companion to Albert the Great. Theology, Philosophy, and the Sciences*, ed. Irven M. Resnick, 269–297. Leiden–Boston: Brill.

Ebbesen, Sten, Christina Thomsen Thörnqvist, and Véronique Decaix. 2015. Questions on "De sensu et sensato", "De memoria" and "De somno et vigilia". A Catalogue. *Bulletin de philosophie médiévale* 57: 59–115.

Eschenbaum, Natalie. 2015. Robert Herrick and the Five (or Six) Senses. In *The Senses in Early Modern England, 1558–1660*, ed. Simon Smith, Jackie Watson, and Amy Kenny, 113–129. Manchester: Manchester University Press.

Federici Vescovini, Graziella. 1974. *Le "Quaestiones de anima" di Biagio Pelacani da Parma*. Firenze: Olschki.

Freeland, Cynthia. 1992. Aristotle on the Sense of Touch. In *Essays on Aristotle's "De Anima"*, ed. Martha C. Nussbaum and Amélie Oksenberg Rorty, 227–248. Oxford: Oxford University Press.

Fulkerson, Matthew. 2016. Touch. In *The Stanford Encyclopedia of Philosophy*. https://plato.stanford.edu/archives/spr2016/entries/touch/. Accessed 12 Sept 2019.

Goldner Steiner, Rebecca. 2018. Aristotle and the Priority of Touch. In *Touch and the Ancient Senses*, ed. Alex Purves, 50–63. London: Routledge.
Grellard, Christophe. 2017. Le philosophe et le glouton. Le plaisir de boire et de manger dans les commentaires de l'*Éthique À Nicomaque* de Jean Buridan et Nicole Oresme. In *Miroir de l'amitié. Mélanges offerts à Joël Biard*, ed. Christophe Grellard, 371–385. Paris: Vrin.
Guido, Francesco. 1970. Cenni biografici su Dino e Tommaso del Garbo. In *Atti del XXI Congresso Internazionale di Storia della Medicina, Siena (Italia), 22–28 settembre 1968*, 156–163. Roma: Arti grafiche Cossidente.
Hamilton, William. 1863. *The Works of Thomas Reid, D.D. Now Fully Collected, with Selections from His Unpublished Letters. Preface, Notes and Supplementary Dissertations by Sir William Hamilton*. Edinburgh: Maclachlan and Stewart.
Hasse, Dag Nikolaus. 2000. *Avicenna's "De anima" in the Latin West. The Formation of a Peripatetic Philosophy of the Soul 1160–1300*. London: The Warburg Institute; Torino: N. Aragno Editore.
———. 2001. Pietro d'Abano's "Conciliator" and the Theory of the Soul in Paris. In *Nach der Verurteilung von 1277. Philosophie und Theologie an der Universität von Paris im letzten Viertel des 13. Jahrhunderts. Studien und Texte*, ed. Jan A. Aertsen, Kent Emery, and Andreas Speer, 635–653. Berlin: Walter de Gruyter.
———. 2010. The Soul's Faculties. In *The Cambridge History of Medieval Philosophy*, ed. Robert Pasnau, vol. 1, 305–319. Cambridge: Cambridge University Press.
Jacquart, Danielle. 1993. La scolastica medica. In *Storia del pensiero medico occidentale, 1. Antichità e Medioevo*, ed. Mirko D. Grmek, 261–322. Roma–Bari: Laterza.
———. 1998. *La médecine médiévale dans le cadre parisien. XIVe–XVe siècle*. Paris: Fayard.
Jacquart, Danielle, and Françoise Micheau. 1990. *La médecine arabe et l'Occident médiéval*. Paris: G.-P. Maisonneuve et Larose.
Jacquart, Danielle, and Claude Thomasset. 1985. *Sexualité et savoir medical au Moyen Age*. Paris: Presses universitaires de France.
Johansen, Thomas Kjeller. 1997. *Aristotle on the Sense-Organs*. Cambridge: Cambridge University Press.
Knuuttila, Simo, and Pekka A. Kärkkäinen. 2013. Sense Perception. Medieval Theories. In *Sourcebook for the History of the Philosophy of Mind: Philosophical Psychology from Plato to Kant*, ed. Simo Knuuttila and Juha Sihvola, 61–81. Dordrecht: Springer.
Lagerlund, Henrik. 2008. Pietro d'Abano and the Anatomy of Perception. In *Theories of Perception in Medieval and Early Modern Philosophy*, ed. Simo Knuuttila and Pekka A. Kärkkäinen, 117–130. Dordrecht: Springer.
Laqueur, Thomas Walter. 1990. *Making Sex: Body and Gender from the Greeks to Freud*. Cambridge, MA–London: Harvard University Press.
Marshall, Peter. 1983. Parisian Psychology in the Mid-Fourteenth Century. *Archives d'histoire doctrinale et littéraire du Moyen Age* 50: 101–193.
Martorelli Vico, Romana. 2000. La medicina scolastica tra galenismo e aristotelismo. *Studi Medievali* 41: 311–338.
Miteva, Evelina. 2018. "Iam ergo patet veritas eius quod dixit Aristoteles, et causa deceptionis Galieni." Philosophers vs. Medics in Albertus Magnus' Account on Conception. In *Irrtum. Error. Erreur*, ed. Andreas Speer and Maxime Mauriège, 107–122. Berlin–Boston: Walter de Gruyter.
Ottosson, Per Gunnar. 1984. *Scholastic Medicine and Philosophy. A Study of Commentaries on Galen's "Tegni" (ca. 1300–1400)*. Napoli: Bibliopolis.
Park, Katharine. 1985. *Doctors and Medicine in Early Renaissance Florence*. Princeton: Princeton University Press.
Pogliano, Claudio. 2015. *Senso lato. Il tatto e la cultura occidentale*. Roma: Carocci Editore.
Scheerer, Eckart. 1987. Muscle Sense and Innervation Feelings: A Chapter in the History of Perception and Action. In *Perspectives on Perception and Action*, ed. Herbert Heuer and Andries Sanders, 171–194. London: Routledge.

Siraisi, Nancy Gillian. 1980. The Medical Learning of Albertus Magnus. In *Albertus Magnus and the Sciences*, ed. James A. Weisheipl, 379–440. Toronto: Pontifical Institute of Mediaeval Studies.

———. 1981. *Taddeo Alderotti and his Pupils. Two Generations of Italian Medical Learning*. Princeton: Princeton University Press.

———. 1990. *Medieval and Early Renaissance Medicine: An Introduction to Knowledge and Practice*. Chicago: The University of Chicago Press.

Smith, Roger. 2011. "The Sixth Sense". Towards a History of Muscular Sensation. *Gesnerus* 68 (1): 218–271.

Sobol, Peter. 2017. John Buridan on External and Internal Sensation. In *Questions on the Soul by John Buridan and Others: A Companion to John Buridan's Philosophy of Mind*, ed. Guyla Klima, 95–106. Dordrecht: Springer.

Vanni Rovighi, Sofia. 1951. *Le "Quaestiones de anima" di Taddeo da Parma*. Milano: Vita e Pensiero.

Yrjönsuuri, Mikko. 2008. Perceiving One's Own Body. In *Theories of Perception in Medieval and Early Modern Philosophy*, ed. Simo Knuuttila and Pekka A. Kärkkäinen, 101–116. Dordrecht: Springer.

Zupko, Jack. 2008. Horse Sense and Human Sense: The Heterogeneity of Sense Perception in Buridan's Philosophical Psychology. In *Theories of Perception in Medieval and Early Modern Philosophy*, ed. Simo Knuuttila and Pekka A. Kärkkäinen, 171–186. Dordrecht: Springer.

———. 2018. John Buridan. In *The Stanford Encyclopedia of Philosophy*. https://plato.stanford.edu/archives/fall2018/entries/buridan/. Accessed 12 Sept 2019.

———. 2003. *John Buridan: Portrait of a Fourteenth-Century Arts Master*. Notre Dame, IN: University of Notre Dame Press.

Chapter 14
Buridan Wycliffised? The Nature of the Intellect in Late Medieval Prague University Disputations

Lukáš Lička

Abstract The paper delves into manuscript sources connected with various disputations held at Prague University from around 1390 to 1420 and singles out a set of hitherto unknown *quaestiones* dealing with the nature of the human intellect and its relation to the body. Prague disputations from around 1400 arguably offer a unique vantage point on late medieval anthropological issues, since they encompass an entanglement of numerous doctrinal influences from Buridanian *De anima* commentaries to John Wyclif's theories. The paper delineates several conceptual tensions regarding the nature of the intellect, e.g., between materialism (entailed by the emphasis on the intellect's inherence in the body) and personal immortality. It presents several strategies Prague masters employed to overcome these tensions. For example, an anonymous participant of the 1409 *quodlibet* develops the Buridanian distinction between the rationally demonstrable materialist tendency and the indemonstrable "catholic truth" about the intellect both inhering in and separable from the body. On the other hand, Wyclif's adherents (Jacob of Mies and another anonymous master) postulate an immortal spirit hypostatically united to each human being beside the human soul educed from the potency of the matter. Yet, the boundaries between the doctrinal standpoints in question seem permeable, whereby a rigid definition of antagonistic groups in late medieval Prague intellectual milieu (e.g., Buridanians vs Wycliffites) is rendered ineffective.

Keywords Intellect · Late medieval Prague university · Medieval disputations · John Buridan · John Wyclif

---

Work on this study received financial support from the Czech Science Foundation (GA ČR) project "Philosophy at the University of Prague around 1409: Matěj of Knín's Quodlibet as a Crossroads of European Medieval Knowledge," grant n. 19-16793S, carried out at the Institute of Philosophy of the Czech Academy of Sciences.

L. Lička (✉)
Czech Academy of Sciences, Prague, Czechia
e-mail: licka@flu.cas.cz

M. Gensler et al. (eds.), *The Embodied Soul*, Historical-Analytical Studies on Nature, Mind and Action 11, https://doi.org/10.1007/978-3-030-99453-2_14

14.1 Introduction

Medieval philosophers, like many others before and after them, invested an immense amount of intellectual energy in investigating the question of what the human being is. The feature they singled out as *the* one unique for human beings, differentiating them from the rest of animate nature, was the intellect, or the capacity of universal conceptual thought. On the other hand, there was a pressing need to distinguish humans from superior, incorporeal beings, such as angels and God himself. The need was assuaged by emphasizing the human intellect's dependence on the material body. The resulting double-nature picture invited medieval thinkers to develop various more or less sophisticated strategies to answer the pivotal anthropological question.

Far from being unified, medieval accounts of the human being and the nature of the intellect evince two tendencies, not easily compatible. One strives for a *unified* explanation of nature, applicable to everything. Aristotelian hylomorphism epitomizes the tendency, treating the human as a kind of material being, in which the material body is in-formed by the soul, which serves as a principle of both bodily and intellectual operations. However, as the soul is merely a part of the composite, the issue of its separability and thus of personal immortality becomes problematic. On the other hand, the Platonic (or generally theological) tendency emphasizes a divine element of human nature *transcending* strictly material nature, potentially self-subsisting and separable and thus enabling at least a part of the human being to survive bodily death. Nevertheless, the tendency contests the unity of the human being, as the composition of two radically different natures calls for a special explanation.[1]

Running after two hares, the bulk of medieval philosophers tried to embrace *both* hylomorphic composition and personal immortality. The hylomorphic model was even declared as the only one consonant with the Catholic faith at the Council of Vienne in 1312. Everybody denying that the intellective soul is *per se* and essentially the form of the human body was to be considered a heretic.[2] Interpreting the exact nature of the intellect-body relation was far from unanimous. The constitution did not calm the dispute over how many substantial forms there are in the human being, an issue arduously debated since the early thirteenth century.[3] Unitarians, like Thomas Aquinas or John Buridan, regarded the intellect as the only substantial form of the human being, inhering directly in the body. Such a claim, however, calls for a special explanation of why the intellect is not like the forms of other material

[1] For the repercussions of taking the intellective soul as a form, albeit potentially separable from the body, see De Boer (2013, 25–36); see also Dales (1995).

[2] See Duba (2012, 175–177) for the quotation of the constitution. Note that personal immortality had to wait until 1513 to be supported by the papal bull *Apostolici regiminis*. See, e.g., Pluta (2010, 85–89).

[3] Despite the assertion of the contrary, often repeated by medieval and modern scholars, the constitution does not seem to aspire to settling down the issue by condemning *pluralism*, on which, see Duba (2012, 171–180).

objects, i.e., educed from matter, inseparable, and corruptible. On the contrary, the *pluralist* perspective (taken, among many others, by Peter Olivi, John Duns Scotus, or John Wyclif) postulates several substantial forms in a single human, often at least two: a sensory soul inhering in the body, which is educed from matter and thus corruptible, and the potentially separable intellect. In such a scenario the unity of the human being becomes problematic.[4]

Although the literature on the issue is vast, the present paper aspires to shed some further light on the issue from an unconventional perspective. Instead of focusing on one or several pivotal figures and presenting their doctrine, the paper aims to reconstruct a more dynamic picture of the issue, focusing on written records of various disputational practices at late medieval Prague University, flourishing around 1400, where the anthropological topics regularly re-emerge. The overwhelming majority of such written records of acute debates of the era are extant in the form of short *quaestiones* (often without determination), drafted for various university acts (quodlibetal or ordinary disputations, determinations of students during exams). Often extant in a single manuscript copy, buried in not properly catalogued codices, they are usually considered too derivative, unimportant, and trifling to be transcribed in full and studied adequately.

The present paper suggests that even texts of this kind are worth studying, as they provide almost immediate access to actual late medieval disputational practices in raw form, not refined by several layers of author's reworking and centuries of textual transmission, which every researcher has to face when dealing with texts by the renowned medieval thinkers. Besides, Prague university milieu around 1400 offers a unique vantage point at anthropological issues, since numerous doctrinal influences are entangled in Prague disputations then, from Buridanian *De anima*-commentaries, ubiquitous at Central European universities, to John Wyclif's theories, studied nowhere so extensively as in Prague during this period. One of the aims of the paper is to suggest that these doctrinal influences were numerous and the boundaries between them were permeable and that, at least in the context of the nature of the intellect, it is not particularly effective to construe rigidly defined antagonistic groups of texts/authors (e.g., Buridanians vs. Wycliffites).

The paper builds upon an idiosyncratic type of textual sources, so tightly connected with medieval scholarly practices that the specific nature of the former is not intelligible without knowledge of the latter. Hence, Sect. 14.2 is devoted to a purely historical account of the institutional context of various types of late medieval university disputations and their manuscript outcomes, both with particular respect to the Prague Faculty of Arts.

Section 14.3 introduces the text base of the paper: a set of Prague *quaestiones* dealing with the nature of the intellect and its relation to the body from different perspectives and doctrinal stances, selected from manuscripts with records of various disputational occasions. Incidentally, several textual discoveries are announced

[4] On the dispute, see, e.g., De Boer (2013, 36–43; 216–227); a metaphysical perspective is provided by Pasnau (2011, 574–596); see also Pasnau (2010).

in Sect. 14.2 and 14.3 (e.g., three Cracow codices preserving two sets of questions, possibly two hitherto unknown Prague quodlibets, or two commentary-like manuscript notes on *De anima*, possibly by John of Borotín and an anonymous Bohemian Wycliffite).

Section 14.4 presents one possible method of utilizing these allegedly second-rate texts. Basing my account on selected arguments harvested from the Prague questions, I delineate the conceptual tensions involved in the medieval articulation of the intellect's nature. I focus on different properties attributed to the intellect, their logical relations, and mutual dependence, but also the incompatibility of some of them. Also, the considerably rich range of philosophical sources exploited by Prague masters is revealed here.

Finally, Sect. 14.5 delineates several main strategies to deal with the nature of the intellect employed by late medieval Prague scholars. While both the Alexandrist materialist position and the Averroist view of a single immortal intellect shared by all people are almost absent in the Prague context, the texts often tend towards Buridanian or Wycliffite positions. A Buridanian view, advocated by an anonymous participant of the 1409 Prague quodlibet, distinguishes between the rationally demonstrable materialist tendency and the indemonstrable "catholic truth" about the intellect both inhering in and separable from the body. On the other hand, Wycliffites (such as an anonymous student or Jacob of Mies) try to overcome the Buridanian position, so prone to the unsatisfying double-truth theory, by a conceptual innovation enriching Aristotelian anthropology by the Christological notion of the hypostatic union. Even if the human intellect were educed from matter, an immortal spirit is hypostatically united with such a composite. The last piece of the picture is a passage from peculiar notes on Aristotle's *De anima*, loosely modelled after Buridan's questions on *De anima*, where Buridan is explicitly Wycliffised.

14.2 Prague University Disputations Around 1400: Scholarly Practices and Manuscript Sources

The first step in investigating the contributions of Prague scholars around 1400 to the medieval debate on the nature of the human intellect is to set the contents and boundaries of the textual corpus. To emphasize the dynamic and controversial aspect of the issue, I deliberately focus almost exclusively on the outcomes of *disputations* held at the Prague Faculty of Arts, which means that I put aside texts designed for lecturing purposes (chiefly *De anima* commentaries) and texts originating at the Faculty of Theology (primarily *Sentences* commentaries). Nevertheless, I do not follow this resolution blindly, as I make use of quotations from some Prague notes on *De anima* to corroborate claims based on the disputations-connected material and treat Jacob of Mies's question, written probably in the theological milieu, as an exemplification of the Wycliffite approach.

Before the textual base can be delineated, several idiosyncrasies of late medieval university disputations and their written outcomes ought to be introduced. Since written manuscript sources are closely related to oral disputation practice, the latter needs to be depicted to make the former more intelligible.

First and foremost, two structural patterns can be recognized in the Prague disputation questions. Both proceed from the nature of the particular type of disputation to which the questions are connected, and, thus, they differ slightly from the traditional *quaestio* format. First, there are questions without any response, designed as a preparation for the disputing master, embracing arguments *contra concedentem* (i.e., arguments to be used against the respondent affirming the question) and *contra negantem* (i.e., those by which a negative answer to the question is repulsed); particular arguments are also often independent of each other. The other kind is the actual determination proposed by the respondent, often called "position" (*positio*) in the manuscripts. It usually comprises of *notabilia* (definitions and distinctions of the terms used and theories implied in the title of the question) and conclusions with corollaries, both supplied with justifications. Solutions of the *opposing* arguments are often missing in the positions; the dialectical development of ideas, typical for the traditional *quaestio* format, must have taken place in the actual disputation between proponent and respondent and was rarely recorded.[5]

The texts investigated here are products of various types of university disputations, each with its specific features.[6] The following sketch focuses primarily on the idiosyncrasies of written outcomes of Prague disputations as preserved in manuscript sources, delineating the corpus of disputation-connected texts, from which I select the sources on the nature of intellect. The most important disputations at the Prague Faculty of Arts were the quodlibetal ones, organized once a year, in which all masters affiliated at the faculty were expected to participate. A *quodlibetarius*, elected several months before to host the session, presided over the disputation and proposed a question to each participant.[7] The textual outcomes of the quodlibets are twofold. First, there are the so-called handbooks or manuals, book-length texts prepared by the *quodlibetarii* and including dozens of questions for particular masters, usually in the form of preparations.[8] Second, in some cases, the positions authored by individual masters are extant, usually dispersed in codices with materials resulting from various university acts.

Until now, research has identified 13 quodlibets taking place at Prague Arts Faculty from the 1390s to 1410s, where at least the name of the presiding master is

[5] See Kejř (1971, 15–16 on preparations, 66–67 on positions); cf. Weijers (2015, 102).

[6] On various kinds of late medieval university disputations, see Weijers (2013, 121–137); for Prague Faculty of Arts, see Kavka (1967, 33–38), Kejř (1960), also Pavlíček and Hanke (2021, 207–210) on the role of sophisms in various Prague disputations.

[7] For the Prague quodlibetal tradition and its highlights, see especially Kejř (1971), Šmahel (2007, 336–386), Weijers (2002, 298–307), Pavlíček (2018, 328–340), Székely (2019, 303–306), and Lička (2021).

[8] However, not exclusively; as, e.g., the handbook of Matthias of Legnica contains positions, as shown below.

known. The manuscripts containing the handbooks are extant merely in seven cases.[9] I assume that this corpus of seven quodlibetal handbooks may be expanded by two further sets of questions extant in three codices originating from Prague but preserved in Cracow. I argue that they may represent material connected with two hitherto unknown Prague quodlibets.

The first set of questions is preserved in two Cracow manuscripts (Biblioteka Jagiellońska [hereafter: BJ] 649, ff. 2v–163v, and BJ 624, ff. 1r–216v) and manifests all the salient features of Prague quodlibetal handbooks. The copy in BJ 649, which seems complete and well-organized, begins with a question which is unnumbered and explicitly designated *quaestio principalis*, on which the *quodlibetarius* himself usually disputed.[10] Additionally, there are 63 numbered questions (in the form of preparations) for other participants of the quodlibets. A so-called *problema* (a further question of minor importance, often curious) is attached to each question, just as in other Prague quodlibetal handbooks. The copy is to be dated to the 1390s according to the watermarks.[11]

The copy in BJ 624, written in a very similar, perhaps identical hand, provides the *quaestio principalis*[12] and 61 other questions (in comparison to BJ 649, two quodlibetal questions are missing in this copy). However, the majority of the questions are unnumbered, and their ordering is confused, both by the scribe and by the binding of the codex.[13] The set includes 15 additional questions, apparently connected with an earlier Prague quodlibet organized by Henry of Ribenicz in the early 1390s, whose handbook is extant in a Leipzig manuscript.[14] The watermarks suggest the 1390s as the date of the copy and that BJ 624 is slightly older than BJ 649.[15] The Ribenicz intrusion corroborates the Prague origin of the codex, propounded by the authors of the catalogue, and suggests the mid-1390s as the earliest date of the quodlibet.

[9] See the list in Šmahel (2007, 377–381). Having sifted through these seven handbooks, I singled out several questions/preparations on the nature of the human intellect from handbooks by Matthias of Knín (Matthias de Knin, handbook in MS Prague, Knihovna metropolitní kapituly L.45, ff. 1r–156v), Matthias of Legnica (Matthias de Legnitz, handbook in MS Stralsund, Stadtarchiv 1067, ff. 207ra–278rb), and Prokop of Kladruby (Procopius de Cladrub, handbook in MS Prague, Knihovna metropolitní kapituly L.27, ff. 1r–132r).

[10] Actually, the position of the *quodlibetarius* is not recorded, only the debate on it.

[11] For the codicological description and list of questions and *problemata*, see Kowalczyk et al. (1980, 4: 398–406).

[12] Here, the principal question is untitled and again encompasses only the debate, slightly differing from the version in BJ 649.

[13] See Kowalczyk et al. (1980, 4: 338–343); the authors try to reconstruct the original ordering of the questions.

[14] Henricus de Ribenicz, *Quodlibet* (MS Leipzig, Universitätsbibliothek 1414, ff. 1v–231v). Although unnoticed by the authors of the catalogue, I have been able to identify variants of the questions 2–4, 6, 9, 12, 13, 15, 17, 19–21, 27, 30–31 of Ribenicz's quodlibet. See Šmahel (2007, 384–386) for the list of questions of Ribenicz's quodlibet, as preserved in the Leipzig codex.

[15] Although the ordering of BJ 624 is messy, at least in the case of the question considered below the codex offers more a plausible reading.

Another set of questions is preserved in Cracow codex BJ 736 on ff. 81r–199r.[16] Although both its Prague origin and its quodlibetal nature are probable in this case, neither is indubitable. Again, the initial question has a special status (it is unnumbered and includes both the position and a discussion of its conclusions), while other 90 questions are numbered preparations (albeit without *problemata*). The authors of the catalogue call the set *quaestiones ordinariae*; yet, they do not look like the outcome of ordinary disputations. In Prague, ordinary disputations took place every Saturday and lasted only until evening (*ad horam vesperorum*). It does not seem possible to discuss 91 questions in one day. Still, the set of questions may be a guide for organizing ordinary disputations, rather than an outcome of a single event or a series of disputations. Nevertheless, a regular element of these sessions was a disputation on sophisms (no more than three per session), in each of which three participants responded (the first by conceding, the second by negating, the third by doubting); yet no sophisms or preparations to a disputation of sophisms are included in BJ 736.[17] In conclusion, the set of questions shares more similarities with other Prague *quodlibetal* handbooks than with the outcomes of other types of disputation.

The Prague origin of the codex is not self-evident either. As the authors of the catalogue assert, at least some parts of the codex originated in Prague. They suggest the handwriting of the part with the set of questions resembles that of BJ 624 and 649; the date of the part seems to be "after 1395" (according to watermarks). The codex was probably bound in Poland, perhaps for its first owner, John of Radochonice (Johannes de Radochoncze), who earned his bachelor's degree in Prague in 1407[18] and later moved to Cracow where he continued in arts and theology studies.[19] He may have procured the quires later bound into BJ 736 (one of which contained a copy of the aforementioned set of questions) during his bachelor's studies in Prague.[20]

---

[16] See Kowalczyk et al. (1980–2012, 5: 236–245).

[17] On ordinary disputations at the Prague Faculty of Arts, see *Statuta facultatis artium universitatis Pragensis redacta anno 1390*, IV.1–2 (ed. Šmahel and Silagi, 2018, 246–247). For a general sketch of medieval ordinary disputations, see Weijers (2013, 122–133); for theological faculties, see Lawn (1993, 13–15).

[18] Tříška (1981, 284).

[19] Knoll (2016, 614).

[20] If the two sets of questions in MSS BJ 649/624 and BJ 736 represent two 1390s handbooks by Prague *quodlibetarii* (a hypothesis, admittedly, slightly more probable in the former case than in the latter one), they might have been authored by Nicolaus Magni de Jawor and Henricus de Homburg, who were elected *quodlibetarii* for the years 1395 and 1396, respectively. See *Liber decanorum facultatis philosophicae Universitatis Pragensis ab anno Christi 1367 usque ad annum 1585* (ed. Dittrich et al., 1830–1832, 1: 292, 303); and Šmahel (2007, 377). I am indebted to Ota Pavlíček for this suggestion. Nevertheless, perhaps I rush to conclusions, as several manuscripts of quodlibetal handbooks ascribe them to masters not mentioned in the *Liber decanorum* as elected *quodlibetarii* (e.g., Henricus de Ribenicz, Matthias de Legnitz, Johannes Arsen de Langenfeld). Hereafter I tentatively refer to these two texts as to anonymous quodlibets: Anonymus, *Quodlibet* (MSS Cracow, BJ 649, ff. 2v–163v, and Cracow, BJ 624, ff. 1r–216v) and Anonymus, *Quodlibet (?)* (MS Cracow, BJ 736, ff. 81r–199r).

Besides annual quodlibetal disputations, several minor types of disputations took place more frequently, some of them weekly, like the already mentioned ordinary disputations, held by masters every Saturday. Furthermore, there were extraordinary disputations of newly incepted masters on Tuesdays and Thursdays[21] and bachelor's disputations on feast day afternoons.[22] Nevertheless, these activities were so frequent and brief that their actual contents (the questions and sophisms determined during these disputations) were scarcely written down. However, some Prague codices include quires with questions (and sophisms) without determination, with blank spaces between the paragraphs, which were, most likely, intended as preparations for these minor disputations.[23]

Finally, more ceremonial disputations took place during promotions of the candidates for bachelor's and master's degrees (*promotio* of bachelors and *inceptio* of masters). On these occasions, a sophism and a question were proposed to the candidate; after his determination, the older master promoting the candidate delivered a speech recommending the latter (*recommendatio*), and the candidate was officially declared bachelor or master.[24] The promotional acts were obviously recorded, as various codices include quires with questions (positions), often intermingled with sophisms and recommendations. The questions in those manuscripts are usually very concise, the reason probably being that the candidates had to speak off the cuff, as the statutes forbade them to read aloud from a sheet of paper or a book (*de carta vel libro*) during the promotion.[25]

---

[21] *Statuta facultatis artium universitatis Pragensis redacta anno 1390*, II.31 (ed. Šmahel and Silagi, 2018, 244).

[22] *Statuta facultatis artium universitatis Pragensis redacta anno 1390*, II.24 (ed. Šmahel and Silagi, 2018, 242).

[23] The set of questions and sophisms in the codex Prague, Knihovna metropolitní kapituly M.92, ff. 88r–172v (of which one question on the intellect is mentioned below) can serve as an example of this type of manuscript evidence. Already Jiří Kejř studied this codex and its relation to the quodlibetal handbooks, on which, see Kejř (1960, 53–54). Several questions (related to the problem of universals) are also listed in Šmahel (1980, 42–43).

[24] For an outline of the entire ceremony, see *Statuta facultatis artium universitatis Pragensis redacta anno 1390*, II.1–25 (ed. Šmahel and Silagi, 2018, 236–245) for bachelors, and II.26–32 (ed. Šmahel and Silagi, 2018, 243–245) for masters; see also Kavka (1967, 21–25).

[25] See *Statuta facultatis artium universitatis Pragensis redacta anno 1390*, II.16 (ed. Šmahel and Silagi, 2018, 241). The material culture of the codices provokes speculation that the candidates did not adhere to the rule universally, as some disputation volumes include little sheets of paper bound into the gatherings with promotional records. See, e.g., MS Uppsala, Universitetsbibliotek, C 639, f. 159r (a "cheat sheet" with a question on accidents not inhering in the substance). This part of the codex originated from Prague, as it includes, e.g., a question determined *Prage sub m<agistro> Rybbeni<cz>* (f. 160r).

14.3 The Intellect Disputed: A Review of Sources

Most of the texts considered here can be dated to the period between 1390 and 1420. A thorough examination of the codices related to various types (and instances) of Prague disputations (kept mostly in Prague, Cracow and Leipzig libraries today) has resulted in a corpus of 10 questions dealing with the nature of the intellect and its relation to the body. They have so far been not only unedited, but also *unstudied*; most of them are extant in a single manuscript copy. Their titles, in a tentative chronological order, are as follows[26]:

1. Anonymus, *Quodlibet*, q. 11: *Utrum anima hominis sit extensa*, MSS Cracow, BJ 649, ff. 34r–35v; BJ 624, ff. 76v–78v. 1390s. Type and structure: Preparation; 8 arguments *pro*, 8 arguments *contra*.
2. Anonymus, *Quodlibet* (?), q. 16: *Utrum animam humanam esse indivisibilem et immortalem sive incorruptibilem possit ostendi naturali ratione*, MS Cracow, BJ 736, ff. 116r–117v. 1390s. Preparation; 8 arg. *contra*, 6 arg. *pro*.
3. Anonymus, *Quodlibet* (?), q. 17: *Utrum sit evidenter probabile naturali ratione quod anima intellectiva sit forma substantialis hominis inhaesive*, MS Cracow, BJ 736, ff. 117v–118v. 1390s. Preparation; 8 arg. *contra*, 7 arg. *pro*.
4. Matthias de Legnitz, *Quodlibet*, q. 52: *Utrum animam intellectivam esse formam hominis possit ratione evidenti ostendi*, MS Stralsund, Stadtarchiv 1067, ff. 253ra–va. 1397.[27] Position; 2 *suppositiones*, 2 *notabilia*, 5 conclusions (with corollaries), 2 arg. *contra conclusiones*.
5. Anonymus, *Utrum anima intellectiva educta de potentia materiae sit corruptibilis*, MS Prague, Národní knihovna VIII.E.5, ff. 55v–56v. 1400s–1410s (possibly 1402).[28] Position; 3 *notabilia*, 3 conclusions (with corollaries).
6. Matthias de Knin, *Quodlibet*, q. 91: *Utrum anima intellectiva sit forma inhaerens corpori humano de potentia materiae per generationem educta*, MS

[26] I have prepared preliminary transcriptions of all the texts listed and plan to publish editions of some of them. For practical reasons, I usually do not quote from these texts *in extenso*. However, I sketch the structure of each text here in order to highlight the differences between preparations and positions.

[27] I prefer this date suggested and argued for by Ota Pavlíček, rather than those maintained by former research: 1394 (Josef Tříška and Vilém Herold), or 1399 (František Šmahel). See Pavlíček (2021, 17–19) for the issue.

[28] The codex Prague, Národní knihovna VIII.E.5 encompasses chiefly positions presented at the promotional acts. The question on the corruptibility of the intellect is involved in the same quire as a question on Aristotle's *Metaphysics* (f. 62v) assigned to Wenceslaus de Egra (Václav of Cheb). Given that the *Metaphysics* was usually part of a master's studies (see the analysis of two lists of lectures attended by Prague students in Šmahel (2007, 316–335, esp. 329–330), and that Egra's *inceptio* took place in 1402 (*Liber decanorum*, ed. Dittrich et al., 1830–1832, 1: 369), it can be inferred that also the other questions in the same quire originated around 1402. See also Kejř's remarks on the codex in Kejř (1960, 54–55), and Kejř (1955–1956, 3: 231, note 98), where he points out Egra's name but gives a wrong date of his inception.

Prague, Knihovna metropolitní kapituly L.45, ff. 100v–101r. 1409. Preparation; 9 arg. *pro*, 4 arg. *contra*.

7. A response to Knín's question by an anonymous master participating in his quodlibet: Anonymus, *Utrum anima intellectiva sit forma inhaerens corpori humano de potentia materiae per generationem educta*, MS Prague, Národní knihovna X.H.18, ff. 40r–42r. 1409. Position; article 1 (8 *notabilia*), article 2 (3 opinions, 2 conclusions with corollaries), article 3 (3 conclusions with corollaries).
8. Procopius de Cladrub, *Quodlibet*, q. 26: *Utrum anima intellectiva* [1] *habens operationem propriam, quam non communicat corpori,* [2] *sit perpetua et immortalis*, MS Prague, Knihovna metropolitní kapituly L.27, ff. 41v–43r. 1417. Preparation; 6 arg. *pro supposito* [1], 10 arg. *contra quaesitum* [2], 6 arg. *contra suppositum* [1], 12 arg. *pro quaesito* [2].
9. Jacobellus de Misa, *Utrum intellectus hominis est homo*, MS Prague, Národní knihovna V.H.13, ff. 140r–141r. 1410s (before 1419). Position; 3 suppositions, 3 conclusions with corollaries.
10. Anonymus, *Utrum intellectus qui est potentia animae rationalis sit potentia organica*, MS Prague, Knihovna metropolitní kapituly M.92, ff. 147v–148r. 1410s–1440s. Preparation; 7 arg. *pro*, 4 arg. *contra*.

Only four of the ten texts have certain authors: the Bohemian masters Matthias of Knín, Prokop of Kladruby, Jacob of Mies, and the Silesian master Matthias of Legnica.[29] The texts are mainly the outcomes of (or preparations for) Prague quodlibetal disputations (no. 1–4, 6–8), but also of promotional disputations (no. 5 and possibly 9), and also of a disputation whose purpose is uncertain (no. 10).

As for the topics of the texts, various properties of the intellect are investigated, especially its relation to the body (whether it is educed from matter, is extended, or even is an organic potency—no. 1, 5–7, 10) and the issue of its (im)mortality and (in)corruptibility (no. 2, 5, 8). Some titles are reminiscent of the Vienne constitution about the intellect as the substantial form of the body (no. 3, 4; no. 9 even asks whether a human being is identical with its intellect); the issue of the intellect's inherence in the body also appears (no. 3, 6, 7). Some questions are rephrased epistemically, asking whether particular properties of the human intellect can be rationally demonstrated (no. 2–4).

Bearing in mind the ephemeral nature of these texts, a question arises: why should we deal with such second-rate sources at all? First of all, they provide a unique insight into how intellectuals at the newly founded universities in Central Europe delved into philosophical issues related to the nature of the intellect. The choice of the Prague university milieu of the 1400s is motivated both philosophically and historically, as it represents a real "melting pot" of various philosophical traditions, but the details of their overlaps and struggles are as yet understudied.

---

[29] For these authors, see Spunar (1985–1995, 1: 214–250 on Mies, and 352–362 on Kladruby), and Tříška (1981, 364) for Knín and Legnica. See also a fresh recent study of Knín's Wycliffite metaphysics in Campi (2020).

Since the studied texts are connected with disputational practices, investigating them offers a picture of what late medieval university scholars were actually doing.[30]

Further, the disputation-connected texts can compensate for the lack of full-fledged Prague *De anima*-commentaries in the period under consideration. The lectors of Aristotle's *De anima* in Prague around 1400 usually made use of Buridan's questions on *De anima* (or its various variants and abbreviations), or else of commentaries from the same philosophical tradition (for example, questions by Lawrence of Lindores). Texts originating directly from Prague are scarce; there are only a few expositions of the authoritative text, some longer, like that by Jenek of Prague (1375); some shorter and more elementary, like those by Henry Totting of Oyta (1360s) and the anonymous exposition of *De anima "secundum Buridanum"* from the 1370s.[31]

I have been able to identify two peculiar materials connected with the *De anima* tradition, which can be added to this scanty list. First, there are *notabilia* related to questions on *De anima* III, written in the hand of the Bohemian master John of Borotín (d. after 1458), most likely around 1410.[32] Second, there are questions on *De anima* II–III, with titles borrowed from Buridan's second (or penultimate) redaction of *De anima* questions,[33] but with contents completely reworked by a Bohemian author of Wycliffite leanings, possibly in the 1410s.[34] I briefly elaborate on both texts in the last section of the paper.

Another shortcoming of the disputational texts is their derivativeness: their authors borrowed many of the arguments incorporated in the questions from the popular treatises of the time. Besides Buridan and Wyclif, they refer to authors such as John Eucles, Thomas of Strasbourg, John of La Rochelle, or John of Jandun. This apparent weakness can turn out to be an advantage, since texts tightly bound to the scholarly practices of Prague masters become a precious document witnessing a clash of philosophical traditions and a gradual transition from Aristotelianism, mostly Buridanian, to Wycliffism, with different positions and accents lurking under the same terminology. Similarly, the occasional reappearance of various arguments in the questions need not be interpreted as plagiarism, but rather as constituting a

[30] That investigating such short-lived texts may be very fruitful is evident from studies on student notebooks with reports of various disputations in fifteenth-century Cologne and Vienna; see, e.g., Hoenen (2011) and Flüeler (2008).

[31] For Jenek of Prague, see Mráz (1982); for a partial edition of Totting's exposition and a complete edition of the 1370s exposition (extant in MS Cracow, BJ 704), see Pluta (1986, 96–97) and Mansfeld (2016), respectively.

[32] Johannes de Borotin (?), *Notabilia super librum III De anima Aristotelis* (MS Prague, Národní knihovna X.H.18, ff. 55r–56v); see Lička (2021, 284), for the list of questions.

[33] For the redactions of Buridan's questions, see note 73 below.

[34] Anonymus, *Quaestiones in libros II–III De anima Aristotelis* (MS Prague, Národní knihovna X.H.17, ff. 157r–180r). In the manuscript, numerous excerpts, notes and definitions are scattered among the questions on *De anima*. These passages deal mostly with theological matters, sometimes being longer than the questions themselves. They seem unrelated to the *De anima* commentary but are written in the same hand. Several Czech words dispersed in both commentary and notes suggest the author was of Czech descent.

terminological and doctrinal framework shared across several generations of Prague scholars. Also, repeating passages present in older materials may corroborate the Prague origin when not fully certain (especially in the questions from BJ 736).[35]

Further, since the disputation context allows (even invites) one to propose bolder claims and push their consequences further than a declaratory text, such as a lecture or a commentary, the questions sometimes encompass sophisticated arguments even for heterodox positions (such as Alexandrism), or for positions explicitly condemned by the Church (e.g., the denial of the claim that an intellect is a substantial form inhering in a body). The environment of late medieval Prague Faculty of Arts might arguably have been liberal. Institutional restrictions prohibiting the *artistae* from discussing theology-related matters (such as the immortality of the human soul), notorious at Paris university, but also in Cologne or Vienna, are not documented in Prague.[36]

Finally, the questions are also philosophically valuable, at least to a certain extent. Despite their sketchiness, one frequently stumbles upon a sharp and compelling piece of reasoning in them. As some arguments are excerpts from massive texts, such as commentaries on Aristotle or the *Sentences*, sometimes unedited, it is possible quickly to focus on the sharpest arguments, which would otherwise remain buried in the source texts among a load of other text. Historians of philosophy may object that especially the preparations are useless for their agenda. How is one to reconstruct a well-argued philosophical position, if there is *no* position maintained in the source text? In the next section, I try to demonstrate that even mere sets of arguments *pro* and *contra* can serve as material for conceptual analysis and for uncovering the framework in which the thinkers conceptualized philosophical issues. The last section will be, by contrast, devoted mainly to questions in the form of position, reconstructing some doctrinal stances of Prague philosophers.

[35] The textual parallels between various quodlibetal handbooks have already been pointed out by Kejř (1971, 49–61) and Šmahel (2007, 339–341, 363–364, 369), and recently further evidenced by Székely (2018) and Lička (2021). The *quodlibetarii* sometimes express their dependence on former handbooks explicitly; see the reference to Arsen's handbook in Matthias of Knín's quodlibet: "Alia argumenta: quaere in M. Io. Artsen circa tale signum." (MS Prague, Knihovna metropolitní kapituly L.45, f. 43v; for Knín's other references to Arsen see Šmahel, 2007, 342.)

[36] See Pavlíček (2018, 340–355, especially 346–351). Admittedly, various restrictions regarding Wyclif took place at early fifteenth-century Prague University (for example, in 1403, an assembly of university masters condemned a list of 45 Wyclif's theses, which had been compiled by John Hübner). See, e.g., Šmahel (2007, 467–489) and Herold (1985, 148–170). Nevertheless, these restrictions did not generally suppress the dynamic evolvement of reformist ideas, but rather provoked it. Further, as practically-oriented (mainly on ecclesiological and Eucharistic matters), these restrictions seem not to have affected the philosophical enterprise other than indirectly.

14.4 The Nature of the Intellect: Conceptual Tensions Between Materialism and Immortality

Once the corpus of the Prague questions on the nature of the intellect has been established, its contents can be examined. This section aims to present how the philosophical problems associated with the nature of the intellect were articulated in late medieval disputational practice. It exploits the arguments, especially from the preparations, to reconstruct the medieval setting of the anthropological issue. This setting, which is complex and not entirely coherent at face value, was a seedbed for the emergence of various *conceptual tensions*. Several properties are ascribed to the intellect, some of which seem to be, at least at first glance, mutually exclusive.

The starting point of several debates was the constitution of the Council of Vienne (1312) proclaiming that the intellect is the substantial form of the human being. The claim itself guided medieval intellectuals to a certain anthropological preunderstanding: they construed the human soul as a form in the Aristotelian sense. Historically, the claim was designed as a safeguard against Averroism and its doctrine of a single intellect only loosely connected with all human individuals.[37] Nevertheless, even such an Averroist single intellect can be understood as a form, albeit not informing the particular humans but merely assisting them in their operations. Similarly, the intelligences, being movers of the celestial spheres, are their forms: although they do not strictly inform them, they still function as the principles of the spheres' motion. To evade the threat of Averroism, late medieval philosophers (especially in the Buridanian tradition) carefully distinguished between two types of forms: those *inhering* in their subjects (intrinsically sustaining them and moving according to the movements of the subjects) and forms merely *assisting* the subjects in operation, present to, adhering to and appropriated by the subjects, yet extrinsic to them.[38] The relation of the intellect to the body must be tighter: it is not only the substantial form of the body (as the Vienne constitution asserts), but a form *inhering* in the body. Matthias of Knín provides a simple Aristotelian argument for such a claim: a property of being *A* evinced by a subject *S* presupposes the form of *A* inher-

---

[37] Cf. Anonymus, *Quodlibet* (?), q. 17: *Utrum sit evidenter probabile naturali ratione quod anima intellectiva sit forma substantialis hominis inhaesive*, pro 5 (MS Cracow, BJ 736, f. 118v).

[38] See, e.g., Johannes Buridanus, *Quaestiones super De anima secundum tertiam sive ultimam lecturam*, III.4, §2 (ed. Hartman et al., forthcoming), Laurentius Lindorius, *Quaestiones in Aristotelis libros De anima*, III.4.1 (ed. Dewender and Pluta, 1997, 208), or Johannes Eucles, *Quaestiones in Aristotelis libros De anima*, III.4.1 (ed. Pluta, 1988, 521), who speaks about forms *coexisting* with the body. See De Libera (2014) for the big picture. In the Bohemian sources, the anonymous participant of Knín's quodlibet borrows the distinction from Eucles (Anonymus, *Utrum anima intellectiva sit forma inhaerens corpori humano de potentia materiae per generationem educta*, a. 1, not. 3, MS Prague, Národní knihovna X.H.18, f. 40r), the anonymous *De anima* commentary speaks about *forma adhaerens*, a terminology which seems idiosyncratic for Bohemian sources; see Anonymus, *Quaestiones in libros II–III De anima Aristotelis*, q. 26 (MS Prague, Národní knihovna X.H.17, f. 176v) and the passage from Knín's quodlibet quoted in note 65 below.

ing in *S*. Since the human body "lives rationally" (*vivere intellective et racionaliter*), it must be so due to an intellective form inhering in the body.[39]

However, the emphasis on inherence creates room for unwanted consequences, balancing on the edge of materialism. How do forms inhere in matter? Medieval scholars distinguished among three opinions on this issue. First, Platonists embrace the opinion that forms are impressed into matter by an external supernatural agent, so-called form giver (*dator formarum*). The second opinion (attributed to Anaxagoras) suggests that all forms are already present in matter, even in actuality; however, they coalesce and lurk in the matter, insensible to us. The third opinion is Aristotelian, according to which forms are in matter merely potentially and are educed or derived from potentiality to actuality (*educuntur de potentia in actum*) by an external mover.[40]

If the question of the origin of the intellect is posed and universal applicability is considered a virtue of the theory, then an Aristotelian-minded thinker should accept the third opinion and account for the intellect as a form educed from the potentiality of the body. Both Matthias of Knín and Prokop of Kladruby present similar articulations of the materialist intuition while asserting that any form of a body must be caused by this body.[41] Such an intuition is corroborated by an argument targeted at the Platonizing notion of the intellect as a form infused into the body by an external principle. Without the intellect educed from matter, univocal generation would not be possible: if the intellect (i.e., the substantial form) were given to the offspring not

[39] Matthias de Knin, *Quodlibet*, q. 91: *Utrum anima intellectiva sit forma inhaerens corpori humano de potentia materiae per generationem educta*, pro 2–5 (MS Prague, Knihovna metropolitní kapituly L.45, f. 100v). Note that the notion of inherence, borrowed likely from Buridanian *De anima* commentaries, plays a significant role in the questions by Legnica, Knín, and the anonymous Knín's respondent. The older quodlibetal question, despite asking whether the intellect is the human substantial form "inhaesive" and arguing extensively against Averroism, does not employ the inherence terminology and speaks about the intellect *informing* the subject; see Anonymus, *Quodlibet* (?), q. 17: *Utrum sit evidenter probabile naturali ratione quod anima intellectiva sit forma substantialis hominis inhaesive*, pro 1–7 (MS Cracow, BJ 736, ff. 118r–v). For the distinction devised by Pierre d'Ailly between *inhering* in matter, pertaining to material forms and always entailing materialist consequences, and *informing* the body, which is typical of immaterial forms, see Klein (2019, 167, 219).

[40] I follow a neat summary by Albertus de Orlamunda, *Philosophia pauperum* I.4 (ed. Borgnet, 1890, 449b). See also Richardson (2011). Note that Johannes Wyclif, *De compositione hominis*, cap. 6 (ed. Beer, 1884, 101–102) is prone to the Aristotelian position, against those who postulate *latentia formarum* or *datores formarum*.

[41] Matthias de Knin, *Quodlibet*, q. 91: *Utrum anima intellectiva sit forma inhaerens corpori humano de potentia materiae per generationem educta*, pro 6 (MS Prague, Knihovna metropolitní kapituly L.45, ff. 100v–101r): "[…] 'corpus animari intellective substantialiter' non potest esse nisi causetur subiective a corpore"; Procopius de Cladrub, *Quodlibet*, q. 26: *Utrum anima intellectiva habens operationem propriam, quam non communicat corpori, sit perpetua et immortalis*, contra quaes. 6 (MS Prague, Knihovna metropolitní kapituly L.27, f. 42r): "[…] anima intellectiva <forma> corporis non potest esse nisi causetur a corpore, ergo anima intellectiva est educta de potentia materiae."

by the parent but by an external principle, the parent and the offspring would not be of the same species.[42]

One finds even bolder materialist claims in the Prague questions. Matthias of Knín and Prokop of Kladruby employ variants of the argument implying that nothing supernatural takes place in the production of the intellect. Built upon the eduction model, the argument assumes that the nobler matter is, and the more harmonious complexion it has, the nobler and ontologically superior the form educed from it is. However, humans are endowed with the noblest matter among all animals, with complexion balanced in the most perfect way. Consequently, a soul superior to the sensory one must originate from such well-balanced matter, which is nothing other than the intellect itself.[43]

In the Buridanian tradition, such materialist tendencies and the terminology of the intellect educed from matter are usually connected with the late ancient Aristotelian scholar Alexander of Aphrodisias.[44] Defining Alexander's position, Buridan asserts that the intellect is a material form, which is extended and educed from matter, and its subtlety results from "the noble complexion of the human body."[45] The Bohemian scholars seem to build upon this sketchy exposition of Alexander's position and supplement it with notions from the medical tradition and

---

[42] Many questions repeat the argument, see Anonymus, *Quodlibet*, q. 11: *Utrum anima hominis sit extensa*, pro 2 (MSS Cracow, BJ 649, f. 34v; BJ 624, f. 77v); Anonymus, *Quodlibet* (?), q. 16: *Utrum animam humanam esse indivisibilem et immortalem sive incorruptibilem possit ostendi naturali ratione*, contra 4 (MS Cracow, BJ 736, f. 116v); Matthias de Knin, *Quodlibet*, q. 91: *Utrum anima intellectiva sit forma inhaerens corpori humano de potentia materiae per generationem educta*, pro 8 (MS Prague, Knihovna metropolitní kapituly L.45, f. 101r); see also Matthias de Knin, *Quodlibet*, q. 40: *Utrum quaelibet anima sit forma indivisibilis quoad molem*, contra 1 (MS Prague, Knihovna metropolitní kapituly L.45, f. 47v). All passages (more or less explicitly) refer to the Aristotelian saying that "man is begotten by man and by the sun." See Aristoteles, *Physica*, II.2, 194b13–14 (ed. Leonina, 1884, 64); cf. Johannes de Fonte, *Auctoritates Aristotelis*, II.65 (ed. Hamesse, 1974, 145): "Homo generat hominem et sol."

[43] Matthias de Knin, *Quodlibet*, q. 91: *Utrum anima intellectiva sit forma inhaerens corpori humano de potentia materiae per generationem educta*, pro 9 (MS Prague, Knihovna metropolitní kapituly L.45, f. 101r); Procopius de Cladrub, *Quodlibet*, q. 26: *Utrum anima intellectiva habens operationem propriam, quam non communicat corpori, sit perpetua et immortalis*, contra quaes. 5 (MS Prague, Knihovna metropolitní kapituly L.27, f. 42r); see also Matthias de Knin, *Quodlibet*, q. 40: *Utrum quaelibet anima sit forma indivisibilis quoad molem*, contra 1 (MS Prague, Knihovna metropolitní kapituly L.45, f. 47v). For the claim that the *sensitive* soul arises from the harmonious mixture of elements and Geoffrey of Aspall's criticism, see Dunne, Chap. 9 in this volume, pp. 163–190.

[44] See Pluta (1994, 2001).

[45] Notice that Buridan speaks about "nobilitas complexionis corporis humani" only in the penultimate redaction of his questions; see Johannes Buridanus, *Quaestiones super De anima (non de ultima lectura)*, III.3 (MS Uppsala, Universitetsbiblioteket, cod. C. 624, f. 193ra–b). The Uppsala manuscript was copied in Prague in 1375. The "noble complexion" is missing in the ultimate redaction (cf. Johannes Buridanus, *Quaestiones super De anima secundum tertiam sive ultimam lecturam*, III.3, §10, ed. Hartman et al., forthcoming).

The anonymous Prague questions on *De anima* also mentions "harmonious complexion" as the source of the intellect: "Alexander namque ponit intellectum esse formam eductam de potentia <materiae> ad actum, quae resultat de complexione harmonica." Anonymus, *Quaestiones in libros II–III De anima Aristotelis*, q. 25 (MS Prague, Národní knihovna X.H.17, f. 176v).

John Wyclif. One version of Knín's argument justifies the claim that the matter of the human body is so noble and harmonious that it is capable of bringing forth the intellect simply by the authority of Alexander.[46] In another instance, he invokes Avicenna's assertion from *Canon* that God himself gave to man the most tempered complexion.[47] Interestingly, Prokop of Kladruby justifies the same claim by a passage borrowed from John Wyclif's *De compositione hominis*, who proposes that human souls are forms educed from matter.[48] In all three instances, the form educed from a human body is identified with the intellective soul.

Both lines of thought corroborating the materialist view of the human soul (the impossibility of univocal generation and the pointer to the most refined nature of human body) are not frequent in late medieval *De anima* commentaries. However, they resemble two articles from the notorious list of theses condemned in Paris in 1277 by Bishop Stephan Tempier.[49]

In light of the fact that many medieval scholars understood Tempier's list as condemning the articles included,[50] the late medieval Prague Faculty of Arts seems to

---

[46] Matthias de Knin, *Quodlibet*, q. 40: *Utrum quaelibet anima sit forma indivisibilis quoad molem*, contra 1 (MS Prague, Knihovna metropolitní kapituly L.45, f. 47v): "[...] probatur tertio *De anima* auctoritate Alexandri."

[47] Matthias de Knin, *Quodlibet*, q. 91: *Utrum anima intellectiva sit forma inhaerens corpori humano de potentia materiae per generationem educta*, pro 9 (MS Prague, Knihovna metropolitní kapituly L.45, f. 101r): "[...] homo inter omnia animalia habet nobiliorem complexionem membrorum, propter quod Avicenna, capitulo *De complexionibus membrorum* dicit quod deus donavit homini complexionem temperatiorem quam in hoc mundo esse est possibile."

[48] Procopius de Cladrub, *Quodlibet*, q. 26: *Utrum anima intellectiva habens operationem propriam, quam non communicat corpori, sit perpetua et immortalis*, contra quaes. 5 (MS Prague, Knihovna metropolitní kapituly L.27, f. 42r): "Omnis anima intellectiva est forma educta de potentia materiae [...] <quod probatur> sic: 'Proportionaliter, ut materia est in commixtione vel complexione perfectior, sufficit nobiliorem ex se principiare et educere formam. Sed constat quod corpus humanum est plus reductum ad temperamentum quam quodcumque corpus sublunare; ergo multo perfectius sufficit formam superadditam principiare.' Si ergo quodcumque corpus bruti ratione suae complexionis educit et principiat animam sensitivam, sequitur quod corpus humanum nobilius complexionatum animam principiabit et educet non viliorem; quae erit intellectiva, dummodo nulla mediat inter illas." The portion of the text designated by inverted commas ('...') is verbatim borrowed from Johannes Wyclif, *De compositione hominis*, cap. 4 (ed. Beer, 1884, 59).

[49] The first argument (for the eduction model of the production of the intellect, without which allegedly univocal generation is not possible) corresponds to article 120: "Quod forma hominis non est ab extrinseco, sed educitur de potentia materiae, quia aliter non esset generatio univoca." (See Hissette, 1977, 195–196.) The second argument (that from the most harmonious matter the noblest soul, i.e., the intellect, arises) resembles article 133, according to which the soul is inseparable and perishes with the disintegration of bodily harmony: "Quod anima est inseparabilis a corpore; et quod ad corruptionem harmoniae corporis corrumpitur anima." (See Hissette, 1977, 212.) Note that Roland Hissette interprets both articles as targeted not against materialism, but against Averroism.

[50] For example, Peter of Abano was allegedly accused of proposing the materialist view on the production of the intellect, on which see Hasse (2001, 635–636). However, see Bianchi (1998, 93–96) for the claim that Tempier's list was probably not intended as a doctrinal *condemnation*, but rather as a "mere" *prohibition* to teach certain theses in order to protect less capable students from error.

have been much more benevolent to discuss the arguments on behalf of heterodox positions. As these arguments occur in a purely dialectical context (in a preparation for disputation, not in a position actually propounded), they might even have been intended as disguised references to the condemned doctrines. In such case the respondent might have been expected to recognize these hints and take special care in dealing with the intricacies of the heterodox doctrines.

The materialist consequences of the hylomorphic framework, which are in an implicit tension with the Christian doctrine of personal immortality, emerge not only in the context of commenting on Aristotle, but also among philosophically trained theologians, namely amidst the disputes regarding unicity vs plurality of forms. The preparation from one of the 1390s quodlibets echoes a particular chapter in the story of these debates, testifying that purely theological literature also gained access to the handbooks of Prague *quodlibetarii*.[51] The anonymous author of asks "whether the human soul is extended," addressing a tension arising from the claim that there is only one substantial form in a human being. If the intellect is the only form of a human, then it should also adopt properties of the lower sensory soul which is, as a material form, educed from the potency of matter and materially extended. However, if the intellect is understood as immaterial, the sensory part of the soul evinces properties incompatible with the properties of the intellect, and, thus, the former cannot be identified with or subsumed within the latter.

The set of arguments for the affirmative answer in this preparation begins with the assumption that the human soul is a form educed from the potency of matter (which is taken for granted), which naturally implies that the soul is extended.[52] However, the most interesting justification of the same claim is a thought experiment about a handicapped man's new hand. Imagine a man deprived of his hand which is, however, miraculously re-created by God. How will the newly acquired hand be animated? A newly created soul will not inform the hand since the man would consequently have two souls, nor does the soul informing the body before the miracle move to the new hand from other parts of the body, leaving them inanimated. Thus, the most plausible option is that the soul *extends* itself into this new part of the body. This picture, however, testifies that the human soul is actually extended.[53] To the objection that it is rather the sensory soul that informs the limbs,

[51] Note that almost all Prague *quodlibetarii* were students of theology while organizing quodlibets at the Faculty of Arts and preparing their handbooks.

[52] Anonymus, *Quodlibet*, q. 11: *Utrum anima hominis sit extensa*, pro 1 (MSS Cracow, BJ 649, ff. 34r–v; BJ 624, f. 77r).

[53] Anonymus, *Quodlibet*, q. 11: *Utrum anima hominis sit extensa*, pro 3 (MSS Cracow, BJ 624, f. 77v; cf. BJ 649, f. 34v): "Capiatur mancus et tribuat sibi deus miraculose manum, tunc quaeritur vel huiusmodi manus animabitur seu informabitur per animam de novo creatam, aut anima praeexistens movebitur ad istam partem [...] dereliquens alias partes quas prius informavit, aut anima praeexistens et manens in partibus istis quas prius informavit extendet se ad istam partem de novo creatam per sufficientem divisionem. Non primum, quia tunc unus homo haberet duas animas realiter distinctas [...]; nec secundum, quia tunc aliae partes corporis [...] fierent inanimatae [...]. Igitur relinquitur tertium et per consequens sequitur intentum, quod ipsa <anima hominis> est vere extensa."

endowing them, for example, with the sense of touch, it may be replied that if unity of forms is taken for granted, it must be ultimately the intellect itself which informs all parts of the body, providing them, among others, with bodily functions.[54]

The Prague master did not design the thought experiment himself; most likely, his ultimate source was the early fourteenth-century Carmelite theologian Gerard of Bologna. Gerard endorsed unitarianism and investigated whether or not this position results in the claim that the intellect, as the sole form of the body, is extended. He gathered several arguments for the claim without taking a determinate position.[55] The anonymous Prague *quodlibetarius*, however, does not seem to have drawn directly on Gerard, but rather on the Augustinian theologian Thomas of Strasbourg, whose commentary on the *Sentences* (written in the 1330s in Paris and later influential at Prague University) includes a version of the argument with the newly-created hand.[56] Altogether, many arguments included in Prague disputation-connected texts imply a surprisingly materialist view of human nature. This view results from various tenets and influences, from the basic principles of Aristotelian hylomorphic physics, the legacy of Alexander of Aphrodisias's provocative view, to the unitarian position proposing that the human being is informed by only one substantial form.

Another crucial tenet emerging in the Prague disputations is the personal immortality of humans. The recurrent claim is that the intellect not only inheres in matter (which accords with its individuality, against Averroism), but at the same time is understood as an incorporeal entity, potentially separable from the body and, thus, immortal (against Alexandrism). Two anonymous *quodlibetarii* argue for the incorporeal nature of the intellect from a Neoplatonic perspective. The intellect evinces a unique property: it is a form capable of reverting upon itself (*super se ipsam conversiva*). Such a property, however, is an unmistakable sign of incorporeal nature, as substantiated by references to Proclus's *Elements of Theology*.[57]

[54] A unitarian response to this threat can be that strictly speaking, *no* form as such is extended. See Klima (2017, 56–57) for this point in Aquinas. A similar argument is also included by the Prague master, see Anonymus, *Quodlibet*, q. 11: *Utrum anima hominis sit extensa*, contra 8 (MSS Cracow, BJ 649, f. 35v; BJ 624, f. 78v).

[55] See Nolan (2013, 37–38).

[56] Although Gerard's *Quodlibets* are extant in three manuscripts entirely and in other five manuscripts fragmentarily, no manuscript is of Central European origin (see Schabel, 2007, 505–514). On the contrary, the dissemination of Thomas of Strasbourg's commentary was apparently wider. A copy of Book I, originating from Prague around 1400, is preserved in MS Prague, Národní knihovna III.B.8, ff. 1ra–190rb. Besides, one of the most eminent pre-Hussite Prague scholars, Conrad of Soltau, was heavily inspired by Strasbourg's commentary while lecturing on the *Sentences* in Prague around 1380, on which see Schabel et al. (2015, 30–33); see also Lička (2021, 258–259) for a further piece of evidence. For Strasbourg's rendering of Gerard's scenario with the newly created hand, see Thomas de Argentina, *Commentaria in IV libros Sententiarum*, I.8.2.1 (Genova 1585, ff. 53vb–54ra; see also MS Prague, Národní knihovna III.B.8, f. 75rb).

[57] Anonymus, *Quodlibet*, q. 11: *Utrum anima hominis sit extensa*, in oppositum (MSS Cracow, BJ 649, f. 34r; BJ 624, ff. 76v–77r); Anonymus, *Quodlibet* (?), q. 16: *Utrum animam humanam esse indivisibilem et immortalem sive incorruptibilem possit ostendi naturali ratione*, in oppositum (MS Cracow, BJ 736, f. 116r). The arguments do not concord literally; the version in the latter source is, however, almost literally borrowed from Thomas of Strasbourg's *Sentences* commentary,

Needless to say, the Aristotelian tradition also emphasizes the unique nature of the intellect. The Prague authors were well aware of the fact that the overwhelming majority of Aristotelians, especially among the Christian thinkers, did not understand the intellect as an ordinary material form, which comes to being by eduction from matter. Instead, they tend to view the intellect as an immaterial form infused into the body by a divine principle. Matthias of Knín introduces an argument asserting that it is precisely the unique immaterial and incorporeal nature of the intellect that makes its eduction from matter impossible. The intellect is created "from above" by the First Cause.[58] After all, Aristotle himself explicitly advocates this conception of the human intellect as something divine in an embryological passage of his *De generatione animalium*: whereas the vegetative and the sensitive souls are gradually derived from the matter provided by the mother and father's semen, the intellect is infused into the foetus from the outside.[59]

If an act of creation (and not simple generation) produces the individual human intellect, it does not perish like an ordinary material form (due to the disintegration of the hylomorphic composite), but is immortal, annihilable only by God himself. The Prague sources sometimes justify the immortality of the intellect by a simple reference to the Christian faith.[60] Prokop of Kladruby provides further reasoning for

---

even from the same article from which the author of the former source takes Gerard of Bologna's arguments. See Thomas de Argentina, *Commentaria in IV libros Sententiarum*, I.8.2.1 (Genova 1585, f. 53va–b); see also MS Prague, Národní knihovna III.B.8, f. 75ra. Both anonymous *quodlibetarii* evidently drew on Strasbourg's commentary (or an excerpt from it). All three sources incorporate explicit references to propositions 15 and 146 of Proclus's *Elements of Theology*. For the first reference, see Proclus, *Elementatio theologica* (ed. Boese, 1987, 11), prop. 15: "Omne quod ad se ipsum conversivum est incorporeum est." The second reference is, however, rather to prop. 186; see ibidem (ed. Boese, 1987, 91): "Omnis anima est incorporea substantia et separabilis a corpore."

58 Matthias de Knin, *Quodlibet*, q. 91: *Utrum anima intellectiva sit forma inhaerens corpori humano de potentia materiae per generationem educta*, contra 4–5 (MS Prague, Knihovna metropolitní kapituly L.45, f. 101r): "Cum ergo anima intellectiva habeat in se esse immateriale et spirituale et incorporeum, sequitur, quod ipsa non potest educi de potentia materiae, sed desuper causari a prima substantia intellectiva."

59 See Aristoteles, *De generatione animalium*, II.3, 736b27–28 (ed. Drossaart Lulofs, 1966, 54); also Johannes de Fonte, *Auctoritates Aristotelis*, IX.13.190 (ed. Hamesse, 1974, 224): "Solus intellectus est in nobis ab extrinseco, quia ipse solus est divinus." In Prague context, the passage is advocated in Anonymus, *Quodlibet*, q. 11: *Utrum anima hominis sit extensa*, contra 1 (MSS Cracow, BJ 649, f. 35r; BJ 624, f. 78r); Anonymus, *Quodlibet* (?), q. 16: *Utrum animam humanam esse indivisibilem et immortalem sive incorruptibilem possit ostendi naturali ratione*, pro 4 (MS Cracow, BJ 736, f. 117r). Medieval thinkers built on the passage already since the thirteenth century; see Dales (1995, 10, 29, 36–37, 187–188).

60 Anonymus, *Quodlibet*, q. 11: *Utrum anima hominis sit extensa*, contra 1 (MSS Cracow, BJ 624, f. 78r; BJ 649, f. 35r): "Si <anima intellectiva> esset extensa, tunc esset educta de potentia materiae, quod est falsum. Falsitas probatur, quia fide credimus quod sit creata ex nihilo." See also the same move in the positions: Matthias de Legnitz, *Quodlibet*, q. 52: *Utrum animam intellectivam esse formam hominis possit ratione evidenti ostendi*, concl. 4 (MS Stralsund, Stadtarchiv 1067, f. 253rb); Jacobellus de Misa, *Utrum intellectus hominis est homo*, sup. 3 (MS Prague, Národní knihovna V.H.13, f. 140r).

personal immortality from an ethical perspective, borrowing a series of arguments for immortality from *Summa de anima* by the early thirteenth-century Franciscan John of La Rochelle.[61] Prokop points out that the post-mortal life of all souls is necessary if rewards and punishments are to be redistributed among good and bad people according to justice.[62]

Various conceptual tensions apparent in medieval conceptualizations of human nature become explicit and burning in the Prague disputation texts. On the one hand, the emphasis on the intellect's inherence in the human body tempted medieval intellectuals to treat the intellect as an ordinary material form—and the Prague masters were willing to develop arguments pushing the Aristotelian principles to the materialist consequences. On the other hand, the emphasis on the unique immaterial and immortal nature of the intellect was ubiquitous. The dialectic nature of the preparations considered in this section demanded sharpening the reasoning but provided no room for harmonizing these tensions. The next section reconstructs the possible doctrinal positions held by Prague intellectuals.

14.5 Doctrinal Sources and Positions in Prague Disputations

Amidst the lively Prague debates on the human intellect, four different strategies to overcome the alleged incoherence between the intellect's inherence and personal immortality crystallized. Two of the options disregard the dilemma and stress only one of its horns. The materialist *Eduction Model* of the human intellect, setting the immortality issue aside and focusing on the inherence claim, takes the intellect to be a form educed from matter. On the contrary, the *Adherence Model* favours the intellect's immortality and postulates a single intellect for all humans, not inhering in each body as its substantial form, but merely assisting, or (to use the terminology common in the Prague sources to describe such an Averroist position) *adhering to* humans in intellectual operations. Third, the *Inherence Model*, the most popular one in late medieval philosophy and also adopted by two of the Prague questions,

[61] Procopius de Cladrub, *Quodlibet*, q. 26: *Utrum anima intellectiva habens operationem propriam, quam non communicat corpori, sit perpetua et immortalis*, pro quaes. 1–12 (MS Prague, Knihovna metropolitní kapituly L.27, ff. 42v–43r). All twelve of Prokop's arguments are found in Johannes de Rupella, *Summa de anima*, I.VII.44 (ed. Bougerol, 1995, 134–139). La Rochelle's *Summa* gained a certain popularity among Bohemian intellectuals, as evinced by two early fifteenth-century copies (MS Prague, Knihovna metropolitní kapituly L.36, ff. 35ra–94ra, around 1415; and MS Prague, Národní knihovna V.G.10, ff. 126r–137r, incomplete, around 1400). Several glosses in Old Czech in both codices indicate Bohemian origin.

[62] Procopius de Cladrub, *Quodlibet*, q. 26: *Utrum anima intellectiva habens operationem propriam, quam non communicat corpori, sit perpetua et immortalis*, pro quaes. 1 (MS Prague, Knihovna metropolitní kapituly L.27, f. 42v). See Anonymus, *Quodlibet* (?), q. 17: *Utrum sit evidenter probabile naturali ratione quod anima intellectiva sit forma substantialis hominis inhaesive*, pro 5 (MS Cracow, BJ 736, f. 118v), which uses the same point against Averroism as failing to secure *personal* immortality.

proposes *both* the intellect's inherence *and* personal immortality. Besides these three positions, notoriously epitomized as "the three famous opinions" in the Buridanian tradition,[63] the *Hypostatic Union Model* is idiosyncratic for the Prague intellectual environment. Derived from John Wyclif, it tries to preserve both inherence and immortality by postulating an immortal and created spirit hypostatically united with the human soul educed from the potency of matter.

Hardly any scholars at medieval universities embraced the Alexandrist Eduction Model, and only a few supported the Averroist Adherence Model (and if so then indirectly). The Prague scholars around 1400 were no exception in this matter. A modern interpretation is at pains to ascribe an Alexandrist leaning to Jenek of Prague on the basis of his *De anima* commentary; nevertheless, Jenek does not subscribe to the crucial claim of all medieval definitions of Alexander's position: that the intellect is derived from the harmonious complexion of the human body.[64]

Nor had the Adherence Model found a devoted advocate among Prague masters. Nevertheless, compared to Alexandrism, Averroist influences are more palpable in the Prague texts. For example, the relation of the intellect to the body, usually illustrated by the metaphor of a helmsman navigating a ship, gets a new expression in Matthias of Knín. An argument against the intellect's inherence in the body compares the body/intellect relation to the relation between a tunic and a body. As the body does not inhere in the tunic, but rather the tunic is attached to the body, so the body is something temporarily attached to an eternal and self-subsistent intellect.[65]

---

[63] The triad of opinions is included in both versions of Buridan's *Questions*, see Johannes Buridanus, *Quaestiones super De anima secundum tertiam sive ultimam lecturam*, III.3, §9–12 (ed. Hartman et al., forthcoming); Buridanus, *Quaestiones super De anima (non de ultima lectura)*, III.3 (MS Uppsala, Universitetsbiblioteket, cod. C. 624, ff. 193ra–va). See Pluta (1986, 38–41), Zupko (2004, 44–49), De Boer (2013, 288–289), Klein (2019, 124–168). The triad became extremely popular in late-medieval *De anima* commentaries, see the extensive list in Andrews (2016, 241–245).

[64] See Mráz (1982, 86–88). According to the passage from Jenek's commentary edited in Mráz (1982, 90–91) (cf. Jenko Wenceslai de Praga, *Commentarius in I–III libros De Anima Aristotelis*, MS Prague, Národní knihovna VIII.G.30, f. 102r), the agent intellect is identified with God and, thus, immortal, perpetual, and separated. On the contrary, the human intellect is passive, corruptible, and unable to perform intellection without the divine agent intellect. Admittedly, this accords with Alexander's position so far. Nevertheless, Jenek also explicitly asserts that the human soul is "from the outside" ("anima intellectiva [...] non est de potentia materiae educta, sed de foris data"). Surprisingly, he adds that it is not (philosophical to say it is) created, nor is it attached to the human in the Averroist sense, but, in his view, produced by *emanation* from the first principle. Mráz dubiously interprets Jenek as proposing "moderate Alexandrism." Certainly, such a view should not be read in an Averroist spirit, as Jenek explicitly refutes the doctrine of the single intellect as heretical, on which see Mráz (1982, 87).

[65] Matthias de Knin, *Quodlibet*, q. 91: *Utrum anima intellectiva sit forma inhaerens corpori humano de potentia materiae per generationem educta*, contra 2 (MS Prague, Knihovna metropolitní kapituly L.45, f. 101r): "[...] sicut corpus non est forma inhaerens tunicae, sed magis tunica adhaeret corpori, sic videtur, quod substantia intellectiva perpetua et incorporea [...] non est forma inhaerens corpori, sed magis corpus adhaeret illi spiritu intellectivo per se existenti." While the tunic metaphor seems original, the "helmsman in the ship" is an example popular since antiquity, by which medieval philosophers often illustrated the Platonic and Averroist conception of the

The most tangible evidence of Averroist leanings occurs in John of Borotín's *Notabilia* on *De anima*, in a passage implying a version of the plurality thesis: there are two substantial forms of the human; one inhering in and educed from the body, constituting an essential unity with the body; the other one providing an operation to the body, but not inhering in it. The inhering form is the vegetative and sensitive soul (*anima cogitativa* in Averroist terminology); the other form, the intellective soul, is merely "appropriate to" the body. Borotín probably borrowed the doctrine from the early fourteenth-century Averroist John of Jandun's questions on *De anima*, which is particularly noteworthy as no extant manuscript of Jandun's text is of Bohemian origin.[66]

Two texts of the present corpus advocate the Inherence Model. One is a question from the codex with materials from Matthias of Legnica's quodlibet, asking whether one may rationally demonstrate that the intellective soul is the form of the human.[67] The position comprises a set of conclusions and corollaries supported by a shallow and clumsy reasoning, defending, step by step, that the intellective soul is a substance, a form, a substantial form, the substantial form of a human being, and,

---

soul as the act of the body, but not in the sense of form but as mover (*motor*). See Aristoteles, *De anima*, II.1, 413a4–10 (ed. Gauthier, 1984, 74b), Averroes, *Commentarium magnum in Aristotelis De anima libros*, II.11 (ed. Crawford, 1953, 147–148), Thomas de Aquino, *Sententia libri De anima*, II.2 (ed. Gauthier, 1984, 76b), Johannes Buridanus, *Quaestiones super De anima secundum tertiam sive ultimam lecturam*, II.1, §23 (ed. Hartman et al., forthcoming).

[66] Johannes de Borotin (?), *Notabilia super librum III De anima Aristotelis* (MS Prague, Národní knihovna X.H.18, f. 56v): "Nota: duplex est forma substantialis, una inhaerens ipsi corpori et dans esse corpori formaliter et facit unum essentialiter cum corpore; alia est forma substantialis dans (dans] *scripsi,* differunt *cod.*) esse et operari corpori, ita quod non est inhaerens corpori, nec perficiens corpus. Et differunt illae formae, nam prima forma est educta de potentia materiae, secunda autem non. [...] Igitur: unius <hominis> erunt plures formae substantiales, videlicet cogitativa et anima intellectiva. Dico quod illud non est inconveniens sed <secundum> intentionem Aristotelis, ita quod una sit inhaerens ipsi, per quam sit generabilis et corruptibilis, et alia, quae sit ipsi appropriata. Et illud fuit de intentione Commentatoris." See Johannes de Janduno, *Super libros Aristotelis De anima subtilissimae quaestiones*, III.12 (Venice, 1552, f. 71ra). On the cogitative and the intellective soul as two substantial forms of a human, see Mahoney (1987, 274–276) and Brenet (2008). John of Borotín possibly had also Jandun's questions on *De sensu* at his disposal, on which, see Lička (2021, 276).

Another piece of evidence for the dissemination of Jandun's commentary among Prague intellectuals is Anonymus, *Utrum intellectus qui est potentia animae rationalis sit potentia organica* (MS Prague, Knihovna metropolitní kapituly M.92, ff. 147v–148r). The question, asking whether the intellect is an organic power, amounts to a compilation of arguments from the question in Jandun bearing a similar title; see Johannes de Janduno, *Super libros Aristotelis De anima subtilissimae quaestiones* III.8 (Venice, 1552, ff. 66ra–67va).

[67] It is not certain who authored the position, whether the *quodlibetarius* himself, who designed it as a storage of arguments for the upcoming disputation, or a master participating in Legnica's quodlibet. (The former option seems less likely, as the structure of the question does not resemble the typical structure of Prague disputation preparations.) A broader question to be determined by future research is whether the Stralsund codex preserves Legnica's handbook of preparations or proceedings from the quodlibet consisting of positions.

finally, that evident reasoning can demonstrate the last claim.[68] Legnica's justification of the Inherence Model is a little simple-minded. Any inhering form makes a subject such-and-such; thus, just as a form of whiteness makes the subject white, a soul inhering in matter makes it animated.[69] Further, humans are *ex definitione* capable of reasoning, which is granted by the intellective soul, i.e., a form inhering in the human body. As this capacity constitutes the essential property of the human being, the form enabling it is a substantial (and not merely an accidental) form.[70] Admittedly, this is only a trivial rendering of the Inherence Model, not conceptualizing its possible shortcomings. The author does not address the issue of the intellect's separability and immortality at all. On the contrary, he emphasizes that the intellect is not separated but *immersed* in the matter.[71]

A more refined version of the Inherence Model, indebted to the Buridanian approach to the issue, is found in the position of an anonymous participant in Matthias of Knín's quodlibet in 1409. Already the title of the question intimates Buridanian framework, as it asks whether the intellective soul is both a form *inhering* in the body and *educed* from the potentiality of matter. The Buridanian influence is not surprising given the immense popularity John Buridan's works (and his questions on *De anima* in particular) enjoyed at the newly-founded Central European universities.[72] Prague University, of course, was no exception.[73]

John Buridan delineates three possible positions in the issue of the nature of the intellect —Alexandrism, Averroism, and the Catholic faith. The first two, being intrinsically coherent but mutually exclusive, are both unacceptable for a medieval Christian, as Alexandrism cannot secure personal immortality and Averroism the intellect's individuality. In the second move, Buridan highlights the epistemic aspect of the issue. In his view, if one sets aside every assumption provided by the faith and

[68] Matthias de Legnitz, *Quodlibet*, q. 52: *Utrum animam intellectivam esse formam hominis possit ratione evidenti ostendi*, concl. 1–5 (MS Stralsund, Stadtarchiv 1067, ff. 253ra–va).

[69] Matthias de Legnitz, *Quodlibet*, q. 52: *Utrum animam intellectivam esse formam hominis possit ratione evidenti ostendi*, concl. 4, cor. 4 (MS Stralsund, Stadtarchiv 1067, f. 253rb).

[70] Matthias de Legnitz, *Quodlibet*, q. 52: *Utrum animam intellectivam esse formam hominis possit ratione evidenti ostendi*, concl. 4 (MS Stralsund, Stadtarchiv 1067, f. 253rb).

[71] Matthias de Legnitz, *Quodlibet*, q. 52: *Utrum animam intellectivam esse formam hominis possit ratione evidenti ostendi*, concl. 4, cor. 5 (MS Stralsund, Stadtarchiv 1067, f. 253rb): "[...] intellectus humanus non est separatus a materia hominis, immo est immersus ei."

[72] For the dissemination of Buridanism in Central Europe, see Michael (1985, 1: 331–365, and especially 334–340 on Prague University). A sketch of the influence of Buridan's *De anima* on Central European universities (with respect to the immortality issue) is provided by Pluta (1986, 45–48).

[73] Interestingly, preliminary research reveals that Prague scholars used *only* the penultimate redaction of Buridan's questions (called "non ultima lectura" or "secunda lectura" nowadays). Prague even seems to be a centre of dissemination of this redaction in Central Europe (of all the manuscripts of this redaction extant today, 50% were copied in Prague and several others may relate to the Bohemian intellectual milieu). On various versions of Buridan's commentary on *De anima*, see Michael (1985, 2: 677–735); for a census of manuscripts of the penultimate redaction and the third or ultimate redaction ("tertia sive ultima lectura") of Buridan's *Questions on De anima*, see Michael (1985, 2: 684–689, 693–704).

follows natural reason alone, then the Alexandrist materialist position is preferable; “a pagan philosopher would hold Alexander’s opinion.”[74] However, a Christian “should hold firmly” (*firmiter tenendum est*) the third position, the Catholic faith, which in Buridan’s account amounts to a combination of both Alexandrist and Averroist claims, proposing *both* the intellect’s inherence in the body and its immortality. Nevertheless, as these properties are incoherent, the position is not demonstrable by natural reason alone, unaided by faith.[75]

Buridan’s framing effectively set a new paradigm of how to deal with the issue of the nature of the intellect.[76] The anonymous participant of Knín’s quodlibet draws on this Buridanian lore, his immediate source being the commentary on *De anima* attributed to an obscure master called John Eucles. The commentary, once ascribed to Albert of Saxony, is dated 1412 and preserved in a single Leipzig manuscript.[77] The date of Knín’s quodlibet and of the anonymous position, whose author draws on Eucles, entitle us to push the *terminus ante quem* of Eucles’s commentary before 1409 and understand the Leipzig manuscript as a later copy of Eucles’s commentary. The commentary apparently gained a certain popularity in Prague; it is even possible that it was composed there.[78]

[74] Johannes Buridanus, *Quaestiones super De anima secundum tertiam sive ultimam lecturam*, III.4, §13 (ed. Hartman et al., forthcoming): “[…] ego puto quod philosophus paganus teneret opinionem Alexandri […].”

[75] Buridan’s position, as expressed in the ultimate redaction of his questions on *De anima*, is scrutinized in Klein (2019, 123–223); see also Pluta (1986, 38–41), Zupko (2004, 44–50; 2007), Beneduce (2019). The scholarship is not unanimous regarding what Buridan’s position exactly was; the interpretations oscillate on a scale between the picture of Buridan as an undercover materialist (e.g., Olaf Pluta) and a reading framing the nature of the intellect as a dialectical question for Buridan, undecidable by pure reason and to be rendered to theology (Jack Zupko). Klein (2019, 214–223), providing an up-to-date investigation of the issue and the interpretative tradition, reads Buridan as compliant with materialism as a position inferred rationally from evident principles, but acknowledging the Christian faith as another source of epistemic certainty. On the contrary, Beneduce (2019) presents a fresh and textually well-argued interpretation, suggesting that the contradiction between inherence and immortality is not insurmountable for Buridan: both tenets are reconciled once the natural reason is aided by the faith, the only epistemic source capable of providing the *complete* truth on the nature of the intellect. However, as the present paper focuses on how Buridan was understood at newly founded universities (rather than what his genuine position was), it is also worth mentioning that some of his late medieval Central European readers plainly put him in the Alexandrist camp, on which, see Pluta (2007, 151–152). For a similar tension in Thomas Wylton’s account of the intellective soul’s relation to the body, see Trifogli, Chap. 10 in this volume, pp. 191–211.

[76] See De Boer (2013, 288–292). For the Buridanian vein in late medieval approaches to the issue, see, e.g., Pluta (1986, 42–49), Dewender and Pluta (1997) (Lawrence of Lindores), Pluta (2010) (John Eucles and Nicholas of Amsterdam), and Andrews (2016, 241–248) (Bero Magni de Ludosia).

[77] Johannes Eucles, *Quaestiones in Aristotelis libros De anima* (MS Leipzig, Universitätsbibliothek 1416, ff. 141ra–234rb); see Pluta (1988) for the contents of the manuscript, list of questions, and partial edition of four questions on the nature of the intellect.

[78] If before 1409, then the commentary was certainly not composed in Leipzig, as the local university was established in 1409; incidentally, mostly by the German-speaking masters who had withdrawn from the Prague University after the Kutná hora decree was issued. Pluta (2010, 93) also

Regarding the nature of the intellect, the Prague master is faithful to the Buridanian framework, textually following Eucles. Nevertheless, in contrast to Eucles, who distinguishes between the materialist and the orthodox approach to the issue very carefully and investigates the two attitudes in separate articles in each question, thus obscuring his own stance,[79] the Prague master's preference for the non-materialist version of the Inherence Model is more palpable. After the mandatory introduction of the three famous opinions,[80] he states that the intellect is *not* a material and extended form (since it does not require a corporeal organ) and that it inheres in the body (if it did not, what would guarantee that the intellect moved with the body which always has the intellect at its disposal?).[81]

Only after that, he points out that there are two ways to deal with the issue. The philosophical way prevents one from postulating anything not based on previous experience or not justified by an argument stronger than the one for its opposite. However, we have no experience confirming the post-mortal existence of the intellect. Further, arguments based on the principles of natural philosophy imply that the human intellect is mortal, rather than the opposite. Therefore, pure natural reason dictates that the intellect is not everlasting, but a corruptible form educed from the potency of matter and even materially extended.[82]

But if the investigation follows the Christian faith, it reaches the conclusion that the human intellect is, in a way, eternal. Being created by God (and not generated naturally like other forms) it came into existence at a certain point of time, but since then it is everlasting (although still annihilable, at least in principle, by God).[83] It is telling that the author labels this conclusion as constituting the response to the question (*conclusio responsalis*), choosing the approach as his ultimate stance. Also, he does not copy the conclusion from Eucles but elaborates on his own, taking into consideration two passages from Thomas Aquinas (explicitly) and Conrad of Ebrach (implicitly).[84]

implies that Eucles could have been a Prague renegade, suggesting that he may be identical with Johannes de Elbing (a Prague master of arts active there especially in the 1370s, but evidenced in Leipzig in 1409).

[79] Pluta (2010, 93–94, 99).

[80] Anonymus, *Utrum anima intellectiva sit forma inhaerens corpori humano de potentia materiae per generationem educta*, a. 2 (MS Prague, Národní knihovna X.H.18, f. 41r).

[81] Anonymus, *Utrum anima intellectiva sit forma inhaerens corpori humano de potentia materiae per generationem educta*, a. 2, concl. 1–2 (MS Prague, Národní knihovna X.H.18, f. 41r). The arguments are derived from the "faith sections" of Eucles's questions (see Johannes Eucles, *Quaestiones in Aristotelis libros De anima*, III.3.2, III.4.1, ed. Pluta, 1988, 518, 522).

[82] Anonymus, *Utrum anima intellectiva sit forma inhaerens corpori humano de potentia materiae per generationem educta*, a. 3, concl. 2 (MS Prague, Národní knihovna X.H.18, ff. 41v–42r). Again, the reasoning is borrowed from Eucles, but this time from a "pure reason section" of his question on whether the intellect is everlasting (see Johannes Eucles, *Quaestiones in Aristotelis libros De anima*, III.6.1, ed. Pluta, 1988, 532–533).

[83] Anonymus, *Utrum anima intellectiva sit forma inhaerens corpori humano de potentia materiae per generationem educta*, a. 3, concl. resp. (MS Prague, Národní knihovna X.H.18, f. 42r).

[84] More on this in the forthcoming critical edition.

The anonymous author responds to the question whether the intellect inheres in and is educed from matter by accepting the first claim while denying the second. Effectively, it amounts to disconnecting the claim about the intellect's inherence in the body from the claim that it is just an ordinary material form, which seems logically connected. The author does not make much effort to justify the disconnection.[85] Ultimately, he does not seem to develop much the inherited conceptual tension between inherence and immortality, but only (not in a particularly original way) acknowledges that there are two possible approaches, of which one is preferable for extraphilosophical reasons.

Finally, some Prague intellectuals were prone to the Hypostatic Union Model of the human being. The source of this conceptualization of the issue, going beyond the framework of the Inherence Model, is John Wyclif, especially his anthropological work *De compositione hominis*.[86] Wyclif seems a little reluctant to use the terminology of forms and inherence[87]; in his view, the human being principally consists of two different natures: a material body and an immaterial created spirit. The unity of a human being is guaranteed by the fact that both natures constitute a single person, being hypostatically united, as Wyclif asserts using Christological terminology.[88] Into this general framework he incorporates Aristotelian terminology, evidently inclining to the pluralist position: the human soul both inheres in the human body and is educed from the potency of matter, but a spiritual, eternal and individual intellect is joined to the animated body.[89] It is not crystal clear what Wyclif means by the "human soul," which he regards as a material form. Probably only the sensory soul, as he seldom speaks about the intellect as educed from matter. The question can hardly be solved here[90]; however, it is worth mentioning that some Bohemian authors inclined to the latter reading, in which even the intellective soul is educed

---

[85] A possible way to temper the tension is advanced by Jack Zupko. In his interpretation of Buridan, the inference from intellect's inherence to materialism is precluded by distinguishing between two different meanings of the notion of inherence: while material forms inhere in their subjects as whole in whole and part in part, the intellect (as an immaterial form) inheres in the body as whole in whole and whole in part. See Zupko (2004, 55–56), and, more minutely, Zupko (1993); however, cf. Klein (2019, 218–220).

[86] Eleven manuscript copies of *De compositione hominis* are extant today (see Thomson, 1983, 36–37), of which eight are of Bohemian origin (listed in Šmahel, 1980, 15–16).

[87] See Pasnau (2011, 565, note 14).

[88] See, e.g., Johannes Wyclif, *De compositione hominis*, cap. 1 and 6 (ed. Beer, 1884, 9, and 102–103, 107–108, 115), Johannes Wyclif, *Trialogus*, II.5 (ed. Lechler, 1869, 91), Michael (2003, 356–358), and Lahey (2006, 169–177).

[89] See Michael (2003, 347, 354–358) for such a reading.

[90] The paucity of modern interpretations of Wyclif's philosophical anthropology does not help to crack this interpretative conundrum. Wyclif's thought has, admittedly, been far from neglected in the last decades—see the collective volumes edited by Levy (2006) and Campi and Simonetta (2020); editorial work is also being conducted, of which see Campi's recent critical edition of Wyclif's *De scientia Dei* (ed. Campi, 2017) and Thakkar's research on Wyclif's logical treatises (see, e.g., Thakkar, 2020 and Thakkar, forthcoming). These efforts notwithstanding, little (if any) research has been done on Wyclif's *De compositione hominis* in the last decades besides Michael (2003) and Lahey (2006).

from the potency of matter and immortality is provided by an immaterial spirit united to the ensouled body.

The first Prague question, implicitly reworking the Wycliffian lore, is the position presented in a promotional act of an anonymous candidate at the Faculty of Arts. The title of the question is suggestive of Alexandrism, as it asks whether the intellective soul, educed from the potency of matter, is corruptible. Yet, the author of the position actually shows himself to be a keen student of Wyclif's *De compositione hominis*, from which he tacitly borrows in his text.[91] He recasts the Aristotelian setting of the question in Wyclif's light. The crucial step in developing the response is to distinguish carefully between "intellective soul" and "human soul." The intellective soul is "a created spirit."[92] As such, it is not educed from matter and does not cease to be with the corruption of its subject.[93] On the contrary, the human soul is, according to the famous Aristotelian definition, an act of the physical body, which cannot exist *per se* without the body. In Wyclif's words, it is nothing else than the fact that an organic essence is animated (*essenciam organicam animari*).[94] Being a corporeal act, the human soul can cease to exist.[95]

The created spirit and the human body's animating principle (*animatio*), albeit entailing different properties, are the same thing as they express two aspects of a single human being.[96] Although humans are mortal, as for their corporeal life, they are also "personally identified" with their incorporeal, spiritual, and created intellects and in this regard immortal.[97] The anonymous author concludes that the question is false in the first sense (as the intellect is neither educed nor corruptible), but

[91] The date of composition around 1402, suggested above, proves to be plausible, as it seems unlikely that a candidate, responding in front of the University board composed of masters of all University nations, would choose to draw on Wyclif after the 1403 condemnation of some of Wyclif's theses (and before 1409, when the Bohemian Wycliffite party prevailed at the University).

[92] According to the introductory distinction between the uncreated spirit (God) and created spirit (human intellect), which the author ascribes to Platonists but borrows from Johannes Wyclif, *De compositione hominis*, cap. 1 (ed. Beer, 1884, 8–9); see Anonymus, *Utrum anima intellectiva educta de potentia materiae sit corruptibilis*, not. 2 (MS Prague, Národní knihovna VIII.E.5, ff. 55v–56r).

[93] Anonymus, *Utrum anima intellectiva educta de potentia materiae sit corruptibilis*, concl. 1 (MS Prague, Národní knihovna VIII.E.5, f. 56r).

[94] Anonymus, *Utrum anima intellectiva educta de potentia materiae sit corruptibilis*, concl. 2 (MS Prague, Národní knihovna VIII.E.5, ff. 56r–v); see Johannes Wyclif, *De compositione hominis*, cap. 4 (ed. Beer, 1884, 55).

[95] Anonymus, *Utrum anima intellectiva educta de potentia materiae sit corruptibilis*, concl. 2, cor. 2 (MS Prague, Národní knihovna VIII.E.5, f. 56v).

[96] Anonymus, *Utrum anima intellectiva educta de potentia materiae sit corruptibilis*, concl. 2, cor. 1 (MS Prague, Národní knihovna VIII.E.5, f. 56v).

[97] Anonymus, *Utrum anima intellectiva educta de potentia materiae sit corruptibilis*, concl. 3, cor. 3 (MS Prague, Národní knihovna VIII.E.5, f. 56v): "Sicut spiritus sive intellectus humanus dicitur perpetuus <et> incorruptibilis, sic homo, qui identificatur cum eodem (eodem] *scripsi*, eadem *cod.*) personaliter, est perpetuus et incorruptibilis, quamvis non secundum utramque vitam, puta spiritualem et corporalem."

true in the second one (since the human soul as the animating principle of the body is both educed and corruptible).[98]

A more sophisticated rendering of Wyclif's position is provided in the question entitled "Whether the human intellect is [identical with] the human," attributed in its single manuscript copy to the famous Bohemian reformist Jacob of Mies (ca. 1370–1429). The text, composed probably in a theological milieu, almost completely omits Aristotelian hylomorphic terminology. Jacob singles out the intellect as the human "essential intrinsic being" (*esse essentiale intrinsecum*), which makes the human what she is.[99] Consequently, he identifies the human being with her intellect, which is immortal and incorruptible.[100] Simultaneously, perhaps to avoid any suspicion of Averroism, he hints that there are many everlasting intellects, one for each human individual.[101] The human body, on the contrary, is mortal and corruptible, and these properties, consequently, also apply to human individuals.[102] The corporeal and incorporeal substances are of the same subject, and the unity of such a twofold composite is saved by the fact that they constitute a single person.[103]

Jacob, regrettably, does not address the self-ruining shortcomings of such robust dualism, e.g., the issue of mental causation (how the ontological gap between body and intellect is to be overcome, or how the intellect draws on the sensory material). Jacob's aim was not to construct a plausible psychological theory; he strives to sketch a metaphysical picture of the human being allowing one to account not only for ordinary humans, but also for Jesus Christ, the only person who is truly God and truly man. Thus, in Jacob's account, anthropology is continuous with Christology, which again resembles Wyclif. Since *every* human is a "personal union" of two different natures, it is easier to explain how God became man.[104]

---

[98] Anonymus, *Utrum anima intellectiva educta de potentia materiae sit corruptibilis*, cor. ult. (MS Prague, Národní knihovna VIII.E.5, f. 56v).

[99] Jacobellus de Misa, *Utrum intellectus hominis est homo*, sup. 1–2 (MS Prague, Národní knihovna V.H.13, f. 140r).

[100] Jacobellus de Misa, *Utrum intellectus hominis est homo*, concl. 1 (MS Prague, Národní knihovna V.H.13, f. 140r).

[101] Jacobellus de Misa, *Utrum intellectus hominis est homo*, concl. 2 (MS Prague, Národní knihovna V.H.13, f. 140r).

[102] Jacobellus de Misa, *Utrum intellectus hominis est homo*, concl. 3 (MS Prague, Národní knihovna V.H.13, ff. 140r–v).

[103] Jacobellus de Misa, *Utrum intellectus hominis est homo*, concl. 3, cor. 1 (MS Prague, Národní knihovna V.H.13, f. 140v): "In eodem homine substantia corporea et incorporea sunt eiusdem suppositi et personae idem suppositaliter vel personaliter, inter quae est unio suppositalis sive personalis sic quod eadem persona est ambo illa simul et quodlibet divisim."

[104] Jacobellus de Misa, *Utrum intellectus hominis est homo*, concl. 3, cor. 5 (MS Prague, Národní knihovna V.H.13, f. 140v); cf. also Johannes Wyclif, *De compositione hominis*, cap. 6 (ed. Beer, 1884, 115–116) on parallels and differences between humans and Christ.

14.6 Conclusion

The paper has delved into some ephemeral texts resulting from late medieval Prague university disputations. As I tried to demonstrate, these minor texts witness, perhaps even more faithfully than works by the honourable grand authors, how vibrant and dynamic exchanges took place in the late medieval market of ideas.

The texts related to the issue of the intellect reveal tangible conceptual tensions between the far-reaching materialist consequences of some Aristotelian principles and the postulate of human immortality, whether religion-driven or a simple component of human dignity. Some texts devise strategies to overcome these tensions. The Buridanian approach, manoeuvring between the Alexandrist position, which was rationally plausible but untenable for medieval Christians, and the position of faith, which was well tenable but rationally implausible for them, left an unpleasant taste of a double-truth theory.

Idiosyncratic to the early fifteenth-century Prague intellectual milieu is the urge to rethink the issue anew from the perspective of John Wyclif. While admittedly not strong in the psychological analysis, Wyclif presents a peculiar metaphysics of the human being with the materiality-immortality tension built in from the beginning. Wyclif's Prague readers were evidently fascinated by this strategy and applied it also in questions framed by Buridanian (or generally Aristotelian) terminology.

The ultimate epitome of this tendency can be the anonymous *Questions on De anima*. Here, in the question on whether the intellect is the substantial form inhering in the body, the notorious Buridanian triad of opinions is reworked significantly. Alexandrism and Averroism remain, but the third option is not labelled "faith" and defined merely as proposing that the intellect is created and infused to the body. More surprisingly, a *fourth* position is added representing an extravagant reading of Wyclif's theory: the human intellect is educed from matter and thus corruptible, but a spirit is also united to the intellect by hypostatic union, becoming thus a true intellective soul and serving as the substantial form inhering in the body.[105]

Texts such as these invite us to reconsider the prevailing historical narrative about philosophy at the early fifteenth-century Prague University, which allegedly abandoned its position of the "stronghold of Buridanism" and went under the full influence of Wyclif.[106] Perhaps the relation between the two traditions is more entangled. Rather than simply dismissed and replaced by Wyclif, Buridan may have been gradually Wycliffised.[107]

---

[105] Anonymus, *Quaestiones in libros II–III De anima Aristotelis*, q. 25 (MS Prague, Národní knihovna X.H.17, f. 176v): "Quarta <opinio> ponit, quod intellectus humanus [...] educitur de potentia materiae ad actum rei corruptibilis, sed addit, quod spiritus, cui convenit animare corpus et actuare, est etiam intellectus per hypostaticam unionem et sic sit vera anima rationalis intellectiva."

[106] See, e.g., Michael (1985, 1: 338–339).

[107] I am particularly grateful to Luigi Campi and Olaf Pluta who commented upon an earlier version of the paper, and to the editors of the volume for their careful reading and manifold suggestions. The text also benefits from numerous discussions with Ota Pavlíček on the paper (and the issue of Prague University disputations in general), as well as from suggestions made by Pavel Blažek and Miroslav Hanke.

References

Primary Literature (Manuscripts & Editions)

Albertus de Orlamunda (= Pseudo-Albertus Magnus). 1890. Philosophia pauperum. In *B. Alberti Magni Opera omnia 5*, ed. Augustus Borgnet, 444–536. Paris: Vivès.

Anonymus. *Quaestiones in libros II–III De anima Aristotelis*. MS Prague, Národní knihovna X.H.17, ff. 157r–180r.

Anonymus. *Quodlibet*. MSS Cracow, BJ 649, ff. 2v–163v, and Cracow, BJ 624, ff. 1r–216v.

Anonymus. *Quodlibet (?)*. MS Cracow, BJ 736, ff. 81r–199r.

Anonymus. *Utrum anima intellectiva educta de potentia materiae sit corruptibilis*. MS Prague, Národní knihovna VIII.E.5, ff. 55v–56v.

Anonymus. *Utrum anima intellectiva sit forma inhaerens corpori humano de potentia materiae per generationem educta*. MS Prague, Národní knihovna X.H.18, ff. 40r–42r.

Anonymus. *Utrum intellectus qui est potentia animae rationalis sit potentia organica*. MS Prague, Knihovna metropolitní kapituly M.92, ff. 147v–148r.

Aristoteles. 1884. Physica: Recensio Guillelmi de Moerbeka. In Thomas de Aquino, *Commentaria in octo libros Physicorum Aristotelis*, ed. Comissio Leonina, 3–455 *passim*. Roma: Commissio Leonina.

———. 1984. De anima: Recensio Guillelmi de Moerbeka. In Thomas de Aquino, *Sentencia libri De anima*, ed. René-Antoine Gauthier, 3–258 *passim*. Roma: Commissio Leonina; Paris: Vrin.

———. 1966. In *De generatione animalium: Translatio Guillelmi de Moerbeka*, ed. H.J. Drossaart Lulofs, 3–179. Bruges–Paris: Desclée de Brouwer.

Averroes. 1953. *Commentarium magnum in Aristotelis De anima libros*, ed. Stuart Crawford. Cambridge: Medieval Academy of America.

Henricus de Ribenicz. *Quodlibet*. MS Leipzig, Universitätsbibliothek 1414, ff. 1v–231v.

Jacobellus de Misa. *Utrum intellectus hominis est homo*. MS Prague, Národní knihovna V.H.13, ff. 140r–141r.

Jenko Wenceslai de Praga. *Commentarius in I–III libros De Anima Aristotelis*. MS Prague, Národní knihovna VIII.G.30, ff. 66v–109r.

Johannes de Borotin (?). *Notabilia super librum III De anima Aristotelis*. MS Prague, Národní knihovna X.H.18, ff 55r–56v.

Johannes Buridanus. *Quaestiones super De anima (non de ultima lectura)*. MS Uppsala, Universitetsbiblioteket C.624, ff. 159ra–205va.

———. Forthcoming. *Quaestiones super De anima secundum tertiam sive ultimam lecturam*, ed. Peter Hartman, Gyula Klima, Peter Sobol, and Jack Zupko. Cham: Springer.

Johannes Eucles. *Quaestiones in Aristotelis libros De anima*. MS Leipzig, Universitätsbibliothek 1416, ff. 141ra–234rb.

———. 1988. *Quaestiones in Aristotelis libros De anima*, III.3–6. In Pluta, Olaf. 1988.

Johannes de Fonte. 1974. Auctoritates Aristotelis. In *Les Auctoritates Aristotelis. Un florilège médiéval. Étude historique et édition critique*, ed. Jacqueline Hamesse. Louvain: Béatrice–Nauwelaerts; Paris: Publications universitaires.

Johannes de Janduno. 1552. *Super libros Aristotelis De anima subtilissimae quaestiones*. Venice: Hieronymus Scotus.

Johannes de Rupella. 1995. In *Summa de anima*, ed. Jacques Guy Bougerol. Paris: Vrin.

Johannes Wyclif. 1869. In *Trialogus, cum Supplemento Trialogi*, ed. Gotthard Lechler. Oxford: Clarendon.

———. 1884. In *De compositione hominis*, ed. Rudolf Beer. London: Trübner.

———. 2017. In *De scientia Dei*, ed. Luigi Campi. Oxford: Oxford University Press.

Matthias de Knin. *Quodlibet*. MS Prague, Knihovna metropolitní kapituly L.45, ff. 1r–156v.

Matthias de Legnitz. *Quodlibet*. MS Stralsund, Stadtarchiv 1067, ff. 207ra–278rb.

Procopius de Cladrub. *Quodlibet*. MS Prague, Knihovna metropolitní kapituly L.27, ff. 1r–132r.

Proclus. 1987. In *Elementatio theologica, translata a Guillelmo de Morbecca*, ed. Helmut Boese. Leuven: Leuven University Press.
Thomas de Aquino. 1984. *Sentencia libri De anima*, ed. René-Antoine Gauthier. Roma: Commissio Leonina; Paris: Vrin.
Thomas de Argentina. *Commentarius super primum librum Sententiarum*. MS Prague, Národní knihovna III.B.8, ff. 1ra–190rb.
———. 1585. *Commentaria in IV libros Sententiarum*. Genova: Antonius Ortrius.

Secondary Literature

Andrews, Robert. 2016. *Bero Magni de Ludosia, Questions on the Soul: A Medieval Swedish Philosopher on Life*. Stockholm: Centre for Medieval Studies, Stockholm University.
Beneduce, Chiara. 2019. On the Intellect. A Reading of John Buridan's *De anima*, III, qq. 3–6. *Forum: Supplement to Acta. Philosophica* 5: 83–105.
Bianchi, Luca. 1998. 1277: A Turning Point in Medieval Philosophy? In *Was ist Philosophie im Mittelalter?* ed. Jan A. Aertsen and Andreas Speer, 90–110. Berlin–New York: De Gruyter.
Brenet, Jean-Baptiste. 2008. Ame intellective, âme cogitative: Jean de Jandun et la "duplex forma propria" de l'homme. *Vivarium* 46: 318–341.
Campi, Luigi. 2020. Is Perfection of this World? A *Quaestio* on Creatures' Perfection in Terms of Propinquity to or Distance from the First Being, Ascribed to Matěj of Knín. *Bulletin de philosophie médiévale* 62: 213–250.
Campi, Luigi, and Stefano Simonetta, eds. 2020. *Before and After Wyclif: Sources and Textual Influences*. Basel: Fédération Internationale des Instituts d'Études Médiévales.
Dales, Richard C. 1995. *The Problem of the Rational Soul in the Thirteenth Century*. Leiden–New York–Köln: Brill.
De Boer, Sander W. 2013. *The Science of the Soul: The Commentary Tradition on Aristotle's De Anima, c. 1260–c. 1360*. Leuven: Leuven University Press.
De Libera, Alain. 2014. Formes assistantes et formes inhérentes. Sur l'union de l'âme et du corps, du Moyen Âge à l'Âge classique. *Archives d'histoire doctrinale et littéraire du Moyen Âge* 81: 197–248.
Dewender, Thomas, and Olaf Pluta. 1997. Lawrence of Lindores on Immortality. An Edition with Analysis of Four of his "Quaestiones in Aristotelis libros De anima". *Bochumer Philosophisches Jahrbuch für Antike und Mittelalter* 2: 187–242.
Dittrich, Antonín, Antonín Spirk, Josef Jungmann, and Ladislav Jandera, eds. 1830–1832. *Liber decanorum facultatis philosophicae Universitatis Pragensis ab anno Christi 1367 usque ad annum 1585*. 2 vols. Praha: Jerzabek.
Duba, William. 2012. The Souls after Vienne: Franciscan Theologians' Views on the Plurality of Forms and the Plurality of Souls, ca. 1315–1330. In *Psychology and Other Disciplines: A Case of Cross-Disciplinary Interaction (1250–1750)*, eds. Paul J.J.M. Bakker, Sander W. de Boer, and Cees Leijenhorst, 171–272. Leiden–Boston: Brill.
Flüeler, Christoph. 2008. Teaching Ethics at the University of Vienna: The Making of a Commentary at the Faculty of Arts (A Case Study). In *Virtue Ethics in the Middle Ages: Commentaries on Aristotle's Nicomachean Ethics, 1200–1500*, ed. István P. Bejczy, 277–346. Leiden–Boston: Brill.
Hasse, Dag N. 2001. Pietro d'Abano's "Conciliator" and the Theory of the Soul in Paris. In *Nach der Verurteilung von 1277. Philosophie und Theologie an der Universität von Paris im letzten Viertel des 13. Jahrhunderts. Studien und Texte*, ed. Jan A. Aertsen, Kent Emery, and Andreas A. Speer, 635–653. Berlin–New York: Walter de Gruyter.
Herold, Vilém. 1985. *Pražská univerzita a Wyclif*. Praha: Univerzita Karlova.
Hissette, Roland. 1977. *Enquête sur les 219 articles condamnés à Paris le 7 mars 1277*. Louvain: Vander-Oyez; Paris: Publications Universitaire.

Hoenen, Maarten J.F.M. 2011. Nominalism in Cologne: The Student Notebook of the Dominican Servatius Fanckel. In *Crossing Boundaries at Medieval Universities*, ed. Spencer E. Young, 85–144. Leiden–Boston: Brill.

Kavka, František. 1967. Organisace studia na pražské artistické fakultě v době předhusitské. *Acta universitatis Carolinae – Historia universitatis Carolinae Pragensis* 8: 7–39.

Kejř, Jiří. 1955–1956. Quodlibetní questie kodexu UK X E 24. 3 parts. *Listy filologické* 78: 216–221, 79: 63–70, and 228–233.

———. 1960. Z disputací na pražské universitě v době Husově a husitské. *Sborník historický* 7: 47–78.

———. 1971. *Kvodlibetní disputace na pražské univerzitě*. Praha: Universita Karlova.

Klein, Martin. 2019. *Philosophie des Geistes im Spätmittelalter: Intellekt, Materie und Intentionalität bei Johannes Buridan*. Leiden–Boston: Brill.

Klima, Gyula. 2017. The Trivia of Materialism, Dualism and Hylomorphism: Some Pointers from John Buridan and Others. In *Questions on the Soul by John Buridan and Others. A Companion to John Buridan's Philosophy of Mind*, ed. Gyula Klima, 45–62. Cham: Springer.

Knoll, Paul. 2016. *"A Pearl of Powerful Learning": The University of Cracow in the Fifteenth Century*. Leiden–Boston: Brill.

Kowalczyk, Maria, et al. 1980–2012. *Catalogus codicum manuscriptorum medii aevi latinorum qui in Bibliotheca Jagellonica Cracoviae asservantur*. 10 vols. Wrocław–Kraków: Ossolineum.

Lahey, Stephen E. 2006. Wyclif's Trinitarian and Christological Theology. In *A Companion to John Wyclif, Late Medieval Theologian*, ed. Ian Christopher Levy, 127–198. Leiden–Boston: Brill.

Lawn, Brian. 1993. *The Rise and Decline of the Scholastic "Quaestio disputata". With Special Emphasis on its Use in the Teaching of Medicine and Science*. Leiden–New York–Köln: Brill.

Levy, Ian Christopher, ed. 2006. *A Companion to John Wyclif, Late Medieval Theologian*. Leiden–Boston: Brill.

Lička, Lukáš. 2021. Studying and Discussing Optics at the Prague Faculty of Arts: Optical Topics and Authorities in Prague Quodlibets and John of Borotín's *Quaestio* on Extramission. In *Studying the Arts in Late Medieval Bohemia: Production, Reception and Transmission of Knowledge*, ed. Ota Pavlíček, 251–303. Turnhout: Brepols.

Mahoney, Edward P. 1987. Themes and Problems in the Psychology of John of Jandun. In *Studies in Medieval Philosophy*, ed. John F. Wippel, 273–288. Washington, DC: The Catholic University of America Press.

Mansfeld, Monika. 2016. Expositio in "De anima" Aristotelis secundum Johannem Buridanum abbreviata. *An Edition. Cahiers de l'institut du Moyen Âge grec et latin* 85: 87–127.

Michael, Bernd. 1985. *Johannes Buridan: Studien zu seinem Leben, seinen Werken und zur Rezeption seiner Theorien im Europa des späten Mittelalters*. 2 vols. Berlin: Freie Universität.

Michael, Emily. 2003. John Wyclif on Body and Mind. *Journal of the History of Ideas* 64: 343–360.

Mráz, Milan. 1982. Commentarium magistri Johannis Wenceslai de Praga super De anima Aristotelis (Der gegenwärtige Zustand der Forschungsarbeit). *Mediaevalia Philosophica Polonorum* 26: 79–91.

Nolan, Simon. 2013. John Baconthorpe on Soul, Body and Extension. *Maynooth Philosophical Papers* 7: 33–45.

Pasnau, Robert. 2010. Form and Matter. In *The Cambridge History of Medieval Philosophy*, ed. Robert Pasnau and Christina Van Dyke, vol. 2, 635–646. Cambridge: Cambridge University Press.

———. 2011. *Metaphysical Themes, 1274–1671*. Oxford: Clarendon Press.

Pavlíček, Ota. 2018. Parisian and Prague "Quodlibeta" Compared: The Transfer of the Quodlibetal Disputation between the Faculties and Jerome of Prague's Struggle against the Thematic Limitations Imposed on the Faculty of Arts. In *What is New in the New Universities? Learning in Central Europe in the Late Middle Ages (1348–1500)*, ed. Elżbieta Jung, 325–356. Warszawa: Wydawnictwo IFiS PAN.

———. 2021. Notes on the Prague Faculty of Arts in 1348–1419. Facts, Sources, Innovations. In *Studying the Arts in Late Medieval Bohemia: Production, Reception and Transmission of Knowledge*, ed. Ota Pavlíček, 13–26. Turnhout: Brepols.

Pavlíček, Ota, and Miroslav Hanke. 2021. Jerome of Prague's and Blasius Lupus's Debate in the Context of Solving Sophism at the Prague Faculty of Arts in the Late Middle Ages with a Re-Edition of the "Argumenta sophistica". In *Studying the Arts in Late Medieval Bohemia: Production, Reception and Transmission of Knowledge*, ed. Ota Pavlíček, 205–233. Turnhout: Brepols.

Pluta, Olaf. 1986. *Kritiker der Unsterblichkeitsdoktrin in Mittelalter und Renaissance*. Amsterdam: Grüner.

———. 1988. Die Diskussion der Frage nach der Unsterblichkeit in einer Leipziger Handschrift des frühen 15. Jahrhunderts. In *Die Philosophie im 14. und 15. Jahrhundert. In memoriam Konstanty Michalski (1879–1947)*, ed. Olaf Pluta, 495–534. Amsterdam: Grüner.

———. 1994. Averroes als Vermittler der Gedanken des Alexander von Aphrodisias. In *Averroismus im Mittelalter und der Renaissance*, ed. Friedrich Niewöhner and Loris Sturlese, 201–221. Zürich: Spur.

———. 2001. The Transformations of Alexander of Aphrodisias' Interpretation of Aristotle's Theory of the Soul. In *Renaissance Readings of the Corpus Aristotelicum*, ed. Marianne Pade, 147–165. Copenhagen: Museum Tusculanum Press.

———. 2007. How Matter Becomes Mind: Late-Medieval Theories of Emergence. In *Forming the Mind. Essays on the Internal Senses and the Mind/Body Problem from Avicenna to the Medical Enlightenment*, ed. Henrik Lagerlund, 149–167. Dordrecht: Springer.

———. 2010. "Sed hoc non videtur verum in lumine naturali": Natural Philosophy's Struggle for the Truth in the Immortality Debate of the Fifteenth Century. In *Kritische Religionsphilosophie: Eine Gedenkschrift für Friedrich Niewöhner*, ed. Wilhelm Schmidt-Biggemann, Georges Tamer, and Catherine Newmark, 85–105. Berlin–New York: Walter de Gruyter.

Richardson, Kara. 2011. Avicenna and Aquinas on Form and Generation. In *The Arabic, Hebrew and Latin Reception of Avicenna's Metaphysics*, ed. Dag N. Hasse and Amos Bertolacci, 251–274. Berlin–Boston: Walter de Gruyter.

Schabel, Christopher. 2007. Carmelite Quodlibeta. In *Theological Quodlibeta in the Middle Ages. The Fourteenth Century*, ed. Christopher Schabel, 493–543. Leiden–Boston: Brill.

Schabel, Chris, Monica Brinzei, and Mihai Maga. 2015. A Golden Age of Theology at Prague: Prague "Sentences" Commentaries from 1375 to 1385, the "terminus post quem" for Evidence of Wycliffism in Bohemia. *Acta universitatis Carolinae – Historia universitatis Carolinae Pragensis* 55: 19–40.

Spunar, Pavel. 1985–1995. *Repertorium auctorum Bohemorum provectum idearum post Universitatem Pragensem conditam illustrans*. 2 vols. Wrocław: Ossolineum.

Székely, Iulia. 2018. Pratiques intellectuelles à l'Université de Prague au XVe siècle: Notes sur un quodlibet de Jan Arsen de Langenfeld (c. 1400). *Przegląd Tomistyczny* 24: 227–254.

———. 2019. The "Liber de causis" in Some Central European Quodlibets. In *Reading Proclus and the Book of Causes*. In *Western Scholarly Networks and Debates*, ed. Dragos Calma, vol. 1, 301–323. Leiden–Boston: Brill.

Šmahel, František. 1980. *Verzeichnis der Quellen zum Prager Universalienstreit 1348–1500*. Wrocław: Ossolineum.

———. 2007. *Die Prager Universität im Mittelalter / Charles University in the Middle Ages*. Leiden–Boston: Brill.

Šmahel, František, and Gabriel Silagi, eds. 2018. Statuta facultatis artium universitatis Pragensis redacta anno 1390. In *Statuta et Acta rectorum Universitatis Carolinae Pragensis 1360–1614*, 227–275. Praha: Karolinum.

Thakkar, Mark. 2020. Wyclif's *Logica* and the *Logica Oxoniensis*. In *Before and After Wyclif: Sources and Textual Influences*, ed. Luigi Campi and Stefano Simonetta, 1–31. Basel: Fédération Internationale des Instituts d'Études Médiévales.

———. Forthcoming. Wyclif, The Black Sheep of the Oxford Calculators. In *Quantifying Aristotle: The Impact, Spread and Decline of the Calculatores Tradition*, eds. Daniel A. Di Liscia, and Edith D. Sylla. Leiden: Brill.

Thomson, Williell R. 1983. *The Latin Writings of John Wyclyf: An Annotated Catalog*. Toronto: Pontfical Institute of Mediaeval Studies.
Tříška, Josef. 1981. *Životopisný slovník předhusitské pražské univerzity 1348–1409 / Repertorium biographicum Universitatis Pragensis praehussiticae 1348–1409*. Praha: Univerzita Karlova.
Weijers, Olga. 2002. *La 'disputatio' dans les Facultés des arts au moyen âge*. Turnhout: Brepols.
———. 2013. *In Search of the Truth: A History of Disputation Techniques from Antiquity to Early Modern Times*. Turnhout: Brepols.
———. 2015. *A Scholar's Paradise: Teaching and Debating in Medieval Paris*. Turnhout: Brepols.
Zupko, Jack. 1993. How Are Souls Related to Bodies? A Study of John Buridan. *The Review of Metaphysics* 46: 575–601.
———. 2004. On Buridan's Alleged Alexandrianism: Heterodoxy and Natural Philosophy in Fourteenth-Century Paris. *Vivarium* 42: 42–57.
———. 2007. John Buridan on the Immateriality of the Intellect. In *Forming the Mind. Essays on the Internal Senses and the Mind/Body Problem from Avicenna to the Medical Enlightenment*, ed. Henrik Lagerlund, 129–147. Dordrecht: Springer.

Author Index

M. Gensler et al. (eds.), *The Embodied Soul*, Historical-Analytical Studies on Nature, Mind and Action 11, https://doi.org/10.1007/978-3-030-99453-2

Subject Index

M. Gensler et al. (eds.), *The Embodied Soul*, Historical-Analytical Studies on Nature, Mind and Action 11, https://doi.org/10.1007/978-3-030-99453-2

The manufacturer's authorised representative in the EU is Springer Nature Customer Service Centre GmbH, Europaplatz 3, 69115 Heidelberg, Germany. If you have any concerns regarding our products, please contact ProductSafety@springernature.com

Printed and bound by CPI Group (UK) Ltd, Croydon, CR0 4YY
30/09/2026
02225497-0015